RR

Biotechnology of
Plant-Microbe
Interactions

The McGraw-Hill Environmental Biotechnology Series

Biotechnology of Plant-Microbe Interactions

WITHDRAWN

James P. Nakas

State University of New York
College of Environmental Science and Forestry
Syracuse, New York

Charles Hagedorn

Department of Plant Pathology
Virginia Polytechnic Institute
and State University
Blacksburg, Virginia

McGraw-Hill Publishing Company
New York   St. Louis   San Francisco   Auckland   Bogotá
Caracas   Hamburg   Lisbon   London   Madrid   Mexico
Milan   Montreal   New Delhi   Oklahoma City
Paris   San Juan   São Paulo   Singapore
Sydney   Tokyo   Toronto

Library of Congress Cataloging-in-Publication Data

Biotechnology of plant-microbe interactions / edited by James P. Nakas
and Charles Hagedorn.
    p. cm.
    Includes bibliographical references and index.
    ISBN 0-07-045867-7
    1. Plant-microbe relationships—Molecular aspects. 2. Microbial
biotechnology.   I. Nakas, James P.   II. Hagedorn, Charles.
QR351.B56   1990   89-13603
576'.15—dc20   CIP

1234567890 DOCDOC 9876543210

ISBN 0-07-045867-7

*The editors for this book were Jennifer Mitchell and Caroline Levine,
the designer was Naomi Auerbach, and the production supervisor was
Dianne L. Walber. It was set in Century Schoolbook by the McGraw-
Hill Publishing Company, Professional and Reference Division
Composition Unit.*

Printed and bound by R. R. Donnelly & Sons Company.

*For more information about other McGraw-Hill materials,
call 1-800-2-MCGRAW in the United States. In other
countries, call your nearest McGraw-Hill office.*

Contents

v

Contributors

Carole L. Cramer *Department of Plant Pathology, Physiology, and Weed Science, Virginia Polytechnic Institute and State University, Blacksburg, Virginia* (Chapter 1. Molecular Biology of Plants)

Thomas G. Currier *Sterling Drug, Inc., Malvern, Pennsylvania* (Chapter 4. Commercial Development of *Bacillus thuringiensis* Bioinsecticide Products*)

F. B. Dazzo *Department of Microbiology and Public Health, Michigan State University, East Lansing, Michigan* (Chapter 6. Microbial Colonization of Plant Roots)

Cynthia Gawron-Burke *Ecogen, Inc., Langhorne, Pennsylvania* (Chapter 4. Commercial Development of *Bacillus thuringiensis* Bioinsecticide Products*)

David Gerhold *Department of Microbiology, University of Tennessee, Knoxville, Tennessee* (Chapter 2. Recent Advances in Molecular Biology Techniques for Studying Phytosymbiotic Microbes)

W. D. Gould *CANMET, Department of Energy, Mines, and Resources, Ottowa, Ontario, Canada* (Chapter 9. Biological Control of Plant Root Diseases by Bacteria)

Charles R. Howell *U.S. Department of Agriculture, Agricultural Research Service, Southern Crops Research Laboratory, College Station, Texas* (Chapter 8. Fungi as Biological Control Agents)

D. A. Klein *Department of Microbiology and Environmental Health, Colorado State University, Fort Collins, Colorado* (Chapter 6. Microbial Colonization of Plant Roots)

James M. Ligon *Ciba-Geigy Corporation, Agricultural Biotechnology Research Unit, Research Triangle Park, North Carolina* (Chapter 5. Molecular Genetics of Nitrogen Fixation in Plant-Bacteria Symbioses)

Steven E. Lindow *Department of Plant Pathology, Hilgard Hall, University of California at Berkeley, Berkeley, California* (Chapter 3. Use of Genetically Altered Bacteria to Achieve Plant Frost Control)

Elizabeth Milewski *U.S. Environmental Protection Agency, Washington, D.C.* (Chapter 10. EPA Regulations Governing Release of Genetically Engineered Microorganisms)

* Copyright © Ecogen Inc. 1989

David N. Radin   *Department of Crop and Soil Environment Sciences, Virginia Polytechnic Institute and State University, Blacksburg, Virginia* (Chapter 1. Molecular Biology of Plants)

J. L. Salzwedel   *Department of Microbiology and Public Health, Michigan State University, East Lansing, Michigan* (Chapter 6. Microbial Colonization of Plant Roots)

Gary Stacey   *Department of Microbiology, University of Tennessee, Knoxville, Tennessee* (Chapter 2. Recent Advances in Molecular Biology Techniques for Studying Phytosymbiotic Microbes)

Hugh E. Wilcox   *Professor Emeritus, Department of Environmental and Forest Biology, State University of New York, College of Environmental Science and Forestry, Syracuse, New York* (Chapter 7. Mycorrhizal Associations)

Preface

During the past ten years, use of the term *biotechnology* has become commonplace in both the scientific as well as the popular literature. Although some consider biotechnology to include only the more recent advances in molecular biology, it is probably more appropriate to include all aspects of applied biology which are geared toward the general improvement of our daily existence. This type of broad interpretation demands a convergence of seemingly unrelated disciplines. In this volume we have attempted to meld the rapidly developing areas of molecular biology with some more traditional approaches in biology to provide an overview of the current status of biotechnology with regard to agronomically significant crops as well as some forestry application. Each author has attempted to include the current applied potential of his or her subject area or, where this is not feasible, the possible future applications or developments resulting from research focused on interactions between plant and microbe.

James P. Nakas
Charles Hagedorn

Biotechnology of
Plant-Microbe
Interactions

Molecular Biology of Plants

Carole L. Cramer

*Department of Plant Pathology, Physiology, and Weed
Science
Virginia Polytechnic Institute and State University
Blacksburg, Virginia 24061-0330*

David N. Radin

*Department of Crop and Soil-Environmental Sciences
Virginia Polytechnic Institute and State University
Blacksburg, Virginia 24061-0404*

Overview

The last five years have marked a period of significant advances in
our understanding of molecular events involved in plant responses
to both symbiotic and pathogenic microbes and in the potential to
directly manipulate these responses. The advances have been due,

in large part, to applications of recombinant DNA technologies and development of techniques for plant transformation and regeneration. Several examples of initial successes in genetically engineering plants for enhanced resistance to specific pathogens, insect pests, or herbicides are currently being tested in the field. Thus, the feasibility of applying the new technologies to problems involving plant-microbe interactions has now been demonstrated. The recent acquisition of many seed companies by agrichemical and biotechnology companies reflects the interest among technology-based industries in using seeds or seed coatings as commercial delivery systems for biotechnology products.

The objectives of this chapter are (1) to summarize recent results in the molecular cloning and characterization of plant genes involved in plant-microbe interactions, (2) to review the current status of recombinant DNA and gene-transfer technologies as applied to plant systems, and (3) to discuss current limitations and future directions in these areas. This task is somewhat daunting due to the tremendous growth of new information in these areas. Accordingly, our approach is to provide an overview of recent results and, where appropriate, refer the reader to several excellent recent reviews covering specific aspects of plant-microbe interactions, plant defense or nodulation genes, and plant gene-transfer technologies.

Identification and Cloning of Plant Genes Involved in Plant-Microbial Interactions

As will become clear in subsequent chapters of this book, plants and microbes have evolved an incredible diversity of interactive relations, both beneficial and detrimental to the host plant. Within the scope of this book, it is the goal of plant genetic engineering to enhance beneficial interactions and to block or minimize damage due to pathogen or pest interactions. Prerequisite to the successful design of genetic engineering strategies for crop improvement is (1) an understanding of the molecular events underlying specific plant-microbe interactions, (2) identification and characterization of genes (bacterial, fungal, or plant) useful for plant genetic engineering, and (3) development of technologies for efficient production of transgenic plants in commercially adapted cultivars of agronomically important crops.

In this section, we focus on recent progress in understanding molecular events of host responses involved in symbiotic interactions with *Rhizobium* nitrogen-fixing bacteria and in expression of host resistance to pathogenic microorganisms. Plant interactions with pathogenic and "disarmed" *Agrobacterium* strains will be discussed under "Plant Gene-Transfer Biotechnologies" below.

Host-symbiote interactions

One of the most extensively studied plant-microbial relationships is the symbiotic association between representatives of the soil bacteria *Rhizobium* and the roots of plants from the family Leguminosae (Long 1989). The importance of this interaction for the participants undoubtedly lies in the nutritional benefits that each derives from it; plants obtain ammoniacal fixed nitrogen via the expression of bacterial nitrogenase genes, and the microbes obtain a rich supply of sugars from the fixation of CO_2 by the photosynthetic machinery of plant chloroplasts (Vincent 1974). The biological and evolutionary significance of this joint arrangement for solving some primary food needs for these organisms is striking. The biological system that evolved to achieve this result depends on an intimate and complex bacterial-plant association through which a series of mutually regulated morphogenetic steps leads eventually to the virtual imprisonment of the bacterium in a nodule-specific intracellular compartment of the plant's root system. Once constructed, this structure serves as the focus of nitrogen fixation and the accompanying nutrient exchange system that confers the mutual benefits in the relationship. The major problem in understanding this result undoubtedly concerns the molecular mechanisms underlying the regulation of this symbiotic process. Equally intriguing is the evolutionary question of how such a complex interactive system between two very distinct organisms could have developed.

The progress of research on this problem also impinges on the practical importance of the symbiotic process as it relates to the agricultural need for fixing nitrogen naturally as an alternative to environmentally harmful and energetically wasteful practices of the past. Absence of a fundamental understanding of how the genes of bacterial symbiotes and their plant hosts coregulate the development and functioning of nitrogen-fixing root nodules hinders rational efforts to engineer greater efficiency and extended host ranges for this function into other adapted crop varieties.

The primary focus of previous research on symbiotic nitrogen fixation has been on the microbial component of this process (Bauer 1981, Djordjevic et al. 1987). These studies have begun to elucidate the function of rhizobia in regulating the development of plant root nodules. This progress has been advanced by sophisticated microbiological and genetic research techniques available for microbial studies.

The relative ease of mutagenesis and gene identification, recombinant DNA cloning, and plasmid genetics, combined with the facility of obtaining *Rhizobium* in fast-growing cultures, has advanced the bacterial research ahead of the plant studies. Results of this work as discussed by many recent reviewers including the authors of Chap. 5 in this volume indicate that *Rhizobium* exercises major control over the development of plant nodules. This has been demonstrated most convincingly by the isolation and analysis, including gene cloning, of several classes of bacterial mutants that appear to affect virtually every aspect of the nodulation process. These groups of bacterial genes, designated *"nod," "exo," "nif,"* and *"fix"* according to function, have also been used to define major functions of the plant (Long 1989).

Recently, with development of more powerful analytical tools in plant molecular and cellular biology, the lack of information on plant function has become less severe, and several reviews of host symbiotic function have appeared (Vance 1983, Bothe et al. 1988, Vance et al. 1988). This discussion, therefore, will be limited to a brief consideration of the plant's role in the symbiotic process and development of approaches for modifying it.

Characterization of nodulins. "Nodulins" have been defined as tissue-specific proteins synthesized by plants during development and function of the root nodules responsible for nitrogen fixation (Delauney and Verma 1988, Van Kammen 1984). Most of the known nitrogen-fixing symbiotic associations occur between *Rhizobium* and legumes through a complex series of developmental steps involving coordinated regulation between both organisms. During this process the plant is called upon to regulate many functions required for nodule formation (Kondorosi and Kondorosi 1986) including cell division, gene expression, primary and secondary metabolism, morphogenesis, signal transduction, and cell wall and membrane biogenesis. In general the roles of nodulins may be classified into three categories (Legocki and Verma 1980): (1) the morphogenesis involved in nodule development, (2) the functional aspects of symbiotic nitrogen fixation including nitrogen assimilation for use by the plant and carbon assimilation for use as a nutrient resource for the bacteroids, and (3) the compatibility, or host range, of the bacterial-plant relationship. Although a number of specific functions within each category are known, it is only recently that detailed biochemical and genetic characterization has begun (Rolfe and Gresshoff 1988). Thus, for example, although over 45 nodulin mutations have been identified in eight different plant species, none of these genes has yet been isolated and no mutations exist for any of the known nodulins (Verma and Nadler 1984).

Plant nodulins are characterized as "early" or "late" according to their function during the time course of nodule development. Early nodulins may be expressed as soon as several days after the start of infection; they function, for the most part, during nodule development before nitrogen fixation has actually begun. Several plant functions that require nodulins include epidermal root-hair curling that facilitates bacterial penetration (Yao and Vincent 1969), meristamatic cell division in the root cortex that is correlated with penetration (Dudley et al. 1987), growth of "infection threads" that serve as channels through which bacteria travel into the plant cortex, and development of intercellular compartments or envelopes in which bacteroids reside during nitrogen fixation. This compartment or envelope structure has been found to be constructed from plant-derived components (Robertson et al. 1978).

Among the early nodulins are those involved in determining compatibility between plant species and bacteria, i.e., host range (Robertson et al. 1985). For example, the infection thread that is constructed by the plant has been found to be filled with a plant-derived glycoprotein matrix that facilitates movement of the bacteria and may also be involved in the recognition process (Newcomb 1981). Growth of the infection thread also depends on cell wall synthesis, which could likewise be a part of a compatibility-screening function (Callaham and Torrey 1981). Finally, the dual chemotaxis that initiates the infection process is undoubtedly part of the compatibility system because small molecules function as cross-species regulators of gene activation that is required to initiate the nodulation process (Kosslak et al. 1987, Long 1989). Although pathways that regulate synthesis of these flavonoid molecules (flavones and flavanones) are crucial for success of nodulation, the flavonoid molecules are not, strictly speaking, nodulins because they are expressed prior to physical initiation of infection (Bergman et al. 1988).

The major intracellular plant structure involved in symbiotic nitrogen fixation is the envelope that houses the nitrogen-fixing bacteroids. This organelle-like structure is surrounded by a plant-derived limiting membrane (Fortin et al. 1985, Bradley et al. 1988) called the "peribacteroid membrane" (PBM). The plant cells that contain these structures are located in undifferentiated root parenchymal tissue that has been stimulated to undergo cell division as part of the nodulation process (Dudley et al. 1987). Plant growth regulators may play a role in regulating this growth although the origin of the regulators—bacterial versus plant—is not known (Long 1989). Bacteria enter these cells via the infection thread and after several more cell divisions cease reproductive growth (Zhou et al. 1985). At this point the nitrogen-fixing apparatus is fully functional within the confines of the peribacteroid envelope. Virtually all of the morphogenetic activity that has produced this result is governed primarily by plant genes, although crucial regulatory signals that affect this expression undoubtedly originate in the bacterium.

As the symbiotic nitrogen-fixing process begins, another group of plant gene products, the late nodulins, are expressed. Leghemoglobin (Appleby 1984) is the best studied of these; it is recognized by the pink color that it confers on nodules. The function of leghemoglobin is to act as an oxygen sink to buffer the exposure of the anaerobic nitrogenase system to exposure to oxygen. Other nodulins also have important roles in assimilating the products of both nitrogen and carbon fixation. Examples include uricase (Bergmann et al. 1983), glutamine synthetase (Cullimore et al. 1984), and sucrose synthetase (Kahn et al. 1985). A relatively stable balance of nutrient production, exchange, and catabolism appears to be established between plant and bacteroid until senescence of the plant begins (Vance et al. 1988). Coordinated regulation of plant and microbial gene expression is essential if the integrity of this process is to be maintained. However, the mechanisms regulating this complex coordination remain unclear.

Prospects for engineering hosts for enhanced N_2 fixation. Recent progress in molecular cloning of a number of plant nodulin genes (Fuller et al. 1983, Franssen et al. 1987, Govers et al. 1987, Delauney and Verma 1988) from several legume species raises the prospect of modifying the efficiency of the nitrogen-fixing process in agronomically adapted crop varieties. Thus far, however, very few actual functions have been identified for these cloned genes; these include leghemoglobin, uricase, sucrose synthetase, and glutamine synthetase (Verma et al. 1986). Although molecular analysis of nodulin genes still lags substantially behind the molecular genetics of *Rhizobium*, we can now begin to focus on certain aspects of the symbiotic process that might be amenable to positive manipulation in the plant (La Rue et al. 1985). For example, the leghemoglobin and nitrogen and carbon assimilation genes could be manipulated by modifying upstream *cis*-acting enhancer elements or other factors that bind activators. The resultant increase in potential for assimilating nutrients from bacterial or plant nitrogen and carbon fixation could be quite beneficial. Likewise, an understanding of molecular factors influencing leghemoglobin regulation could lead to increased capacity of the plant to modulate the use of oxygen and help protect the nitrogenase system from excessive stress. Jensen et al. (1988), recently reported the discovery of a nodule-specific *trans*-acting factor that affects leghemoglobin gene function in soybean. Other parts of the symbiotic process that could be modified by manipulating plant genes include the early morphogenetic steps required for nodule development. These events, including root-hair curling, growth of infection tubes, and development of the PBM, are central for establishment of symbiosis and could potentially be made more efficient by modifying the plant genome.

Another function that could eventually be enhanced is the compatibility between particularly effective *Rhizobium* lines and legume varieties that are highly adapted to specific growing areas. A particularly attractive prospect is the possibility of playing mix and match between bacterial lines and adapted crop varieties by using genetic engineering to alter one or both partners of the symbiotic system. It should be noted, however, that such speculations are quite premature considering the minimal levels of progress that have been achieved in understanding the functioning of the plant. Perhaps the most promising prospect at the present time is the possibility of a much broader future research effort aimed at understanding regulation of the symbiotic process in plants on a molecular level.

Host-pathogen interactions

Like symbiotic microorganisms, plant-pathogenic microbes generally display a relatively limited host range. That plant disease, especially in wild species, is the exception rather than the rule (Lamb et al. 1989) suggests that plants have evolved very effective mechanisms to defend against most pathogenic agents. Plants, being sessile, utilize a variety of morphological, structural, and chemical defenses to prevent or limit predation and disease. Recently, substantial research efforts have been focused on elucidating the molecular mechanisms involved in host defenses and in the ability of pathogens to overcome or avoid these defenses. Molecular studies on the plant component of these relationships have concentrated primarily on "inducible" resistance, i.e., the ability of the plant to recognize pathogen invasion and rapidly trigger transcriptional activation of defense-response genes leading to the production of both structural and chemical defense compounds (Ebel 1986, Dixon 1986, Dixon et al. 1986). In bean (*Phaseolus vulgaris*), the defense response represents a significant alteration in gene expression with more than 70 genes showing elevated expression and at least 10 down-regulated during the early part of the response (Cramer et al. 1985b). This response is often associated with host-cell death localized to the site of infection, termed a "hypersensitive resistance" (HR) response (Bell 1981, Sequeira 1983). The association of HR responses with resistance to a wide variety of pathogens (including viruses, bacteria, fungi, and nematodes) has fueled the drive to identify and analyze host genes and gene products involved in plant defense behavior.

Identification and characterization of defense-response genes.   Inducible host responses include (1) accumulation of low-molecular-weight antibiotics termed "phytoalexins," (2) deposition of cell wall materials and calloses thought to serve as structural barriers for limiting pathogen penetration, and (3) production of hydrolytic enzymes effective against cell walls of invading microbes (reviewed in Sequeira 1983, Dixon 1986, Boller 1987). In addition, proteinase inhibitors (effective, for example, as feeding deterrents against insects) are induced both locally and systemically after wounding or pathogenic attack (Ryan and An 1988). Strategies that have been utilized to identify and clone host genes involved in these responses are based either on knowledge of specific defense products (or required biosynthetic enzymes) or on differential-screening approaches to identify cDNA clones of unknown function. The latter have been termed "pathogenesis-related" (PR) proteins, which are induced in response to pathogenic attack or chemical elicitors (Collinge and Slusarenko 1987). Defense-related genes that have been cloned and characterized by these methods are listed in Table 1 and are described briefly below.

The critical role of phytoalexins in resistance has now been demonstrated in a number of host-pathogen interactions (reviewed in Bailey 1987, Darvil and Albersheim 1984, Kuc and Rush 1985):

1. Moesta and Grisebach (1982) demonstrated that when soybean seedlings of a *Phytopthora megasperma*–resistant cultivar were pretreated with specific inhibitors blocking biosynthesis of phenylpropanoid phytoalexins, they were rendered more susceptible to infection upon subsequent inoculation with *P. megasperma*.

2. Hahn et al. (1985) used specific antibodies directed against glyceollin, the major soybean phytoalexin, and against *P. megasperma* hyphal cell walls to monitor the relationship between phytoalexin accumulation and fungal penetration in compatible (host susceptible) and incompatible (host resistant) interactions. In the incompatible interaction, glyceollin concentration exceeded the 50% kill dose in the plant tissues adjacent to the growing hyphal tip. In contrast, compatible interactions showed glyceollin accumulation to a lesser extent and in tissues substantially behind the growing tips.

3. In several host-pathogen systems [e.g., pea–*Nectria haemetococca* (Kistler and Van Etten 1984) and tomato–*Gibberella pulicaris* (Desjardins and Gardner 1989) interactions], the degree of pathogenicity has been directly correlated with the ability of the pathogen to detoxify the major host phytoalexins—for example, pisitin in pea and rishitin in tomato.

TABLE 1   Cloned Host Defense-Response Genes

Defense-response genes	Host species	Reference[a]
Phytoalexin biosynthesis:		
Phenylpropanoid phytoalexins		
Phenylalanine ammonia lyase	*Phaseolus*	Edwards et al. 1985
	Petroselinum	Kuhn et al. 1984
4-Coumarate CoA ligase	*Petroselinum*	Kuhn et al. 1984
Chalcone synthase	*Petroselinum*	Kreuzaler et al. 1983
	Phaseolus	Ryder et al. 1984
Chalcone isomerase	*Phaseolus*	Mehdy and Lamb 1987
Resveratrol synthase	*Arachis*	Schroder et al. 1988
Terpenoid phytoalexins		
3-Hydroxy-3-methylglutaryl CoA reductase	*Lycopersicon*	Park et al. 1989
Casbene synthetase	*Ricinus*	Moesta and West 1985[b]
Cell wall components:		
Lignin/lignification		
Phenylalanine ammonia lyase	see above	
Cinnamyl alcohol dehydrogenase	*Phaseolus*	Walters et al. 1988
Lignin-forming peroxidase	*Nicotiana*	Lagrimini et al. 1987a,b
Hydroxyproline-rich glycoproteins (HRGPs)	*Phaseolus*	Corbin et al. 1987
Hydrolytic enzymes:		
Chitinase	*Nicotiana*	LeGrand et al. 1987
	Phaseolus	Hedrick et al. 1988
Glucanases	*Nicotiana*	Kauffmann et al. 1987
Others:		
Proteinase inhibitors	*Lycopersicon*	Graham et al. 1985a,b
Thaumatin-like protein	*Nicotiana*	Cornelissen et al. 1986
	Zea	Richardson et al. 1987
PR proteins	*Nicotiana*	Hooft van Huijsduijnen et al. 1986, Krombrink 1988
	Petroselinum	Somssich et al. 1986
Superoxide dismutase	*Nicotiana*	Bowler et al. 1989
Thionins	*Hordeum*	Bohlmann et al. 1988

[a]References are limited to initial cloning and/or defense-related expression in host cells or plants.

[b]Casbene synthetase cDNAs have recently been cloned and characterized (A. Lois and C. A. West, unpublished results), and results of these studies will be published in 1989 (C. A. West, University of California, Los Angeles, personal communication).

Progress has also been made at the molecular level in isolating and characterizing genes involved in biosynthesis of isoflavonoid phytoalexins, the major class of phytoalexins in legumes (see Table 1). Cloned marker genes now include phenylalanine ammonia lyase (PAL) and 4-coumarate CoA ligase (4CL) in the primary pathway, chalcone synthase (CHS) and chalcone isomerase (CHI) in the branch pathway leading to phytoalexin biosynthesis and cinnamyl alcohol dehydrogenase (CAD) leading to lignin synthesis (reviewed in Ebel 1986, Dixon 1986). These genes are rapidly induced in bean and soybean in response to fungal inoculation and in cultured cells in response to fungal cell wall elicitors. In all cases analyzed, the induction of enzyme activity results from de novo synthesis based on a transient increase in mRNA levels. Nuclear run-off experiments have shown that elicitor-induced increases in mRNA result from transcriptional activation of PAL, 4CL, CHS, CHI, and CAD genes (Lawton and Lamb 1987, Templeton and Lamb 1988). In bean cell cultures, an initial increase in gene transcription of these defense genes is seen within 2 to 3 minutes of elicitor addition (Lawton and Lamb 1987). This very rapid transcriptional activation represents the shortest lag for gene activation in response to an external signal reported in plants and is comparable to rapid activational events in animal cells (Templeton and Lamb 1988).

Regarding incompatible (host resistant) responses, the phytoalexin biosynthetic enzymes are rapidly induced at the gene level in tissues adjacent to infection. This induction is transient and precedes the onset of visible HR symptoms (Bell et al. 1986, Cramer et al. 1985a). These genes are also activated in compatible (host susceptible) responses, but at a substantially later period after other disease symptoms are visible. It is believed that this later response represents an attempt at lesion limitation by the plant and is triggered by signals originating in damaged plant cells rather than by specific pathogen recognition events (Bell et al. 1986, Templeton and Lamb 1988). These studies and those of Hahn et al., discussed above, underscore the critical importance of analyzing both temporal and spatial expression patterns when attempting to associate defense-induced compounds with disease resistance.

Substantially less information is available concerning the regulation of genes involved in biosynthesis of other major classes of phytoalexins, the terpenoids and polyacetylenes. The major phytoalexins of the Solanaceae and some agronomically important graminaceous species (e.g., rice) are derived from the terpenoid pathway. Recent progress in molecular cloning of genes encoding several terpenoid biosynthetic enzymes should eventually provide significant inroads in understanding the regulation of this important class of defense compounds. 3-Hydroxy-3-methylglutaryl CoA reductase (HMGR) mediates a key rate-limiting step catalyzing the production of mevalonic acid, precursor of the five-carbon isoprene building block for terpenoid synthesis. HMGR enzyme activity is rapidly elevated in potato tubers in response to elicitors (arachidonic acid) and infection (Stermer and Bostock 1987). We have recently cloned several HMGR genes from tomato and demonstrated a marked induction of HMGR mRNA levels in tomato cells treated with cell wall elicitor from *Verticillium albo atrum* or *Fusarium oxysporum* (Park, Denbow, and Cramer, unpublished results). Preliminary results suggest that the HMGR isogene involved in biosynthesis of phytoalexins (e.g., rishitin) is regulated independently from those leading to sterol biosynthesis (Cramer, unpublished). In potato tuber slices, treatment with arachidonic acid or inoculation with the soft-rotting bacterium *Erwinia carotovora* results in induction of both HMGR mRNA levels and enzyme activity (Yang, Cramer, and Lacy, unpublished results). Interestingly, the HMGR gene induced by *Erwinia* infection is distinct from that induced by wounding (Yang, Lacy, and Cramer, unpublished). West and coworkers have recently isolated a cDNA clone for another gene involved in terpenoid phytoalexin biosynthesis, casbene synthetase (C. A. West, University of California, personal communication). Casbene synthetase is greatly elevated in castor bean (*Ricinus communis*) treated with fungal elicitors, leading to accumulation of the diterpene phytoalexin casbene (Bruce and West 1982). Although analyses of these genes are still in the early stages, evidence suggests that control of terpenoid phytoalexin biosynthesis will function by mechanisms very similar to those directing the accumulation of phenylpropanoid phytoalexins in legumes.

Elicitor- and infection-induced gene activation of other defense-related genes, including genes for hydroxyproline-rich glycoproteins (HRGPs), chitinase, proteinase inhibitors, superoxide dismutase, and certain PR proteins has also been shown (Bowler et al. 1989, Corbin et al. 1987, Hedrick et al. 1988, Ryan and An 1988, Somssich et al. 1986). Interestingly, in bean, the most studied system, elicitor induction of different defense-related genes follows several kinetic patterns. Chitinase and CAD are induced very rapidly, with mRNA levels peaking after 1.5 to 2 hours (Hedrick et al. 1988, Walter et al. 1988). PAL, CHS, and chalcone isomerase are expressed coordinately, with maximal mRNA levels occuring 3 to 4 hours after elicitor treatment (Cramer et al. 1985a, Mehdy and Lamb 1987, Ryder et al. 1984). In contrast, HRGP genes show no transcriptional activation until 2 hours after elicitor treatment, and maximal RNA accumulation takes place after 12 hours (Corbin et al. 1987, Lawton and Lamb 1987). In elicitor-treated tomato cell cultures, PAL mRNA (involved in lignin, but not phytoalexin biosynthesis) is induced with kinetics similar to those shown in bean (Cramer, unpublished results). HMGR mRNA, however, is induced more slowly, with maximal accumulation occurring at about 9 hours (Denbow and Cramer, unpublished results). Preliminary results in potato tubers suggest that one HMGR isozyme is very rapidly induced by wounding (peaking at 30 minutes after excision), while a second is induced more slowly in response to arachidonic acid or infection by *Erwinia carotovora* (Yang, Lacy, and Cramer, unpublished results). The induction kinetics of proteinase inhibitor genes from both potato and tomato suggest that the slower pattern of induction follows leaf wounding (Graham et al. 1986). The differences in induction patterns for these distinct defense responses may reflect separate cellular signals or possibly a regulatory cascade. Interestingly, two genes showing the slower pattern of induction (HRGPs and proteinase inhibitors I and II) are also systemically induced (elevated in regions of the plant far removed from the site of injury or infection). This suggests a regulatory circuitry distinct from that of the phytoalexin pathways.

Two new technologies are being applied to the analysis of defense-related gene activation: in situ hybridization analyses (Meyerowitz 1987) and transgenic plants (discussed below). In the first approach, expression of specific genes is monitored by hybridization of complementary RNA to mRNA directly on thin sections of inoculated tissues. In the second approach, promoters of defense-related genes are fused to reporter genes and transferred into plant cells for transient assays (Dron et al. 1988, Dangl et al. 1987) or for regeneration of transgenic plants. Expression of these transgenes in response to wounding or infection can then be monitored by sensitive enzymatic assays. These approaches have been important in defining key regulatory regions of defense-related genes (Dron et al. 1988, Logemann et al. 1989, Thornburg et al. 1987) and in determining their spatial and temporal expression in relation to the pathogen (Cuypers et al. 1988, Somssich et al. 1988). Both approaches should be important in defining the regulatory circuitry controlling specific aspects of the host's defense responses and directing efforts to delineate regulatory signals and signal transduction pathways in disease resistance.

Signal transduction in triggering host responses.   The determinants underlying the specificity of compatible versus incompatible host-pathogen interactions and the sequence of molecular events involved in pathogen recognition, signal transduction, and activation of host genes remain unknown. However, progress is being made in identifying components that are involved in the recognition–signal transduction pathway as discussed in a recent review by Lamb et al. (1989). Pathogen-derived compounds have been identified that trigger host responses. Best characterized is a β-linked heptaglucan from race 3 of *Phytophthora megasperma* f. sp. *glycinea*. Nanomolar amounts of this glucan effectively elicit glyceollin phytoalexin accumulation in soybean (Sharp et al. 1984). Other elicitor-active compounds include chitosan, arachidonic acid, other glucans, and a variety of glycoproteins (reviewed in Darvill and Albersheim 1984). Although most purified elicitors do not maintain the race-to-cultivar specificity of the pathogen source (Dixon 1986), a glycoprotein from *Colletotrichum lindemuthianum* and a necrosis-inducing polypeptide from *Cladosporium fulvum* are reported to trigger appropriate race-to-cultivar responses on bean and tomato, respectively (Tepper and Anderson 1986, De Wit et al 1988). An elicitor binding site for glucan elicitors isolated from *P. megasperma* has been identified in membrane fractions of soybean (Schmidt and Ebel 1987). Such receptors are candidates for the gene products of genetically defined dominant resistance genes (R-genes), discussed below. Hopefully, further studies of elicitor binding sites and molecular cloning of R-genes, as well as identification of fungal and bacterial determinants, will lead to an understanding of the initial recognition events.

As discussed above, elicitor treatment of cell cultures triggers host-defense-related gene transcriptional activation within 2 to 3 minutes. This suggests that only a limited number of intervening steps can be involved in transducing the signal from receptor to transcriptional activation (Lamb et al. 1989). Many components involved in signal transduction in other systems, for example, Ca^{2+}, cAMP, and protein phosphorylation, have been identified in plant defense responses. Polyphosphoinositides and cAMP do not appear to be signals in *P. megasperma*–soybean interactions (Hahn and Grisebach 1983, Strasser et al. 1986). However, recent experiments on elicitor induction of carrot cells suggest a role for Ca^{2+} and cAMP (Kurosaki et al. 1987). Ca^{2+} fluxes have also been implicated in mediating elicitor induction of isoflavonoid phytoalexin accumulation and activation of callose synthase in soybean cells (Kohle et al. 1985, Stab and Ebel 1987) and terpenoid phytoalexin accumulation in potato (Zook et al. 1987). In addition, both elicitor induction and wound induction have been associated with rapid phosphorylation of specific host proteins (C. Lamb, Salk Institute; C. Ryan, Washington State University, personal communications). In potato, a specific 34-kDa membrane protein is rapidly phosphorylated following wound induction (C. Ryan, personal communication). The identity of these proteins and their role in activation of host defense responses is currently under investigation. Progress has been made in molecular cloning of plant protein kinases (Lawton et al. 1988, Lawton and Lamb, Salk Institute, unpublished results) that may aid in elucidating possible roles of phosphorylation

in triggering defense responses. Hydroxyl radicals have been implicated in the abiotic elicitation of phytoalexins in legumes (Epperlein
et al. 1986). Induction of superoxide dismutases during infection
(Bowler et al. 1989) and induction of defense-related genes by
glutathione (Wingate et al. 1988) suggest that free radicals and/or oxygen stress may play a role in triggering defense responses or producing the hypersensitive localized cell death.

Activation of host defense genes also occurs in adjacent tissues not
in direct contact with the pathogen or wound. For example, HRGPs
and proteinase inhibitors have been shown to be induced systemically
(Bell et al. 1986, Ryan 1987, Corbin et al. 1987). Glutathione,
ethylene, and plant cell wall fragments have all been implicated as
intercellular stress signals (Corbin et al. 1987, Darvill and
Albersheim 1984, Lamb et al. 1989, Ryan 1987, Wingate et al. 1988).
Polygalacturonic acid is released from plant pectin substances upon
wounding and induces proteinase inhibitors in potato and tomato.
However, this substance, if placed on a wound, is not transported
through the plant, suggesting it cannot play a role in systemic induction (Baydoun and Fry 1985, Ryan, personal communication).
Ethylene has been shown to induce many defense-related genes, but
its role in signaling host defense responses is not clear. Mauch et al.
(1984) reported that AVG, an inhibitor of ethylene synthesis, blocked
the induction of the chitinase involved in senescence but did not interfere with defense-related chitinase induction of pea pods in response to *P. megasperma* infection. This suggests that ethylene is not
critical in triggering the disease response in this interaction. Final
identification of intermediate signal molecules and intercellular
stress signals will ultimately depend on identifying receptors, R-gene
products, and defense-gene-binding proteins and on showing specific
interactions of presumptive signal molecules or second messengers.

Significant progress is being made in identifying key *cis*-regulatory
sequences mediating transcriptional activation of defense-related
genes in response to elicitors, infection, or wounding (Dron et al. 1988,
Lawton et al. 1988, Logemann et al. 1989, Sanchez-Serrano et al.
1987, Thornburg et al. 1987). For bean CHS genes, a combination of
deletion–transgenic expression assays and DNase footprinting analyses have identified two promoter regions important in transcriptional
activation and elicitor-specific regulation (Dron et al. 1988, Lamb et
al. 1989, Lawton et al. 1988). These regions are also conserved in bean
PAL promoters showing similar patterns of regulation (Cramer et al.
1989, Liang et al. 1989). These conserved regulatory sequences are being used to identify and clone specific DNA-binding proteins (Lawton
et al. 1988, Lamb et al. personal communication). Recent studies indicate that the signals and *cis*-acting regulatory sequences responding

to induction by elicitor, infection, and wounding are not identical (Liang et al. 1989). Thus, while considerable progress has been made in understanding key events in pathogen attack and susceptible or resistant host responses, there is still much to be learned before we have a thorough understanding of molecular events involved in pathogen recognition and successful triggering of active defense responses.

Efforts to isolate and clone host "resistance" genes.   Classical genetic studies have identified dominant genetic loci, termed "resistance genes" ("R-genes"), that confer resistance to specific pathogens or physiological races of a pathogen. Loci have been identified that are effective against specific viruses, bacteria, fungi, or nematodes and have served as major breeding tools for disease resistance in many crop species. Products encoded by these loci have not been identified, and molecular cloning of a resistance gene has not yet been accomplished. However, based on the fact that R-genes exhibit a dominant resistance genotype and appear to direct a very rapid and localized induction of plant defense-response genes, it is likely that they encode proteins involved in specific recognition of pathogens, such as membrane receptors, or early regulatory or signal transduction steps triggering successful host responses. The potential of these genes for use in genetic engineering of disease resistance is obvious, and there are significant research efforts being focused on identification and molecular cloning of specific R-genes.

Shotgun cloning strategies have been successful in isolating bacterial pathogenicity genes. Genomic libraries of avirulent (the predominant phenotype in bacteria) strains have been transconjugated into recipient virulent races and screened for cultivar-specific patterns of avirulence (Staskawicz et al. 1984, Gabriel et al. 1986, Daniels et al. 1984). Analogous strategies for cloning plant resistance genes are not feasible currently due to the large genome size of most crop species. Such approaches in tomato, for example, would require construction of a genomic library containing DNA from a cultivar containing a specific (or multiple) R-gene(s) in binary vectors (Simoens et al. 1986) for *Agrobacterium*-mediated transformation of a susceptible cultivar. Over 100,000 transgenic seedlings might have to be screened for a shift in disease phenotype to ensure identification of a single copy gene. Similar calculations for flax, which has a relatively small genome, suggest that at least 10,000 independent transformants must be screened to ensure identification of an R-gene starting with a cultivar containing at least four R-genes (Jones et al. 1985). Obviously this approach requires very efficient transformation and regeneration techniques, simple and accurate disease-screening procedures, and considerable greenhouse space. Thus, several alternative approaches are being taken to reduce the number of transformed plants to be screened. These include transposon tagging, restriction fragment length polymorphism (RFLP) mapping combined with production of subgenomic libraries or "chromosome walking," and development of disease-screening methods that do not require use of fully developed whole plants.

Molecular cloning of specific genes, including regulatory genes, by transposon tagging, has been successful in maize (Fedoroff et al. 1984, Schmidt et al. 1987, Hake et al. 1989) and *Antirrhinum* (Martins et al. 1985). In addition, the maize transposable elements, *Spm* and *Ac*, and *Antirrhinum* element *Tam3* have now been introduced successfully into tobacco, tomato, carrot, and *Arabidopsis* (Baker et al. 1987, Van Sluys et al. 1987). This suggests that transposon-tagging approaches may soon be widely applicable to agronomically important crops. However, this methodology is not without some drawbacks. The strategy (reviewed in Weinand and Saedler 1987) involves mutating a gene of interest (e.g., an R-gene) by the insertion of a transposon introduced by a genetic cross or transformation. A genomic library is constructed from total DNA of the mutant plant, and the mutant gene is then isolated based on sequence cross-homology with the transposon. Regions flanking the transposon may be used subsequently to isolate the wild-type gene from a second library. One major drawback to this approach is that transposons generally exist in multiple copies in the genome, and identification of the specific copy inserted into the gene of interest is required. This can be done by establishing genetic linkage of a particular element with the mutant phenotype, by the isolation of "active elements" based on methylation patterns (reviewed in Banks et al. 1988, Fedoroff 1989), or by the simultaneous insertion of two different elements (O'Reilly et al. 1985). In transgenic plants such as tobacco, where a unique element has been introduced through transformation, the copy number appears to stay relatively low, facilitating gene-tagging approaches. Transposons can jump to alternate chromosomal sites, and, thus, the mutant phenotype may be unstable. Additionally, transposons tend to jump to closely linked regions of the chromosome, which limits their utility as a random mutagenic agent. Recent analyses of the regulation of copy number and the isolation of defective elements that reduce mutability of specific elements (Walbot and Warren 1988, Cuypers et al. 1988) may provide future strategies for directly controlling both copy number and stability.

Several research groups are actively involved in the application of transposon-tagging strategies to clone the *RP1* locus in maize. This locus is well-characterized genetically with 14 alleles conferring rust resistance to specific *Puccinia sorghi* races. However, the locus displays a spontaneous reversion rate (Pryor 1987b) substantially higher than the transposon-induced mutation frequency seen in most maize element systems (e.g., Schmidt et al. 1987). This high mutation rate (up to 8×10^{-3}) back to susceptibility makes identification of transposon-induced mutants exceedingly difficult (Bennetzen et al. 1988). These analyses, however, suggest that the *Rp1* locus may be a complex locus consisting of multiple copies in an allelic series (Bennetzen et al. 1988). Such an arrangement provides potential for multiple gene conversion events and genetic rearrangements that could generate diversity in pathogen recognition. A similar mechanism may be involved in regulation of self-incompatibility in plants (reviewed in Ebert et al. 1989). Specific diversity-generating mechanisms for surface antigens, receptors, and recognition components (e.g., antibodies) exist in many organisms and may provide advantages for mediating host-pathogen interactions. If complex resistance loci are the rule, rather than the exception, molecular cloning based on transposon tagging may not be the strategy of choice.

The identification of RFLP markers closely linked to specific R-genes provides an alternate strategy for cloning these genes. This has now been successfully applied to the human genome to clone a number of disease-related genes (reviewed in White and Lalouel 1988). Significant progress has been made in producing detailed RFLP maps of a number of plants including maize, tomato, *Arabidopsis*, soybean, rice, and several *Brassica* species (reviewed in Burr et al. 1988, Tanksley et al. 1987). Several research groups are focusing on the identification of markers very closely linked to specific resistance loci in a number of crops (S. Tanksley, Cornell University, S. Figdore, Native Plants, Inc., Salt Lake City, Utah and M. Saghai-Maroof, Virginia Polytechnic Institute and State University, personal communications; Michaelmore et al. 1987). RFLP markers within 1 to 2 centiMorgans (several million base pairs) flanking a resistance locus could be used to isolate large genomic fragments for construction of subgenomic libraries. Recent advances in the electrophoretic resolution of large DNA fragments (up to 5 million bp) using pulse-field and related systems should facilitate these approaches. Future technical advances in DNA resolution, isolation, and analysis should be a windfall to the plant sciences from the human genome project. Use of subgenomic libraries based on flanking RFLP markers could potentially reduce the number of transgenic plants needed for screening to several hundred, an experimentally feasible number. Alternatively, RFLP markers, closely linked to a gene of interest, could be used as the starting point for "chromosome walking" or "chromosome jumping" approaches (Michaelmore et al. 1987). Final determination of the identity of an R-gene, however, is not a trivial exercise since the only available assay requires transformation of a susceptible cultivar and screening of transgenic plants for a transformation-dependent resistance phenotype.

Technologies that would allow screening of disease phenotypes in transgenic plant cells or tissues grown in tissue cultures would significantly reduce the time and space required for genomic or subgenomic library shotgun experiments and would broaden available approaches in those plant varieties for which efficient regeneration of transgenic whole plants is a problem. The identification of "R-gene-responsive" promoters, which could be fused with selectable markers, could provide a system for direct selection of R-gene-containing clones in culture. However, specific resistance phenotypes of a host are often lost in cultured cells (Darvill and Albersheim 1984, Templeton and Lamb 1988), suggesting that differentiated tissue may be required for activation of a true resistance response. Approaches involving the rapid in vitro production of differentiated target tissue may provide a feasible middle-ground approach. The authors have exploited thin-cell-layer (TCL) explants (Tran Thanh Van 1973, 1981) for the in vitro production of tomato roots for analysis of tomato–root-knot-nematode (*Meloidogyne incognita*) interactions. TCL-derived roots from susceptible (mi^-/mi^-) and resistant (Mi^+) tomato lines show respective galling or hypersensitive resistance responses after infection with *M. incognita* (D. Radin and J. Eisenback, unpublished results). TCL explants from *Brassica napus* have been successfully transformed using *Agrobacterium*-mediated approaches (Charest et al. 1988). Thus a strategy involving direct transformation of TCL explants, induction of rooting on selective medium (e.g., kanamycin), and infection of transformed roots may provide a relatively rapid and effective screen for clone identification. Current technology in tobacco and tomato allows specific induction of roots, shoots, or flowering, and subsequent propagation or regeneration of selected tissues (Tran Thanh Van 1981, Tran Thanh Van et al. 1985, Compton and Veilleux, Virginia Polytechnic Institute and State University, personal communication).

Although significant gaps still exist in our understanding of microbe recognition, signal transduction, and activation of host defense-response genes, substantial progress has been made during the last five years. With the growing research activity in this field, it is probable that several resistance genes will be cloned and characterized within the next few years, and the major components of signal transduction and defense gene regulation will be identified and utilized in genetic engineering strategies.

Plant Gene-Transfer Biotechnologies

The introduction and expression of foreign DNA into stably transformed plants is now routine for several plant species (e.g., tobacco, tomato, petunia). Table 2 lists more than two dozen plant species for

TABLE 2   Transgenic Plant Species

Alfalfa	Cucumber	Peas	Sunflower
Arabidopsis	Flax	Petunia	Tobacco
Asparagus	French bean	Poplar	Tomato
Cabbage	Horseradish	Potato	Walnut
Carrot	Lettuce	Rice[a]	White clover
Celery	Lotus	Rye[a]	
Corn[a]	Oilseed rape	Soybean	
Cotton	Pear	Sugar beet	

[a]Transformation based on direct gene transfer and not on *Agrobacterium*-mediated processes.

which transgenic plants have been reported. These include a number of important crops [soybean (Hinchee et al. 1988, McCabe et al. 1988), rice (Toriyama et al. 1988), cotton (Umbeck et al. 1987), potato (Eckes et al. 1986), oilseed rape (Guerche et al. 1987), rye (de la Pena et al. 1987)] and several woody perennials [walnut (McGranahan et al. 1988) and poplar (Fillatti et al. 1987)]. Successful transformation of the chloroplast genome has also been reported (Blowers et al. 1989, de Block et al. 1985). Production of transgenic plants depends on several critical components:

1. Introduction and stable integration of DNA into plant chromosomes

2. Regeneration of a reproductively competent transgenic plant

3. Appropriate expression of introduced transgenes in the transformed plant and its progeny

Reviews covering various aspects of this process have recently been published (Fraley et al. 1985, Goodman et al. 1987, Klee et al. 1987, Schell 1987, Weising et al. 1988). The reader is referred to these for comprehensive discussions of the transfer and expression of foreign DNA in plants and its application to both basic and applied problems. Discussion here will be limited to recent developments; methods, vectors, and gene markers currently in greatest use; and limitations of current technologies.

Agrobacterium-mediated gene transfer

Agrobacterium–host-cell interactions.   A major factor in the recent successes in plant transformation is the exploitation of a natural plant genetic engineering system discovered in several related *Agrobacterium* plant pathogens, especially *Agrobacterium tumefaciens* and *Agrobacterium rhizogenes*. The pathogenicity of these organisms depends on the insertion of a specific region, the transfer DNA

(T-DNA) of the *Agrobacterium* plasmid into the genome of infected host cells. The inserted DNA then directs the synthesis of plant growth regulators producing tumorous growth (crown gall) and opines, which serve as a bacterial nutrient source. Significant progress has been made in understanding the molecular events occurring within the bacteria during infection (reviewed in Zambryski 1988, Zambryski et al. 1989). However, mechanisms involved in T-DNA transport within plant cells and integration into host chromosomes are presently unknown.

Three regions of the *Agrobacterium tumefaciens* genome are critical for gene transfer into plants: (1) chromosomal DNA loci associated with production of bacterial surface components; (2) the *vir*, or virulence, region of the tumor-inducing plasmid (Ti-plasmid), involved in recognition of susceptible plant cells, excision of T-DNA from the Ti-plasmid, and transfer of T-DNA to the host cell; and (3) the T-DNA, also contained on the Ti-plasmid (reviewed in Weising et al. 1988, Zambryski 1988, Zambryski et al. 1989). *Agrobacterium* is chemotactic toward wounded host cells, and host wound metabolites, especially phenylpropanoid-derived compounds (e.g., acetosyringone), are involved in the induction of the *vir* region (Ashby et al. 1987, Stachel et al. 1985). Products of *virA* and *virG* are involved in recognition of acetosyringone and transcriptional activation of the *vir* loci *B, C, D*, and *E* (Stachel and Zambryski 1986*a,b*), which direct transfer and may regulate integration of T-DNA in plant cells (Weising et al. 1988). Thus, unlike transposable elements, the functions involved in excision and transfer of the T-DNA are not encoded within the DNA that is integrated in the plant chromosome. The only requirements (in *cis*) of the T-DNA itself for efficient transfer and integration are the border regions consisting of highly conserved 25-bp direct repeats (Horsch and Klee 1986, Peralta and Ream 1985, Wang et al. 1984). This 25-bp border region is the target sequence for the endonuclease [*virD* product (Stachel et al. 1987)] involved in excision of T-strand from the Ti-plasmid (Zambryski et al. 1989). Although excised T-DNAs have been identified within *Agrobacterium* as single-stranded DNA, double-stranded DNA, and circular DNA, the strongest evidence suggests that excision involves release of a single-stranded T-strand that is bound by a single-stranded binding protein (product of the *virE* locus; Citovsky et al. 1989). Sequence alterations within the 25-bp border region greatly influence the efficiency of excision and subsequent transfer to plant cells (Wang et al. 1987). The form of T-DNA that is transferred to the plant cell and the site of complementary strand synthesis has not yet been determined. However, the process appears to be quite analogous to bacterial conjugation (Zambryski et al. 1989), suggesting that only the T-strand is transferred. The right border is critical for T-DNA excision and transfer, and also appears to be more frequently conserved in T-DNA integrated in the plant genome, suggesting directionality in both T-strand excision and integration. Elucidation of bacterial components essential for T-DNA delivery within the plant cell and integration within the host genome is of prime importance and may provide valuable tools for plant genetic engineering by other delivery systems as well.

The evolution of *Agrobacterium*–host-cell interactions is remarkable both with respect to the efficient delivery of DNA through an apparent conjugational process and the expression of bacterial genes in a plant host. Genes encoded within the T-DNA have all the appropriate signals for efficient transcription and translation in their eukaryotic host. These genes encode enzymes involved in synthesis of plant growth regulators (indole acetic acid, cytokinin) and opines (especially nopaline or octopine) which serve as nitrogen and carbon sources for the bacterium. For example, the nopaline-type T-DNA is about 23 kb and contains 13 genes (Willmitzer et al. 1983, Holsters et al. 1980). The discovery that these genes are not essential for transfer or integration was critical for the current success in applying *Agrobacterium*-mediated transformation to plant genetic engineering. In "disarmed" Ti-plasmid-derived vectors currently in general use for transgenic plant production, these genes have been removed, thus eliminating the oncogenetic phenotype. Foreign DNA, inserted within intact right- and left-border repeats of an appropriate Ti-plasmid, is efficiently transferred and inserted into the host genome (Fraley et al. 1986, Klee et al. 1987, Weising et al. 1988). Intact nopaline T-DNAs exceed 20 kb in length and are effectively integrated into host chromosomes. Thus, *Agrobacterium*-mediated transformation systems have the potential of delivering large, multigenic segments of DNA into plants, a distinct advantage over alternative delivery systems based on plant viruses, for example.

Once transferred to the plant, one to several copies of the T-DNA are covalently inserted at single or multiple sites in host chromosomes (Ursic et al. 1983, Deroles and Gardner 1988). Rearrangements or truncation of both T-DNA and plant target sequences occur at some frequency (Deroles and Gardner 1988, Gheysen et al. 1987, Jones et al. 1987, Jorgensen et al. 1987). However, the foreign DNA, once integrated, assumes characteristics of typical eukaryotic chromatin (Coates et al. 1987) and is stably inherited (Deroles and Gardner 1988, Muller et al. 1987).

Vectors for transgene delivery, selection, and screening. The structure of both the T-DNA and Ti-plasmids have been extensively modified to enhance their usefulness as plant transformation tools (e.g., Rogers et al. 1987). As mentioned above, the only requirement in *cis* for T-DNA delivery is the 25-bp border repeats, and only the right border is absolutely required (Wang et al. 1984). In addition, in octopine-type (but not nopaline-type) Ti-plasmids, a second 24-bp transmission enhancer region termed "overdrive" is also required for efficient excision and transfer (Peralta et al. 1986). Thus, the T-DNA region has been engineered to eliminate the tumor-inducing genes and to contain a variety of antibiotic markers selectable in plant cells, multiple cloning sites for efficient insertion of foreign DNA, and specific 5', coding, or 3'-flanking sequences allowing expression and assay of inserted promoters or coding sequences in transformed plant cells. Ti-plasmids have been engineered to permit efficient plasmid replication and selection in both *Agrobacterium* and *E. coli* hosts, greatly facilitating gene constructions and genetic manipulations prior to plant transformation steps. In general, two plasmid strategies are used:

1. *Cointegrating vectors:* Genes to be transferred are integrated within the T-DNA of a resident Ti-plasmid.

2. *Binary vectors:* The engineered T-DNA region is on an autonomously replicating plasmid distinct from that carrying the *vir* genes necessary for transformation (reviewed in Weising et al. 1988, Draper et al. 1988).

Cointegrating vectors.   Cointegrate vectors rely on a region of homology between the Ti-plasmid (within the T-DNA border repeats) and the vector plasmid allowing for genetic manipulation of foreign DNA in *E. coli* vectors followed by insertion into Ti-plasmid T-DNA. The *E. coli* plasmid vector contains a bacterial selection marker in addition to the foreign DNA to be inserted, but no replication origin appropriate for propagation in *Agrobacterium*. Following forced matings between the *E. coli* and *Agrobacterium* hosts, agrobacteria selected for antibiotic resistance also contain the foreign gene cointegrated within the T-DNA borders due to a single recombination event in the region of homology.

Two cointegrating vector systems are commonly used for plant transformation. One utilizes the disarmed *Agrobacterium* Ti-plasmid pGV3850, in which the genes encoding plant growth regulators have been replaced with pBR322 sequences. This allows cointegration of any plasmid containing pBR322 sequences within the T-DNA borders (Zambryski et al. 1983). The second approach, the "split-end-vector" (SEV) system, initially carries the T-DNA border repeats on separate plasmids (Fraley et al. 1985). In this system, the left T-DNA border and part of the original T-DNA, termed the "limited internal homology" (LIH) region, remain intact on the avirulent plasmid pTiB6S3SE.

The vector to be introduced into *Agrobacterium*, pMON200, contains the right T-DNA border, the LIH region, the *nos* (nopaline synthase) gene, a multiple cloning site for insertion of foreign DNA, and selectable markers for both bacteria and plants. pMON200 does not contain a replication origin for *Agrobacterium*, and thus selection for the bacterial resistance marker will identify cointegrates that have recombined in the LIH region creating a functional T-DNA. Once the cointegrate plasmid is formed, it is stable in *Agrobacterium* even in the absence of selection. The major disadvantage of cointegrate vectors is the relatively low frequency of cointegrate formation due to low frequency of homologous recombination. For the pMON200 system, this frequency ranges between 10^{-5} and 10^{-7} (Fraley et al. 1985).

Binary vectors. The fact that the Ti-plasmid *vir* region could function to effectively transfer T-DNA when placed on a separate plasmid (Hoekema et al. 1983) or on the bacterial chromosome (Hoekema et al. 1984) led to the development of binary vector systems. Gene transfer in these systems involves two plasmids, both able to replicate in *Agrobacterium*. One plasmid is a modified (or intact) Ti-plasmid, which provides the *vir* functions in *trans*. The second plasmid is a broad-host-range plasmid that can replicate both in *E. coli* and *Agrobacterium* (based in most cases on the RK-2 origin of replication) and contains T-DNA borders flanking the gene construct for insertion and a plant-selectable marker. These systems have now been optimized for a variety of functions allowing plant gene constructs to be maintained, manipulated, replicated, and shuttled back and forth between *E. coli* and *A. tumefaciens* and efficiently transferred into a variety of plant species (An et al. 1985, 1986, Bevan 1984, De Framond et al. 1983, Klee et al. 1985, Matzke and Matzke 1986, Simoens et al. 1986, Simpson et al. 1986, Velten and Schell 1985). Binary vectors have several advantages over cointegrate vectors including (1) the frequency of introduction into *Agrobacterium* (10^{-1}) (Fraley et al. 1985, Klee et al. 1987) and (2) the lack of dependence on specifically engineered Ti-plasmids. In fact, binary vectors work efficiently in a variety of *A. tumefaciens* and *A. rhizogenes* strains, effectively increasing the host range for efficient transformation (Simpson et al. 1986).

Several modifications have been incorporated into binary vectors to facilitate specific tasks of interest in transgenic plants. For example, the bacteriophage lambda *cos* sequences have been incorporated into binary vectors, creating a cosmid system capable of high-frequency cloning of plant genomic sequences up to 40 kb (Prosen and Simpson 1987, Simoens et al. 1986) for shotgun cloning approaches in plants. Specific plant gene cassettes have been constructed that allow easy generation of plant expression vectors for specific transcriptional or translational fusion products (Topfer et al. 1987), selection of

endogenous plant promoters (based on transfer of promoterless selectable marker genes; Teeri et al. 1986), manipulation of promoters inserted in vitro (An 1986, Velten and Schell 1985), or random *Tn5*-
insertion mutagenesis of plant sequences in *E. coli* prior to transfer
and expression in transgenic plants (Koncz et al. 1987*a*).

Markers for selection and screening in transgenic plants. Transformed plant cells were initially identified by tumorous gall formation
or by hormone-independent growth in culture. Disarming of the T-
DNA required the development of alternative selectable markers that
would be expressed in plants for identification of transformed cells or
tissues. Several selectable markers have proved successful in
transgenic plants and are reviewed elsewhere (Fraley et al. 1986,
Weising et al. 1988). The *Tn5*-derived *nptII* gene conferring resistance
to kanamycin continues to be the most widely used and broadly effective antibiotic marker. Kanamycin is inhibitory to plant cell growth,
and *nptII* expression provides a high level of resistance to transformed
cells, plants, and seeds. In addition, sensitive assays for neomycin
phosphotransferase activity are available for crude cell extracts (Reiss
et al. 1984). Kanamycin resistance is not equally effective as a selectable marker in all plant systems, and several alternatives are currently receiving attention. The hygromycin phosphotransferase (*hpt*)
gene from *E. coli* has been effective as a selectable marker in
Arabidopsis thaliana (Lloyd et al. 1986) and other species (Waldron et
al. 1985, Van den Elzen et al. 1985). Gentamicin acetotransferase
genes confer resistance to gentamicin (Hayford et al. 1988), and a mutated mouse dihydrofolate reductase gene confers resistance to
methotrexate (Eichholtz et al. 1987) in transformed plants. Thus, antibiotic resistance genes, inserted within the T-DNA border repeats,
have proved effective in selection of transformed plant cells.

Another group of scorable genes has also been developed to facilitate assaying promoter expression in transgenic plants. The products
of these "reporter genes" can be assayed by very sensitive methods in
crude extracts or even in situ and have no or very low endogenous activities in nontransformed plants. Currently, the most commonly used
reporter genes include chloramphenicol acetyltransferase (*cat*) gene
(e.g., De Block et al. 1984, Dron et al. 1988), the bacterial or firefly
luciferase genes (Koncz et al. 1987*b*, Ow et al. 1986), and the bacterial
β-glucuronidase (*gus*) gene (Jefferson 1987, Jefferson et al. 1986).
Binary-vector expression cassettes allowing easy insertion of cloned
promoter sequences directly upstream of the *cat* or *gus* coding sequences with appropriate polyadenylation sequences for expression in
plants are now widely available and can even be purchased commercially (e.g., Clontech Laboratories, Inc., Palo Alto, California). These
reporter genes have been instrumental in analyses of plant *cis*-

regulatory sequences in transgenic plants or plant cells, in determining inheritance patterns of foreign DNA inserts, and in demonstrating promoter-directed tissue- or cell-specific expression (reviewed in Schell 1987, Kuhlemeier et al. 1987).

Procedures for *Agrobacterium*-mediated transformation.   Early transformation studies with *Agrobacterium* were based initially on gall formation on wounded plant stems or leaves and subsequently on plant protoplasts. Infection of intact plants creates only local transformation and is not easily monitored with disarmed vectors. *Agrobacterium* cocultivation with plant protoplasts is still used in some plants where regeneration from protoplasts is highly successful. The most common approaches for transgenic plant production now involve transformation of tissue explants. Many dicotyledonous plants can be regenerated from explants, and the time from inoculation to mature transgenic plants is significantly faster with explants than by protoplast transformation. The optimum tissue type and age for transformation and regeneration vary with plant species and even cultivars within a species but the technique has been successful with leaf sections (Horsch et al. 1985), stems (Lloyd et al. 1986), cotyledons (McCormick et al. 1986), or thin cell layers from floral stems (Charest et al. 1988, Trinh et al. 1987). In general, the "leaf-disc transformation" (Horsch et al. 1985) involves the following steps:

1. Leaf or cotyledon sections or discs are excised from aseptically grown seedlings.
2. Explants are cultivated in a suspension of *Agrobacterium* carrying appropriately engineered plasmids.
3. Inoculated explants are placed on a nurse culture plate (optional, generally tobacco cell cultures) for several days.
4. Explants are transferred to selective medium that favors shoot formation on plates containing carbenicillin to select against *Agrobacterium* and kanamycin to select against nontransformed plant cells.
5. Shoots are transferred to "rooting" medium containing kanamycin.
6. Plantlets are transferred to soil.

This approach has been successful with many plant species, requires minimal tissue culture expertise, and is more rapid (plantlets in 6 to 8 weeks with some species) than protoplast cocultivation techniques. An alternative that has been successful with several plants that have not been regenerated from protoplasts or leaf discs involves transformation of somatic embryos derived from embryogenic culture lines. For example, in walnuts (*Juglans regia* L.), proliferating somatic embryos were cocultivated with disarmed *Agrobacterium* strains, induced to produce secondary embryos (on medium containing kanamycin), and these embryos were germinated to produce transgenic plants (McGranahan et al. 1988). Thus, a key focus of current research involves optimization of *Agrobacterium* strains, vectors, selectable markers, and regeneration procedures for new crop species and agronomically important cultivars.

The major limitations of *Agrobacterium*-mediated transformation continue to be host range and the requirement for whole-plant regeneration. Host range is being expanded by utilizing different *Agrobacterium* strains. However, transformation of most monocots and especially important cereal crops has not been successful using *Agrobacterium*. A number of alternatives involving direct DNA delivery may be more promising for these crops.

Direct DNA-transfer technologies

Limitations of *Agrobacterium* host range and the requirement for plant regeneration have fueled the search for alternative methods to *Agrobacterium*-mediated transformation to stably transfer genes into plants. Most monocots—including cereals, the world's most important group of crop plants—cannot be infected by *Agrobacterium*. Current research efforts to develop direct gene-transfer technologies are directed primarily toward transforming these important crop species. Maize, rice, wheat, and sugarcane have now been transiently (Fromm et al. 1985, Junker et al. 1987, Ou-Lee et al. 1986, Klein et al. 1987) or stably (Chen et al. 1987, Fromm et al. 1986, Lorz et al. 1985, Paszkowski et al. 1984, Potrykus et al. 1985, Uchimiya et al. 1986) transformed by direct gene-transfer technologies (reviewed in Cocking and Davey 1987). In addition, a variety of dicots have been successfully transformed either stably or transiently by direct gene transfer.

Direct gene-transfer techniques include the introduction of foreign DNA into plant protoplasts mediated by (1) polyethylene glycol (PEG) and poly-L-ornithine (Draper et al. 1982), (2) calcium phosphate coprecipitation (Krens et al. 1982), (3) electrical current (electroporation, Fromm et al. 1985, 1986, Prols et al. 1988), (4) liposome fusion (Deshayes et al. 1985), and (5) particle bombardment (Klein et al. 1986). Transformation efficiencies have been greatly improved due to the development of plasmid vectors containing selectable marker genes, optimization of media and conditions for protoplast isolation and transformation, and synchronization of cultures for transformation during mitosis (Balazs et al. 1985, Hain et al. 1985, Meyer et al. 1985, Negrutiu et al. 1987, Okada et al. 1986, Paszkowski and Saul

1986). While these techniques have been successful in producing transgenic plants in a few species, they are limited to those plant species in which efficient regeneration of plants from protoplasts is possible. Regeneration from protoplasts generally is a time-consuming and labor-intensive process that can result in the generation of somaclonal varients. It is also not applicable to all species or all lines within a species. The use of embryogenic cultures has increased the efficiency of regeneration in several crops and is a significant factor in initial reports of transgenic rice (Abdullah et al. 1986) and maize plants (Rhodes et al. 1988). Unfortunately, the first transgenic maize plants were infertile (Rhodes et al. 1988). It is too early to know if this will be a widespread problem in transgenic plants produced via these techniques. Electroporation has been widely used for studying transient expression of introduced DNA and RNA and has resulted in some stable integration (Fromm et al. 1985, Fromm et al. 1986). Electroporation triggers an increase in DNA synthesis (Rech et al. 1988) and enhances cell division and plant regeneration in some species (Rech et al. 1987, Ochatt et al. 1988), and these properties may further applications of this technique in production of intact transgenic plants. Thus, while progress is being made in increasing regeneration efficiencies and in developing protoplast regeneration systems in other crops (e.g., Luhrs and Lorz 1987), these factors continue to limit widespread usefulness of gene-transfer methods in protoplasts. In addition, somaclonal variation (i.e., variation generated during in vitro culturing phases) may present significant problems in producing transgenic plants in some species (e.g., Lee and Phillips 1987).

Microinjection of DNA directly into immobilized plant protoplasts has been successful in several systems including alfalfa protoplasts (Reich et al. 1986) and tobacco protoplasts (Crossway et al. 1986, Lawrence and Davies 1985). Microinjection has the advantage of relatively high transformation frequencies (6 to 26%) and the potential to deliver molecules to a nucleus or organelle of choice. However, this technology is also limited by plant regeneration techniques from protoplasts.

Several alternative methods for gene transfer directly into intact plant cells or tissues have been successful in generating transiently transformed tissues and, recently, stable transformants. For example, Topfer et al. (1989) report the uptake and expression of vector DNA carrying kanamycin resistance by imbibing seeds in a DNA solution. Uptake occurred only during the early imbibition of dry seeds and was facilitated by the presence of dimethyl sulfoxide (DMSO) in the DNA solution. The technique was successful with wheat, oat, rye, triticale, and several legumes, although the levels of expression varied (Topfer et al. 1989). de la Pena et al. (1987) produced transgenic rye plants by injecting the floral tillers (prior to meiosis) with plasmid DNA carrying kanamycin resistance (*nptII*). Some of the progeny expressed kanamycin resistance. The process was not very efficient, with 2 transformants isolated from 3,000 seedlings derived from 100 injected and cross-pollinated plants (de la Pena et al. 1987). However, the procedure is relatively simple and may have application to a wide variety of plants. Transformation of rice has also been reported via the "pollen-tube pathway" by placing a drop of DNA solution (containing *nptII*) onto severed styles of rice florets and testing the resulting seeds for the presence and expression of *nptII* (Luo and Wu 1988).

The use of "microprojectiles," or "particle bombardment," is another technique that has received substantial attention. This procedure involves coating small gold or tungsten beads with plasmid DNA and propelling the beads into intact cells, embryos, or differentiated tissues using high-velocity particle "guns" or electrical discharges (Klein et al. 1987). Microprojectiles have successfully delivered DNA into the nucleus and mitochondria of yeast, nuclei and chloroplasts of *Chlamydomonas*, epidermal tissues of *Allium cepa*, and suspension cultures of maize (Blowers et al. 1989, reviewed in Klein et al. 1988). Klein et al. (1988) have established optimal conditions for efficient transient expression of microprojectile-mediated transformants of maize. In a recent meeting (American Association for Advancement of Science Annual Meeting, January 14–19, 1989, San Francisco), Winston Brill of Agricetus reported on the stable (i.e., integrated into the plant chromosome) transformation of soybean using microprojectiles delivered by an electric-discharge particle accelerator. Immature meristems growing from seeds were bombarded with constructs carrying a GUS reporter gene, and chimeric plants were regenerated. Some of these plants produced transformed seed, and the time from transformation to field testing was substantially reduced using the particle-accelerated delivery compared to *Agrobacterium*-mediated transformation (McCabe et al. 1988). Brill noted that conditions that favor transient expression do not favor stable integration. A major limitation on the widespread use of microprojectile approaches stems from the limited number of particle guns available and the cost of using them or building them. It is encouraging to see the development of new, lower-cost technologies such as the electrical-discharge accelerator that may bring these methods into more labs. Recent successes in producing transgenic plants with this procedure suggest that this may be the best approach for producing transgenic plants in many species recalcitrant to efficient plant regeneration approaches.

One problem with foreign DNA delivered through direct gene transfer is the high frequency of genetic rearrangement. Truncation, concatemerization, and complex rearrangements of vector DNA may occur prior to or during integration at frequencies significantly higher than those seen with *Agrobacterium*-mediated transfer (reviewed in Weising et al. 1988). The integration pattern is simplified by the use of supercoiled DNA, but transformation frequencies are reduced (Meyer et al. 1988). Meyer et al. (1988) recently identified a petunia DNA fragment that leads to a 20-fold increase in transformation frequencies using supercoiled plasmids and simple integration patterns. The mechanism for enhancing transformation conferred by this 2-kb sequence, named "TBS" for "transformation booster sequence," is currently unknown.

Once integrated, foreign DNA introduced by all of these direct-transfer approaches appears to be stable and inherited in typical Mendelian fashion. In summary, these technologies serve as a valuable alternative to *Agrobacterium*-mediated transfer because they provide rapid transient-expression assays and a major hope for reliable transformation of agronomically important monocots.

Advances in transgene construction and expression

Regulation of transgene expression. As plant transformation becomes more sophisticated, our expectations for precisely engineered transgenic plants grow. One is no longer limited to the random incorporation and expression of specific genes. Biotechnologists now have the potential to direct the expression of transgenes in particular tissues, at specific times in development, or in response to particular environmental signals or stresses. The recent advances in delineating *cis*-acting control regions of plant promoters are due in large part to the application of plant transformation, both in transgenic plants, and where applicable, in transient assays of DNA delivered into plant protoplasts by electroporation. Until recently, these analyses have been based on determining the effects of relatively large deletions on expression (generally using reporter genes) with respect to level, inducibility, or tissue specificity. Recent advances in DNA amplification using the polymerase chain reaction (PCR) technology (Saiki et al. 1988) should accelerate these analyses, providing mechanisms for rapidly constructing very specific alterations in promoter regions to delineate key sequences involved in regulation. The in vivo approaches have been complemented by several in vitro techniques including DNase-sensitivity, DNA footprinting, and gel-retardation assays to identify specific 5'-flanking regions involved in DNA-protein interactions. These studies should lead into the next level of analysis: the identification and cloning of DNA-binding proteins involved in specific gene regulation. Regulatory regions from monocots or dicots are recognized and expressed in both systems, although with differing efficiencies (Keith and Chua 1986), suggesting substantial conservation of the signal transduction and DNA-binding components in plants.

The recent progress in understanding mechanisms involved in regulation of plant gene expression has been substantial. A recent review by Kuhlemeier et al. (1987) describes components involved in gene regulation by light, heat shock, hormones, wounding and biotic stress, and phytohormones as well as expression in specific cell types or at particular stages in development. Sequences involved in correctly targeting gene products have been identified and have resulted recently in the gene products of transgenic constructs being correctly inserted into the proper chloroplast compartments (de Boer et al. 1988, Kuntz et al. 1986). Weising et al. (1988) contains an extensive list of genes from bacterial, plant, and animal sources that have been transferred and expressed in plant cells or transgenic plants. Several of these systems are of particular interest to biotechnologists interested in manipulating plant-microbe interactions.

The majority of plant-microbe interactions discussed in other chapters of this book involve interactions occurring in the rhizosphere. Thus, it is anticipated that genes engineered to enhance these interactions need to be directed to the root and preferably the root surface. Recent analyses of the three phenylalanine ammonia lyase (PAL) genes in bean show high levels of expression of all three genes in roots (Liang et al. 1989). This expression is probably associated with several functions within the roots such as lignification of the vascular system and suberization. The suberization function is likely to occur at the root surface, and PAL genes involved in this response may provide valuable promoters for specific gene expression in these cells. Transgenic tobacco plants containing both PAL2 and PAL3 promoters fused to a GUS reporter gene show GUS activity in the roots, but the cell-specific location of this activity has not yet been determined (Liang et al. 1989, Liang and Lamb, Salk Institute, personal communication).

To affect plant responses during nodulation, expression based on nodulin promoters seems an obvious choice. However, analysis of nodulin promoters has been relatively slow due to the limited ability to efficiently transform legume species. Due to the complex nature of this interaction, experimental manipulation of expression of nodulins will require transformed whole plants. Recent progress in producing transgenic soybean plants (Hinchee et al. 1988, McCabe et al. 1988) and transforming the wild legume *Lotus cornatus* (Jensen and Marcker 1986) should provide systems for detailed analyses of nodulin gene promoters. A leghemoglobin-CAT construct has been transformed into *Lotus* via *Agrobacterium rhizogenes*, and *Rhizobium loti*–induced nodules were shown to express CAT activity in the correct pattern for leghemoglobin (Jensen and Marcker 1986). These breakthroughs in transgene analysis should greatly facilitate analysis of the *cis*-regulatory regions of the nodulin genes and provide strategies for engineering enhanced interactions.

Some success in engineering plants for increased resistance to pathogenic agents and pests has been reported but is currently not based on tissue-specific expression. Insect tolerance has been increased by producing transgenic tobacco or tomato plants expressing either a toxin gene from *Bacillus thuringiensis* (see Chap. 4) or proteinase inhibitor gene from cowpea (Fischoff et al. 1987, Hilder et al. 1987, Vaeck et al. 1987). Transgenic tobacco or tomato plants with enhanced resistance to several RNA viruses have been engineered based on insertion of the viral coat protein gene into the plant genome with appropriate regulatory sequences for expression in plants. This strategy is effective with a number of different viruses including tobacco mosaic virus (Powell Abel et al. 1986), cucumber mosaic virus (Cuozzo et al. 1988, Rezainan et al. 1988), alfalfa mosaic virus (Tumer et al. 1987, Loesch-Fries et al. 1987), tobacco streak virus (van Dun et al. 1987), and potato virus X (Hemenway et al. 1988). The mechanism of resistance afforded by the expressed viral coat protein is unknown but is dependent on the presence of the protein and not the RNA and thus is thought to involve either interference or prevention of "challenge virus" uncoating and disassembly or competition for viral binding sites on the inner plasmalemma (Cuozzo et al. 1988, Hemenway et al. 1988). Two other strategies have shown partial success in providing viral resistance; transformation and expression of antisense viral RNA (Hemenway et al. 1988; discussed further below) and of viral satellite RNA (Harrison et al. 1987, Gerlach et al. 1987). Both approaches are thought to provide protection either by binding the viral replication origin or by competition for limited replicase (viral RNA polymerase).

Resistance genes and promoters of defense-related genes should provide valuable tools for engineering host resistance to pathogens. As discussed in previous sections, expression of inducible host defense responses is precisely regulated in plants. Because at least part of the defense involves localized plant cell death, indiscriminate triggering of defense responses in engineered plants would likely be counterproductive. Thus, utilization of promoters activated in response to wounding or pathogen recognition is critical for enhancing resistance through genetic engineering. Many defense-related genes characterized to date are regulated in this manner. Recent experiments identifying elicitor- and wound-responsive *cis*-acting regulatory elements (Dron et al. 1988, Lawton et al. 1988, Logemann et al. 1989, Ryan and

An 1988, Sanchez-Sarrano et al. 1987, Thornburg et al. 1987) and regulatory consensus sequences in similarly regulated genes (Cramer et al. 1989, Liang et al. 1989) should define required regulatory elements. As a word of caution, however, some wound-inducible genes have regulatory sequences both 5' and 3' of the coding region that are required for efficient gene expression (An et al. 1989, Thornburg et al. 1987). Recent analysis suggests that the mechanism of increased expression conferred by 3' terminator sequences of potato proteinase inhibitor II involves mRNA stabilization (An et al. 1989). These terminator sequences also increase expression of genes driven by the cauliflower mosaic virus 35S promoter (An et al. 1989), suggesting potential applications in transgene constructs where high levels of RNA expression are required.

Several promoter and terminator sequences derived from viral genomes or *Agrobacterium* T-DNA have been extremely useful for efficient transgene expression in plants. The cauliflower mosaic virus (CaMV) 35S is broadly used as a promoter, resulting in high constitutive levels of expression in most plant tissues (Fang et al. 1989, J. Jones et al. 1985, Lawton et al. 1987). This promoter has been extensively analyzed (Fang et al. 1989, Odell et al. 1985, Ow et al. 1987) and contains three regions critical for its transcriptional activity: a proximal region containing the TATAA box, a region about 80 bp from the transcription start containing a CAAT-like box, and a distal region with sequences sharing homology to the SV40 core enhancer element. Duplication of the entire 35S promoter or of the distal enhancer increases levels of expression compared with the wild-type 35S (Kay et al. 1987, Ow et al. 1987). In addition, fusion of the 35S distal enhancer region with other promoters often confers increased expression but under the tissue-specific or inducer-specific expression pattern of the initial promoter. T-DNA-derived *nos* and *ocs* promoters have also been used (An 1986, Koncz et al. 1983, Shaw et al. 1984) but not in as strong promoters as the 35S promoter (Sanders et al. 1987). Other components directly influencing the efficiency of transgene expression in plants include copy number, DNA methylation within the plant, various *trans*-acting factors, and position effects due to location of chromosomal integration.

Antisense and ribozyme strategies.   The potential for control of gene expression in vivo through expression of antisense RNA is currently receiving much attention. A recent meeting, entitled "Discoveries in Antisense Nucleic Acids and Genetic Engineering for Commercial Applications in Agriculture and Human Therapy," organized by International Business Communications Inc. was held in New York City (January 24–25, 1989) to focus on new developments in this area. A recent review by van der Krol et al. (1988*b*) described the applications of antisense technologies to both prokaryotic and eukaryotic systems and delineates the current understanding of mechanisms of RNA antisense modulation of gene expression. The basic strategy in antisense applications is to block expression of a particular gene product by expression of a transgenic construct containing the gene or part of the gene with the transcript in reverse orientation with respect to the promoter. This results in a complementary RNA rather than the normal-sense RNA. The antisense RNA binds with its homologous sense RNA, preventing translation and/or facilitating degradation. Using this approach, one can reduce gene product accumulation by 90 to 99%, thus creating a phenotypic mutant. This technology is of particular importance in plants in which mutational analyses are hampered by high ploidy levels or presence of multigene families. Researchers have been successful in blocking the expression of several plant genes in transgenic plants expressing antisense genes (Cuozzo et al. 1988, Ecker and Davis 1986, Hemenway et al. 1988, Rodermel et al. 1988, van der Krol et al. 1988*a*). Several researchers have inserted an antisense construct of polygalacturonase (PG) into tomato plants (Sheehy et al. 1988, Smith et al. 1988), a result with potential commercial application. Transgenic plants expressing high levels of antisense RNA have only 10% of wild-type levels of PG activity and concomitantly show a significant delay in fruit softening. This could lengthen shelf life and allow ripening to take place in the field prior to picking.

Strategies involving antisense RNA to plant viral RNA sequences have been shown to enhance viral resistance in transgenic plants (e.g., Hemenway et al. 1988). This resistance, however, was not as effective as the protection derived from transformation with viral coat protein gene and could be overcome by higher inoculum of challenge virus (Cuozzo et al. 1988, Hemenway et al. 1988). This finding underscores the major limitation in antisense approaches. To be effective in reducing target gene expression, high ratios of antisense to sense RNA are required (Ecker and Davis 1986, Izant and Weintraub 1985, van der Krol et al. 1988*b*). To prevent viral expression, antisense RNAs have to be present in sufficient amounts in the cytosol to prevent effective viral RNA translation and replication. However, even this barrier can be overcome by very high infection titers. Most antisense constructs used in plants have utilized the cauliflower mosaic virus (CaMV) 35S promoter to increase antisense expression above that of the endogenous sense gene. However, significant variation is seen between transgenic regenerants, which suggests that position effects may be critical in successful modulation of target gene expression by antisense. Antisense oligonucleotides have been effective in blocking gene expression in animal cell cultures but have not yet been applied to plants (van der Krol et al. 1988b). Antisense genes have been effective in reducing expression of nonhomologous genes (20% mismatch, van der Krol et al. 1988b), suggesting that genes cloned from one plant species can be effectively applied to modulate gene expression in another. Antisense strategies for modifying plant-microbe interactions have not yet been applied to systems other than viral interactions. However, these approaches should be valuable in identifying specific genes, and even particular isogenes of multigene families that may play key roles in disease resistance or enhancing symbiotic interactions.

New developments in understanding the mechanisms involved in autocatalytic RNA cleavage may lead to a new class of antisense genes termed "ribozymes." Haseloff and Gerlach (1988) determined the RNA sequence and structure required for self-cleavage of the satellite tobacco ringspot virus (TRV) RNA. Based on this study, they were able to design and synthesize RNA oligonucleotides capable of efficient cleavage of a target RNA sequence(s), in this case CAT template RNA (Haseloff and Gerlach 1988). The requirements of the catalytic RNA or ribozyme were (1) the catalytic core of TRV satellite and (2) flanking regions complementary to target sequence flanking the conserved GUC target cleavage site. Thus, the CAT antisense regions base-paired with the target CAT mRNA and positioned the catalytic core for cleavage. Theoretically, if the transcribed sequence of a target gene is known, one or more ribozymes against GUC sequences within that transcript could be targeted for cleavage (see Knight 1988). Specific destruction of targeted RNAs has, thus far, been shown only in vitro although the TRV satellite cleavage normally functions in vivo. The conditions and extent of base pairing required for ribozyme activity of synthesized catalytic RNA are now being elucidated. If these experiments are successful, ribozymes may provide a valuable tool for genetic engineering and gene therapy.

Limitations of current technologies

Significant progress has been made in identifying and characterizing genes important in mediating plant-microbe interactions, in understanding mechanisms regulating their temporal and spatial patterns of expression, and in successfully delivering and expressing engineered genes in plants. Transformation and analyses of transgenes is now routine in an ever growing list of plant species, and promoter regions have now been identified that allow expression of transferred DNA in specific organs or organelles and at specific times in development. However, as discussed above, current gene-transfer technologies have a number of limitations. These include (1) limited host range of *Agrobacterium* strains, (2) difficulties in transforming cells and regenerating mature plants, (3) difficulties in direct DNA delivery into tissue that will result in stable germ-line transformation, (4) vector rearrangements and truncation preceding stable integration, and (5) variations in transgene expression based on differing sites of integration in the host chromosome. The first four limitations have been discussed in the preceding sections. However, we have not yet discussed position effects noted in most plant systems for which sufficient numbers of transgenic plants are produced to allow valid comparisons. Researchers have reported wide variations (up to 200-fold) of transgene expression between independently derived transgenic plants transformed with identical vectors (J. Jones et al. 1985, An 1986). These effects may be due to location of the inserted DNA with respect to endogenous promoters, enhancers, or silencer regions or to chromatin structure effects. Position effects may be reduced by using vectors with a selectable marker tightly linked to the gene of interest (Velten and Schell 1985) but not eliminated (Eckes et al. 1986, J. Jones et al. 1985). Methylation may play a role in the expression levels of foreign DNA since increased methylation has been associated with reduced gene activity in plants and T-DNA has been shown to be methylated once it is integrated within host chromosomes (Amasino et al. 1984). Copy number of the inserted DNA has been positively correlated with levels of transgene expression in some systems (Scott and Draper 1987, Stockhaus et al. 1987) but shows no correlation in others (Crossway et al. 1986, Jones et al. 1987, Sanders et al. 1987). Development of a homologous recombination system in plants may provide mechanisms for targeting engineered DNA into specific regions of the plant chromosome, thus avoiding major position effects. However, homologous recombination is not a general mechanism of DNA integration in plants. Paszkowski et al. (1988) recently reported directed integration into a predicted chromosomal location based on complementation of a partial *nptII* chromosomal gene with the missing region carried on vector sequences. Transformants were selected on kanamycin such that only those cells undergoing homologous recombination would be rescued. Although the frequency of successful gene targeting was low, these experiments represent an important step toward more precise modification and correction of plant genes through gene-replacement strategies that have been so powerfully employed in bacterial and yeast systems.

In summary, direct manipulation of the plant genome to specifically impact plant-microbe interactions has now been successful in several plant systems, yielding transgenic plants with enhanced resistance to insects and viruses. The advances in gene-transfer technology and in understanding host responses to microbes clearly indicate that genetic engineering of host plants is a viable approach both for basic studies of plant-microbe interactions and for developing commercial crop varieties with enhanced microbial interactions or disease resistance.

References

Abdullah, R., Cocking, E. C., and Thompson, J. A. 1986. Efficient plant regeneration from rice protoplasts through somatic embryogenesis, *Bio/Technology* 4:1087–1090.

Amasino, R. M., Powell, A. L. T., and Gordon, M. P. 1984. Changes in T-DNA methylation and expression are associated with phenotypic variation and plant regeneration in a crown gall tumor line, *Mol. Gen. Genet.* 197:437–446.

An, G. 1986. Development of plant promoter expression vectors and their use for analysis of differential activity of nopaline synthase promoter in transformed tobacco cells, *Plant Physiol.* 81:86–91.

An, G., Watson, B. D., Stachel, S. E., Gordon, M. P., and Nester, E. W. 1985. New cloning vehicles for transformation of higher plants, *EMBO J.* 4:277–284.

An, G., Watson, B. D., and Chiang, C. C. 1986. Transformation of tobacco, tomato, potato, and *Arabidopsis thaliana* using a binary Ti vector system, *Plant Physiol.* 81:301–305.

An, G., Mitra, A., Choi, H. K., Costa, M. A., An, K., Thornburg, R. W., and Ryan, C. A. 1989. Functional analysis of the 3' control region of the potato wound-inducible proteinase inhibitor II gene, *Plant Cell* 1:115–122.

Appleby, C. A. 1984. Leghemoglobin and *Rhizobium* respiration, *Annu. Rev. Plant Physiol.* 35:443–478.

Ashby, A. M., Watson, M. D., and Shaw, C. H. 1987. A Ti plasmid determined function is responsible for chemotaxis of *A. tumefaciens* towards the plant wound compound acetosyringone, *FEMS Microbiol. Lett.* 41:189–192.

Bailey, J. A. 1987. Phytoalexins: a genetic view of their significance. In *Genetics and Plant Pathogenesis*, P. R. Day and G. J. Jellis (eds.). Blackwell Scientific, Oxford, pp. 233–244.

Baker, B., Coupland, G., Fedoroff, N., Starlinger, P., and Schell, J. 1987. Phenotypic assay for excision of the maize controlling element *Ac* in tobacco, *EMBO J.* 6:1547–1554.

Balazs, E., Bouzoubaa, S., Guilley, H., Jonard, G., Paszkowski, J., and Richards, K. 1985. Chimeric vector construction for higher plant transformation, *Gene* 40:343–348.

Banks, J. A., Masson, P., and Fedoroff, N. 1988. Molecular mechanisms in the developmental regulation of the maize *Suppressor-mutator* transposable element, *Genes Devel.* 2:1364–1380.

Bauer, W. D. 1981. Infection of legumes by rhizobia, *Annu. Rev. Plant Physiol.* 32:407–449.

Baydoun, E. A. H., and Fry, S. C. 1985. The immobility of pectic substances in injured tomato leaves and its bearing on the identity of the wound hormone, *Planta* 165:269–276.

Bell, A. A. 1981. Biochemical mechanisms of disease resistance, *Annu. Rev. Plant Physiol.* 32:21–81.

Bell, J. N., Ryder, T. B., Wingate, V. P. M., Bailey, J. A., and Lamb, C. J. 1986. Differential accumulation of plant defense gene transcripts in a compatible and incompatible plant-pathogen interaction, *Mol. Cell. Biol.* 6:1615–1623.

Bennetzen, J. L., Quin, M. M., Ingels, S., and Ellingboe, A. H. 1988. Allele-specific and Mutator-associated instability at the Rp1 disease-resistance locus of maize, *Nature* 332:369–370.

Bergman, K., Gulash-Hofee, M., Hovestadt, R. E., Larosilliere, R. C., Ronco, P. G., and Su, L. 1988. Physiology of behavioral mutants of *Rhizobium meliloti*: evidence for a dual chemotaxis pathway, *J. Bacteriol.* 170:3249–3254.

Bergmann, H., Preddie, E., and Verma, P. D. S. 1983. Nodulin-35: a subunit of specific uricase (uricase II) induced and localized in the uninfected cells of soybean nodules, *EMBO J.* 2:2333–2339.

Bevan, M. 1984. Binary *Agrobacterium* vectors for plant transformation, *Nucleic Acids Res.* 12:8711–8721.

Blowers, A. D., Bogorad, L., Shark, K. B., and Sanford, J. C. 1989. Studies of *Chlamydomonas* chloroplast transformation: foreign DNA can be stably maintained in the chromosome, *Plant Cell* 1:123–132.

Bohlmann, H., Clausen, S., Behnke, S., Giese, H., Hiller, C., Reimann-Philipp, U., Schrader, G., Barkholt, V., and Apel, K. 1988. Leaf-specific thionins of barley—a novel class of cell wall proteins toxic to plant-pathogenic fungi and possibly involved in the defence mechanism of plants, *EMBO J.* 7:1559–1565.

Boller, T. 1987. Hydrolytic enzymes in plant disease resistance. In *Plant-Microbe Interactions: Molecular and Genetic Perspectives*, T. Kosuge and E. W. Nester (eds.), vol. 2. Macmillan, New York, pp. 385–413.

Bothe, H., deBruijn, F. J., and Newton, W. E. (eds.). 1988. *Nitrogen Fixation: Hundred Years After*. Gustave Fisher, Stuttgart.

Bowler, C., Alliotte, T., De Loose, M., Van Montagu, M., and Inze, D. 1989. The induction of manganese superoxide dismutase in response to stress in *Nicotiana plumbaginifolia*, *EMBO J.* 8:31–38.

Bradley, D. J., Wood, E. A., Larkins, A. P., Galfre, G., Butcher, G. W., and Brewin, N. J. 1988. Isolation of monoclonal antibodies reacting with peribacteroid membranes and other components of pea root nodules containing *Rhizobium leguminosarum*, *Planta* 173:149–160.

Bruce, R. J., and West, C. A. 1982. Elicitation of casbene synthetase activity in castor bean, *Plant Physiol.* 69:1181–1188.

Burr, B., Evola, S. V., and Burr, F. A. 1988. The application of restriction length polyporphism to plant breeding. In *Genetic Engineering: Principles and Method*, J. K. Setlow and A. Hollaender (eds.), vol. 5. Plenum, New York, pp. 45–59.

Callaham, D. A., and Torrey, J. G. 1981. The structural basis for infection of root hairs of *Trifolium repens* by *Rhizobium*, *Can. J. Bot.* 59:1647–1664.

Charest, P. J., Holbrook, L. A., Gabard, J., Iyer, V. N., and Miki, B. L. 1988. *Agrobacterium*-mediated transformation of thin cell layer explants from *Brassica napus* L., *Theor. Appl. Genet.* 75:438–445.

Chen, W. H., Gartland, K. M. A., Davey, M. R., Sotak, R., Gartland, J. S., et al. 1987. Transformation of sugarcane protoplasts by direct uptake of a selectable chimeric gene, *Plant Cell Rep.* 6:297–301.

Citovsky, V., Wong, M. L., and Zambryski, P. 1989. Cooperative interaction of *Agrobacterium* VirE2 protein with single-stranded DNA: implications for the T-DNA transfer process, *Proc. Natl. Acad. Sci. USA*, in press.

Coates, D., Taliercio, E. W., and Gelvin, S. B. 1987. Chromatin structure of integrated T-DNA in crown gall tumors, *Plant Mol. Biol.* 8:159–168.

Cocking, E. C., and Davey, M. R. 1987. Gene transfer in cereals, *Science* 236:1259–1262.

Collinge, D. B., and Slusarenko, A. J. 1987. Plant gene expression in response to pathogens, *Plant Mol. Biol.* 9:389–410.

Corbin, D. R., Sauer, N., and Lamb, C. J. 1987. Differential regulation of a hydroxyproline-rich glycoprotein gene family in wounded and infected plants, *Mol. Cell. Biol.* 7:6337–6344.

Cornelissen, B. J. C., Hooft van Huijsduijnen, R. A. M., and Bol, J. F. 1986. A tobacco mosaic virus-induced protein is homologous to the sweet-tasting protein thaumatin, *Nature* 321:531–532.

Cramer, C. L., Bell, J. N., Ryder, T. B., Bailey, J. A., Schuch, W., Bolwell, G. P., Robbins, M. P., Dixon, R. A., and Lamb, C. J. 1985a. Coordinated synthesis of phytoalexin biosynthetic enzymes in biologically-stressed cells of bean (*Phaseolus vulgaris* L.), *EMBO J.* 4:285–289.

Cramer, C. L., Ryder, T. B., Bell, J. N., and Lamb, C. J. 1985b. Rapid switching of plant gene expression induced by fungal elicitor, *Science* 227:1240–1243.

Cramer, C. L., Edwards, K., Dron, M., Liang, X., Dildine, S. L., Schuch, W., Dixon, R. A., and Lamb, C. J. 1989. Phenylalanine ammonia-lyase gene organization and structure, *Plant Mol. Biol.*, 12:367–385.

Crossway, A., Oakes, J. V., Irvine, J. M., Ward, B., Knauf, V. C., and Shewmaker, L. K. 1986. Integration of foreign DNA following microinjection of tobacco mesophyll protoplasts, *Mol. Gen. Genet.* 202:179–185.

Cullimore, J. V., Gebhardt, C., Saarelainen, R., Miflin, B. J., Idler, K. B., and Barker, R. F. 1984. Glutamine synthetase of *Phaseolus vulgaris* L.: organ-specific expression of a multigene family, *J. Molec. Appl. Genet.* 2:589–599.

Cuozzo, M., O'Connell, K. M., Kaniewski, W., Fang, R.-X., Chua, N.-H., and Tumer, N. E. 1988. Viral protection in transgenic tobacco plants expressing the cucumber mosaic virus coat protein or its antisense RNA, *Bio/Technology* 6:549–557.

Cuypers, B., Schmelzer, E., and Hahlbrock, K. 1988. *In situ* localization of rapidly accumulated phenylalanine ammonia-lyase mRNA around penetration sites of *Phytophthora infestans* in potato leaves, *Mol. Plant-Microbe Interact.* 1:157–160.

Cuypers, H., Dash, S., Peterson, P. A., Saedler, H., and Gierl, A. 1988. The defective En-1102 element encodes a product reducing the mutability of the En/Smp transposable element system of *Zea mays*, *EMBO J.* 7:2953–2960.

Daniels, M. J., Barber, C. E., Turner, P. C., Sawczyc, M. K., Byrde, R. J. W., and Fielding, A. H. 1984. Cloning of genes involved in pathogenicity of *Xanthomonas campestris* pv. *campestris* using the broad host range cosmid pLARF1, *EMBO J.* 3:3323–3328.

Dangl, J. L., Hanfle, K. D., Lipphard, S., Hahlbrock, K., and Scheel, D. 1987. Parsley protoplasts retain differential responsiveness to u.v. light and fungal elicitor, *EMBO J.* 6:2551–2556.

Darvill, A. G., and Albersheim, P. 1984. Phytoalexins and their elicitors—a defense against microbial infection in plants, *Annu. Rev. Plant Physiol.* 35:243–298.

De Block, M., Herrera-Estrella, L., Van Montagu, M., Schell, J., and Zambryski, P. 1984. Expression of foreign genes in regenerated plants and in their progeny, *EMBO J.* 3:1681–1689.

De Block, M., Schell, J., and Van Montagu, M. 1985. Chloroplast transformation by *Agrobacterium tumefaciens*, *EMBO J.* 4:1367–1372.

de Boer, D., Cremers, F., Teertstra, R., Smits, L., Hille, J., Smeekens, S., and Weisbeek, P. 1988. *In vivo* import of plastocyanin and a fusion protein into developmentally different plastids of transgenic plants, *EMBO J.* 7:2631–2635.

De Framond, A. J., Barton, K. A., and Chilton, M. D. 1983. Mini-Ti: a new vector strategy for plant genetic engineering, *Biotechnology* 1:262–269.

de la Pena, A., Lorz, H., and Schell, J. 1987. Transgenic rye plants obtained by injecting DNA into young floral tillers, *Nature* 325:274–276.

Delauney, A. J., and Verma, D. P. S. 1988. Cloned nodulin genes for symbiotic nitrogen fixation, *Plant Mol. Biol. Rep.* 6:279–285.

Deroles, S. C., and Gardner, R. C. 1988. Analysis of the T-DNA structure in a large number of transgenic petunias generated by *Agrobacterium*-mediated transformation, *Plant Mol. Biol.* 11:365–377.

Deshayes, A., Herrera-Estrella, L., and Caboche, M. 1985. Liposome-mediated transformation of tobacco mesophyll protoplasts by an *Escherichia coli* plasmid, *EMBO J.* 4:2731–2737.

Desjardins, A. E., and Gardner, H. W. 1989. Genetic analysis in *Gibberella pulicaris*: rishitin tolerance, rishitin metabolism, and virulence on potato tubers, *Mol. Plant-Microbe Interact.* 2:26–34.

De Wit, P. J. G. M., Toma, I. M. J., and Joosten, M. H. A. J. 1988. Race-specific elicitors and pathogenicity factors in the *Cladosporium fulvum*-tomato interaction. In *Physiology and Biochemistry of Plant-Microbial Interactions*, N. T. Keen, T. Kosuge, and L. L. Walling (eds.). American Society of Plant Physiologists, Rockville, Maryland, pp. 111–119.

Dixon, R. A. 1986. The phytoalexin response: elicitation, signalling and the control of host gene expression, *Biol. Rev.* 61:239–291.

Dixon, R. A., Bailey, J. A., Bell, J. N., Bolwell, G. P., Cramer, C. L., Edwards, K., Hamden, M. A. M. S., Lamb, C. J., Robbins, M. P., Ryder, T. B., and Schuch, W. 1986. Rapid changes in gene expression in response to microbial elicitation, *Phil. Trans. R. Soc. Lond. B* 314:411–426.

Djordjevic, M. A., Gabriel, D. W., and Long, S. W. 1987. Microscopic studies of plant cell division induced in alfalfa roots by *Rhizobium*—the refined parasite of legumes, *Annu. Rev. Phytopathol.* 25:145–168.

Draper, J., Davey, M. R., Freeman, J. P., Cocking, E. C., and Cox, B. J. 1982. Ti plasmid homologous sequences present in tissues from *Agrobacterium* plasmid-transformed Petunia protoplasts, *Plant Cell Physiol.* 23:451–458.

Draper, J., Scott, R., Armitage, P., and Walden, R. (eds.). 1988. *Plant Genetic Transformation and Gene Expression. A Laboratory Manual.* Blackwell Scientific, Oxford, U.K.

Dron, M., Clouse, S. D., Lawton, M. A., Dixon, R. A., and Lamb, C. J. 1988. Glutathione and fungal elicitor regulation of a plant defense gene promoter in electroporated protoplasts, *Proc. Natl. Acad. Sci. USA* 85:6738–6742.

Dudley, M. E., and Long, S. R. 1989. A non-nodulating mutant displays neither root hair curling nor early cell division in response to *Rhizobium meliloti*, *Plant Cell* 1:65–72.

Dudley, M. E., Jacobs, T. W., and Long, S. R. 1987. Microscopic studies of cell division induced in alfalfa roots by *Rhizobium meliloti*, *Planta* 171:289–301.

Ebel, J. 1986. Phytoalexin synthesis: The biochemical analysis of the induction process, *Annu. Rev. Phytopathol.* 24:235–264.

Ebert, P. R., Anderson, M. A., Bernatzky, R., Altschuler, M., and Clarke, A. E. 1989. Genetic polymorphism of self-incompatibility in flowering plants, *Cell* 56:255–262.

Ecker, J. R., and Davis, R. W. 1986. Inhibition of gene expression in plant cells by expression of antisense RNA, *Proc. Natl. Acad. Sci. USA* 83:5372–5376.

Eckes, P., Rosahl, S., Schell, J., and Willmitzer, L. 1986. Isolation and characterization of a light-inducible, organ-specific gene from potato and its analysis of expression after tagging and transfer into tobacco and potato shoots, *Mol. Gen. Genet.* 205:14–22.

Edwards, K., Cramer, C. L., Bolwell, G. P., Dixon, R. A., Schuch, W., and Lamb, C. J. 1985. Rapid transient induction of phenylalanine ammonia-lyase mRNA in elicitor-treated bean cells, *Proc. Natl. Acad. Sci. USA* 82:6731–6735.

Eichholtz, D. A., Rogers, S. G., Horsch, R. B., Klee, H. J., Hayford, M., Hoffmann, N. L., Bradford, S. B., Fink, C., Flick, J., and O'Connell, K. M. 1987. Expression of mouse dihydrofolate reductase gene confers methotrexate resistance in transgenic petunia plants, *Somat. Cell. Mol. Genet.* 13:67–76.

Ellis, J. G., Lawrence, G. J., Peacock, W. J., and Pryor, A. J. 1988. Approaches to cloning plant genes conferring resistance to fungal pathogens, *Annu. Rev. Phytopathol.* 26:245–263.

Epperlein, M. M., Noronha-Dulta, A. A., and Strange, R. N. 1986. Involvement of the hydroxyl radical in the abiotic elicitation of phytoalexins in legumes, *Physiol. Mol. Plant Pathol.* 28:67–77.

Fang, R.-X., Nagy, F., Sivasubramaniam, S., and Chua, N.-H. 1989. Multiple *cis* regulatory elements for maximal expression of the cauliflower mosaic virus 35S promoter in transgenic plants, *Plant Cell* 1:141–150.

Fedoroff, N. V. 1989. About maize transposable elements and development, *Cell* 56:181–191.

Fedoroff, N. V., Furtek, D. B., and Nelson, O. E. 1984. Cloning of the bronze locus in maize by a simple and generalizable procedure using the transposable element *Activator (Ac)*, *Proc. Natl. Acad. Sci. USA* 81:3825–3829.

Fillatti, J. J., Sellmer, J., McCown, B., Hassig, B., and Comai, L. 1987. *Agrobacterium* mediated transformation and regeneration of *Populus*, *Mol. Gen. Genet.* 206:192–199.

Fischoff, D. A., Bowdish, K. S., Perlak, F. T., Marrone, P. G., McCormick, S. M., Niedermeyer, J. G., Dean, D. A., Kusano-Kretzmer, K., Mayer, E. J., Rochester, D. E., Rogers, S. G., and Fraley, R. T. 1987. Insect tolerant transgenic tomato plants, *Bio/Technology* 5:807–813.

Fortin, M. A., Zelechowska, M., and Verma, D. P. S. 1985. Specific targeting of membrane nodulins to the bacteroid enclosing compartment in soybean nodules, *EMBO J.* 4:3041–3046.

Fraley, R. T., Rogers, S. G., Horsch, R. B., Eichholts, D. A., Flick, J. S., Fink, C. L., Hoffmann, N. L., and Sanders, P. R. 1985. The *SEV* system: a new disarmed Ti plasmid vector for plant transformation, *Bio/Technology* 3:629–635.

Fraley, R. T., Rogers, S. G., and Horsch, R. B. 1986. Genetic transformation in higher plants, *CRC Crit. Rev. Plant Sci.* 4:1–46.

Franssen, H. J., Nap, J. P., Gloudemans, T., Stiekema, W., Van Dam, H., Govers, F., Louwerse, J., and Bisseling, T. 1987. Characterization of cDNA for nodulin-75 of soybean: a gene product involved in early stages of root nodule development, *Proc. Natl. Acad. Sci. USA* 84:495–499.

Fromm, M., Taylor, L. P., and Walbot, V. 1985. Expression of genes transferred into monocot and dicot plant cells by electroporation, *Proc. Natl. Acad. Sci. USA* 82:5824–5828.

Fromm, M., Taylor, L. P., and Walbot, V. 1986. Stable transformation of maize after gene transfer by electroporation, *Nature* 319:791–793.

Fuller, F., Kunstner, P. W., Nguyen, T., and Verma, D. P. S. 1983. Soybean nodulin genes: analysis of cDNA clones reveals several major tissue specific sequences in nitrogen-fixing root nodules, *Proc. Natl. Acad. Sci USA* 80:2594–2598.

Gabriel, D., Burges, A., and Lazo, G. 1986. Gene-for-gene interactions of five cloned avirulence genes from *Xanthomonas campestris* pv. *malvacearum* with specific resistance genes in cotton, *Proc. Natl. Acad. Sci. USA* 83:6415–6419.

Gerlach, W. L., Llewellyn, D., and Haseloff, J. 1987. Construction of a plant disease resistance gene from the satellite RNA of tobacco ringspot virus, *Nature* 328:802–805.

Gheysen, G., Van Montagu, M., and Zambryski, P. 1987. Integration of *Agrobacterium tumefaciens* transfer DNA (T-DNA) involves rearrangements of target plant DNA sequences, *Proc. Natl. Acad. Sci. USA* 84:6169–6173.

Goodman, R. M., Hauptli, H., Crossway, A., and Knauf, V. C. 1987. Gene transfer in crop improvement, *Science* 236:48–54.

Govers, F., Nap, J. P., Moerman, M., Franssen, H. J., Van Kammen, A., and Bisseling, T. 1987. cDNA cloning and developmental expression of pea nodulin genes, *Plant Mol. Biol.* 8:425–435.

Graham, J. S., Pearce, G., Merryweathers, J., Titani, K., Ericsson, L. H., and Ryan, C. A. 1985a. Wound-induced proteinase inhibitors from tomato leaves I. The cDNA-deduced primary structure of pre-inhibitor I and its post-translational processing, *J. Biol. Chem.* 260:6555–6560.

Graham, J. S., Pearce, G., Merryweathers, J., Titani, K., Ericsson, L. H., and Ryan, C. A. 1985b. Wound-induced proteinase inhibitors from tomato leaves II. The cDNA-deduced primary structure of pre-inhibitor II, *J. Biol. Chem.* 260:6561–6566.

Graham, J. S., Hall, G., Pearce, G., and Ryan, C. A. 1986. Regulation of synthesis of proteinase inhibitors I and II mRNAs in leaves of wounded tomato plants, *Planta* 169:399–405.

Guerche, P., Jouanin, L., Tepfer, D., and Pelletier, G. 1987. Genetic transformation of oilseed rape (*Brassica napus*) by the Ri T-DNA of *Agrobacterium rhizogenes* and analysis of inheritance of the transformed phenotype, *Mol. Gen. Genet.* 206:328–334.

Hahn, M. G., and Griesbach, H. 1983. Cyclic AMP is not involved as a second messenger in the response of soybean to infection by *Phytophthora megasperma* f.sp. *glycinea*, *Z. Naturforsch.* 38C:578–582.

Hahn, M. G., Banhoff, A., and Grisebach, H. 1985. Quantitative localization of the phytoalexin glyceollin I in relation to fungal hyphae in soybean roots infected with *Phytophthora megasperma* f.sp. *glycinea*, *Plant Physiol.* 77:591–601.

Hain, R., Stabel, P., Czernilofsky, A. P., and Steinbiss, H. H. 1985. Uptake, integration, expression and genetic transmission of a selectable chimeric gene by plant protoplasts, *Mol. Gen. Genet.* 199:161–168.

Hake, S., Vollbrecht, E., and Freeling, M. 1989. Cloning *Knotted*, the dominant morphological mutant in maize using *Ds2* as a transposon tag, *EMBO J.* 8:15–22.

Harrison, B. D., Mayo, M. A., and Baulcombe, D. C. 1987. Virus resistance in transgenic plants that express cucumber mosaic virus satallite RNA, *Nature* 328:799–820.

Haseloff, J., and Gerlach, W. L. 1988. Simple RNA enzymes with new and highly specific endoribonuclease activities, *Nature* 334:585–591.

Hayford, M. B., Medford, J. I., Hoffman, N. L., Rogers, S. G., and Klee, H. J. 1988. Development of a plant transformation selection system based on expression of genes encoding gentamicin acetyltransferases, *Plant Physiol.* 86:1216–1222.

Hedrick, S. A., Bell, J. N., Boller, T., and Lamb, C. J. 1988. Chitinase cDNA cloning and mRNA induction by fungal elicitor, wounding and infection, *Plant Physiol.* 86:182–186.

Hemenway, C., Fang, R.-X., Chua, N.-H., and Tumer, N. E. 1988. Analysis of the mechanisms of protection in transgenic plants expressing the potato virus X coat protein or its antisense RNA, *EMBO J.* 7:1273–1280.

Hilder, V. A., Gatehouse, A. M. R., and Sheerman, S. E. 1987. A novel mechanism of insect resistance engineered into tobacco, *Nature* 330:160–163.

Hinchee, M. A. W., Connor-Ward, D. V., Newell, C. A., McDonnell, R. E., Sato, S. J., Gasser, C. S., Fischoff, D. A., Re, D. B., Fraley, R. T., and Horsch, R. B. 1988. Production of transgenic soybean plants using *Agrobacterium*-mediated DNA transfer, *Bio/Technology* 6:915–922.

Hoekema, A., Hirsch, P. R., Hooykaas, P. J. J., and Schilperoot, R. A. 1983. A binary plant vector strategy based on separation of *vir*- and T-regions of the *Agrobacterium tumefaciens* Ti plasmid, *Nature* 303:179–180.

Hoekema, A., Roelvink, P. W., Hooykaas, P. J. J., and Schilperoot, R. A. 1984. Delivery of T-DNA from the *Agrobacterium tumefaciens* chromosome into plant cells, *EMBO J.* 3:2485–2490.

Holsters, M., Silva, B., Van Vliet, F., Genetello, C., DeBlock, M., Dhaese, P., Depicker, A., Inze, D., Engler, G., Villarroel, R., Van Montagu, M., and Schell, J. 1980. The functional organization of the nopaline *A. tumefaciens* plasmid pTiC58, *Plasmid* 3:212–230.

Hooft van Huijsduijnen, R. A. M., Van Loon, L. C., and Bol, J. F. 1986. cDNA cloning of six mRNAs induced by TMV infection of tobacco and a characterization of their translation products, *EMBO J.* 5:2057–2061.

Horsch, R. B., and Klee, H. J. 1986. Rapid assay of foreign gene expression in leaf discs transformed by *Agrobacterium tumefaciens*: role of T-DNA borders in the transfer process, *Proc. Natl. Acad. Sci. USA* 83:4428–4432.

Horsch, R. B., Fry, J. E., Hoffmann, N. L., Eichholtz, D., Rogers, S. G., and Fraley, R. T. 1985. A simple and general method for transferring genes into plants, *Science* 227:1229–1231.

Izant, J. G., and Weintraub, H. 1985. Constitutive and conditional suppression of exogenous and endogenous genes by anti-sense RNA, *Science* 229:345–352.

Jefferson, R. A. 1987. Assaying chimeric genes in plants: the GUS gene fusion system, *Plant Mol. Biol. Rep.* 5:387–405.

Jefferson, R. A., Burgess, S. M., and Hirsh, D. 1986. Beta-Glucuronidase from *Escherichia coli* as a gene-fusion marker, *Proc. Natl. Acad. Sci. USA* 83:8447–8451.

Jensen, E. O., Marcker, K. A., Schell, J., and deBruijn, F. J. 1988. Interaction of a nodule specific trans-acting factor with distinct DNA elements in the soybean leghaemoglobin lbc_3 5' upstream region, *EMBO J.* 7:1265–1271.

Jones, D. A., Zhan, X.-C., and Kerr, A. 1985. To clone a gene from flax (*Linum usitatissimum*) for resistance to flax rust (*Melampsora lini*). In *Current Communications in Molecular Biology: Plant Cell/Cell Interactions*, I. Sussex, A. Ellingboe, M. Crouch, and R. Malmberg (eds.). Cold Spring Harbor Laboratories, New York, pp. 115–119.

Jones, J. D. G., Dunsmuir, P., and Bedbrook, J. 1985. High level expression of introduced chimaeric genes in regenerated transformed plants, *EMBO J.* 4:2411–2418.

Jones, J. D. G., Gilbert, D. E., Grady, K. L., and Jorgensen, R. A. 1987. T-DNA structure and gene expression in petunia plants transformed by *Agrobacterium tumefaciens* C58 derivatives, *Mol. Gen. Genet.* 207:478–485.

Jorgensen, R., Snyder, C., and Jones, J. D. G. 1987. T-DNA is organized predominantly in inverted repeat structures in plants transformed with *Agrobacterium tumefaciens* C58 derivatives, *Mol. Gen. Genet.* 207:471–477.

Junker, B., Zimny, J., Luhrs, R., and Lorz, H. 1987. Transient expression of chimaeric genes in dividing and non-dividing cereal protoplasts after PEG-induced DNA uptake, *Plant Cell Rep.* 6:329–332.

Kahn, M., Kraus, J., and Somerville, J. E. 1985. A model of nutrient exchange in the *Rhizobium*-legume symbiosis. In *Nitrogen Fixation Research Progress*, H. J. Evans, P. J. Bottomly, and W. E. Newton (eds.). Martinus Nijhoff, Dordrecht, The Netherlands, pp. 193–199.

Kauffmann, S., Legrand, M., Geoffroy, P., and Fritig, B. 1987. Biological function of "pathogenesis-related" proteins: four PR proteins of tobacco have 1,3-beta-glucanase activity, *EMBO J.* 6:3209–3212.

Kay, R., Chan, A., Daly, M., and McPherson, J. 1987. Duplication of CaMV 35S promoter sequences creates a strong enhancer for plant genes, *Science* 236:1299–1302.

Keith, B., and Chua, N.-H. 1986. Monocot and dicot pre-mRNAs are processed with different efficiencies in transgenic tobacco, *EMBO J.* 5:2419–2425.

Kistler, H. C., and VanEtten, H. D. 1984. Regulation of pisatin demethylation in *Nectria haematococca* and its influence on pisatin tolerance and virulence, *J. Gen. Microbiol.* 130:2605–2613.

Klee, H. J., Yanofsky, M., and Nester, E. W. 1985. Vectors for transformation of higher plants, *Bio/Technology* 3:637–642.

Klee, H. J., Horsch, R., and Rogers, S. 1987. *Agrobacterium*-mediated plant transformation and its further application to plant biology, *Annu. Rev. Plant Physiol.* 38:467–486.

Klein, T. M., Wolf, E. D., Wu, R., and Sanford, J. C. 1987. High-velocity microprojectiles for delivering nucleic acids into living cells, *Nature* 327:70–73.

Klein, T. M., Gradziel, T., Fromm, M. E., and Sanford, J. C. 1988. Factors influencing gene delivery into *Zea mays* cells by high-velocity microprojectiles, *Bio/Technology* 6:559–563.

Knight, P. 1988. Biocatalysts by design, *Bio/Technology* 6:826–827.

Kohle, H., Jeblick, W., Poten, F., Blaschek, W., and Klauss, H. 1985. Chitosan-elicited callose synthesis in soybean cells as a Ca^{+2} dependent process, *Plant Physiol.* 77:544–581.

Kombrink, E., Schroder, M., and Hahlbrock, K. 1988. Several "pathogenesis-related" proteins in potato are 1,3-beta-glucanases and chitinases, *Proc. Natl. Acad. Sci. USA* 85:782–786.

Koncz, C., and Schell, J. 1986. The promoter of TL-DNA gene 5 controls the tissue-specific expression of chimeric genes carried by a novel type of *Agrobacterium* binary vector, *Mol. Gen. Genet.* 204:383–396.

Koncz, C., De Greve, H., Andre, D., Deboeck, F., Van Montagu, M., and Schell, J. 1983. The opine synthase genes carried by the Ti plasmids contain all signals necessary for expression in plants, *EMBO J.* 2:1597–1603.

Koncz, C., Koncz-Kalman, Z., and Schell, J. 1987a. Transposon Tn5 mediated gene transfer into plants, *Mol. Gen. Genet.* 207:99–105.

Koncz, C., Olsson, O., Langridge, W. H. R., Schell, J., and Szalay, A. A. 1987b. Expression and assembly of functional bacterial luciferase in plants, *Proc. Natl. Acad. Sci. USA* 84:131–135.

Konderosi, E., and Konderosi, A. 1986. Nodule induction on plant roots by *Rhizobium*, *Trends Biochem. Sci.* 11:296–299.

Kosslak, R. M., Bookland, R., Barkei, J., Paaren, H. E., and Applebaum, E. R. 1987. Induction of *Bradyrhizobium japonicum* common *nod* genes by isoflavones isolated from *Glycine max.*, *Proc. Natl. Acad. Sci. USA* 84:7428–7432.

Krens, F. A., Molendijk, L., Wullems, G. J., and Schilperoort, R. A. 1982. In vitro transformation of plant protoplasts with Ti-plasmid DNA, *Nature* 296:72–74.

Kreuzaler, F., Ragg, H., Fautz, E., Kuhn, D. N., and Hahlbrock, K. 1983. UV-induction of chalcone synthase mRNA in cell suspension cultures of *Petroselinum hortense*, *Proc. Natl. Acad. Sci. USA* 80:2591–2593.

Kuc, J., and Rush, J. S. 1985. Phytoalexins, *Arch. Biochem. Biophys.* 236:455–472.

Kuhlemeier, C., Green, P. J., and Chua, N. -H. 1987. Regulation of gene expression in higher plants, *Annu. Rev. Plant Physiol.* 38:221–257.

Kuhn, D. N., Chappell, J., Boudet, A., and Hahlbrock, K. 1984. Induction of phenylalanine ammonia-lyase and 4-coumarate:CoA ligase in cultured plant cells by UV light or fungal elicitor, *Proc. Natl. Acad. Sci. USA* 81:1102–1106.

Kuntz, M., Simons, A., Schell, J., and Scheier, P. H. 1986. Targeting of protein to chloroplasts in transgenic tobacco by fusion to mutated transit peptide, *Mol. Gen. Genet.* 205:454–460.

Kurosaki, F., Tsurusawa, Y., and Nishi, A. 1987. The elicitation of phytoalexins by Ca^{2+} and cyclic AMP in carrot cells, *Phytochemistry* 26:1919–1923.

Lagrimini, L. M., and Rothstein, S. 1987. Tissue specificity of tobacco peroxidase isozymes and their induction by wounding and tobacco mosaic virus infection, *Plant Physiol.* 84:438–442.

Lagrimini, L. M., Burkhart, W., Moyer, M., and Rothstein, S. 1987. Molecular cloning of complementary DNA encoding the lignin-forming peroxidase from tobacco: molecular analysis and tissue specific expression, *Proc. Natl. Acad. Sci. USA* 84:7542–7546.

LaRue, T. A., Kneen, B. E., and Gartside, E. 1985. Plant mutants defective in symbiotic nitrogen fixation. In *Analysis of Plant Genes Involved in the Legume*-Rhizobium *Symbiosis*, R. Martcellin (ed.). OECD Publications, Paris, pp. 39–48.

Lamb, C. J., Lawton, M. A., Dron, M., and Dixon, R. A. 1989. Signals and transduction mechanisms for activation of plant defenses against microbial attack, *Cell* 56:215–224.

Lawrence, W. A., and Davies, D. R. 1985. A method for the microinjection and culture of protoplasts at very low densities, *Plant Cell Rep.* 4:33–35.

Lawton, M. A., Kragh, K., Jenkins, S. M., Dron, M., Clouse, S. D., Dixon, R. A., and Lamb, C. J. 1988. Nuclear protein binding to a cis-acting silencer in a defense gene promoter, *J. Cell Biol.* 12c(suppl.):222.

Lawton, M. A., and Lamb, C. J. 1987. Transcriptional activation of plant defense genes by fungal elicitor, wounding, and infection, *Mol. Cell. Biol.* 7:335–341.

Lawton, M. A., Tierney, M. A., Nakamura, I., Anderson, E., Komeda, Y., Dube, P., Hoffman, N., Fraley, R. T., and Beachy, R. N. 1987. Expression of a soybean beta-conclycinin gene under the control of the Cauliflower Mosaic Virus 35S and 19S promoters in transformed petunia tissues, *Plant Mol. Biol.* 9:315–324.

Lee, M., and Phillips, R. L. 1987. Genetic variants in progeny of regenerated maize plants, *Genome* 29:834–838.

Legocki, R. P., and Verma, D. P. S. 1980. Identification of nodule specific host-proteins (nodulins) involved in the development of *Rhizobium*-legume symbiosis, *Cell* 20:153–163.

Legrand, M., Kauffmann, S., Geoffroy, P., and Fritig, B. 1987. Biological function of pathogenesis-related proteins: four tobacco pathogenesis-related proteins are chitinases, *Proc. Natl. Acad. Sci. USA* 84:6750–6754.

Liang, X., Dron, M., Cramer, C. L., Dixon, R. A., and Lamb, C. J. 1989. Differential regulation of phenylalanine ammonia-lyase genes during plant development and by environmental cues, *J. Biol. Chem* 264:14486–14492.

Lloyd, A. M., Barnason, A. R., Rogers, S. G., Byrne, M. C., Fraley, R. T., and Horsch, R. B. 1986. Transformation of *Arabidopsis thaliana* with *Agrobacterium tumefaciens*, *Science* 234:464–466.

Loesch-Fries, L. S., Merlo, D., Zinnen, T., Burhop, L., Hill, K., Krahn, K., Jarvis, N., Nelson, S., and Halk, E. 1987. Expression of alfalfa mosaic virus RNA4 in transgenic plants confers virus resistance, *EMBO J.* 6:1845–1851.

Logemann, J., Lipphardt, S., Lorz, H., Hauser, I., Willmitzer, L., and Schell, J. 1989. 5'-Upstream sequences from the *wun*1 gene are responsible for gene activation by wounding in transgenic plants, *Plant Cell* 1:151–158.

Long, S. R. 1989. *Rhizobium*-legume nodulation: life together in the underground, *Cell* 56:203–214.

Lorz, H., Gobel, E., and Brown, P. T. H. 1985. Advances in tissue culture and progress toward genetic transformation of cereals, *Plant Breeding* 100:1–25.

Luhrs, R., and Lorz, H. 1987. Plant regeneration in vitro from embryogenic cultures of spring- and winter-type barley (*Hordeum vulgare* L.) varieties, *Theor. Appl. Genet.* 75:16–25.

Luo, Z., and Wu, R. 1988. A simple method for the transformation of rice via the pollen-tube pathway, *Plant Mol. Biol. Rep.* 6:165–174.

Martins, C., Carpenter, R., Sommer, H., Seadler, H., and Coen, E. S. 1985. Molecular analysis of instability in flower pigmentation of *Antirrhinum majus*, following isolation of the *pallisa* locus by transposon tagging, *EMBO J.* 4:1625–1630.

Matzke, A. J. M., and Matzke, M. A. 1986. A set of novel Ti-plasmid derived vectors for production of transgenic plants, *Plant Mol. Biol.* 7:357–365.

Mauch, F., Hadwiger, L. A., and Boller, T. 1984. Ethylene: symptom, not signal for the induction of chitinase and β-1,3-glucanase in pea pods by pathogens and elicitors, *Plant Physiol.* 76:607–611.

McCabe, D. E., Swain, W. F., Martinell, B. J., and Christou, P. 1988. Stable transformation of soybean (*Glycine max*) by particle acceleration, *Bio/Technology* 6:923–926.

McCormick, S., Niedermeyer, J., Fry, J., Barnason, A., Horsch, R., and Fraley, R. 1986. Leaf disc transformation of cultivated tomato (*L. esculentum*) using *Agrobacterium tumefaciens, Plant Cell Rep.* 5:81–84.

McGranahan, G. H., Leslie, C. A., Uratsu, S. L., Martin, L. A., and Dandekar, A. M. 1988. *Agrobacterium*-mediated transformation of walnut somatic embryos and regeneration of transgenic plants, *Bio/Technology* 6:800–804.

Mehdy, M. C., and Lamb, C. J. 1987. Chalcone isomerase cDNA cloning and mRNA induction by fungal elicitor, wounding and infection, *EMBO J.* 6:1527–1533.

Meyer, P., Walgenbach, E., Bussmann, K., Hombrecher, G., and Saedler, H. 1985. Synchronized tobacco protoplasts are efficiently transformed by DNA, *Mol. Gen. Genet.* 201:513–518.

Meyer, P., Kartzke, S., Niedenhof, I., Heidmann, I., Bussmann, K., and Saedler, H. 1988. A genomic DNA segment from *Petunia hybrida* leads to increased transformation frequencies and simple integration patterns, *Proc. Natl. Acad. Sci. USA* 85:8568–8572.

Meyerowitz, E. M. 1987. *In situ* hybridization to RNA in plant tissue, *Plant Mol. Biol. Rep.* 5:242–250.

Michaelmore, R. W., Hulbert, S. H., Landry, B. S., and Leung, H. 1987. Toward a molecular understanding of lettuce downy mildew. In *Genetics and Plant Pathogenesis,* P. R. Day and G. J. Jellis (eds.). Blackwell Scientific, Oxford, U.K., pp. 220–223.

Moesta, P., and Grisebach, H. 1982. L-2-Aminooxy-3-phenylproprionic acid inhibits phytoalexin accumulation in soybean with comcomitant loss of resistance against *Phytopthora megasperma* f. sp. *glycinea, Physiol. Plant Pathol.* 21:65–70.

Moesta, P., and West, C. A. 1985. Casbene synthetase: regulation of phytoalexin biosynthesis in *Ricinus communis* L. seedlings. Purification of casbene synthetase and regulation of its biosynthesis during elicitation, *Arch. Biochem. Biophys.* 238:325–333.

Muller, A. J., Mendel, R. R., Schiemann, J., Simoens, C., and Inze, D. 1987. High meiotic stability of a foreign gene introduced into tobacco by *Agrobacterium*-mediated transformation, *Mol. Gen. Genet.* 207:171–175.

Negrutiu, I., Shillito, R. D., Potrykus, I., Biasini, G., and Sala, F. 1987. Hybrid genes in the analysis of transformation conditions. I. Setting up a simple method for direct gene transfer in plant protoplasts, *Plant Mol. Biol.* 8:363–373.

Newcomb, W. 1981. Nodule morphogenesis. In *International Review of Cytology,* suppl. 13, G. H. Bourne and J. F. Danielli (eds.), pp. 246–298.

Ochatt, S. J., Chand, P. K., Rech, E. L., Davey, M. R., and Power, J. B. 1988. Electroporation-mediated stimulation of plant regeneration from Colt cherry (*Prunus avium* × *psuedocerasus*) protoplasts, *Plant Sci.* 54:165–169.

Odell, J. T., Nagy, F., and Chua, N. -H. 1985. Identification of DNA sequences required for activity of the cauliflower mosaic virus 35S promoter, *Nature* 313:810–812.

Okada, K., Takebe, I., and Nagata, T. 1986. Expression and integration of genes introduced into highly synchronized plant protoplasts, *Mol. Gen. Genet.* 205:398–403.

O'Reilly, C., Shepherd, N. S., Pereira, A., Schwarz-Sommer, Z., Bertram, I., Robertson, D. S., Peterson, P. A. and Saedler, H. 1985. Molecular cloning of the *a1* locus of *Zea mays* using the transposable elements *En* and *Mu1, EMBO J.* 4:877–882.

Ou-Lee, T. -M., Turgeon, R., and Wu, R. 1986. Expression of a foreign gene linked to either a plant virus or a *Drosophila* promoter, after electroporation of protoplasts of rice, wheat and sorghum, *Proc. Natl. Acad. Sci. USA* 83:6815–6819.

Ow, D. W., Wood, K. V., DeLuca, M., de Wet, J. R., Helinski, D. R., Howell, S. H. 1986. Transient and stable expression of the firefly luciferase gene in plant cells and transgenic plants, *Science* 234:856–860.

Ow, D. W., Jacobs, J. D., and Howell, S. H. 1987. Functional regions of the cauliflower mosaic virus 35S RNA promoter determined by use of the firefly luciferase gene as a reporter of promoter activity, *Proc. Natl. Acad. Sci. USA* 84:4870–4874.

Park, H. -S., Denbow, C. J., and Cramer, C. L. 1989. Molecular cloning and defense-related expression of HMG-CoA reductase genes in tomato cells, manuscript in preparation.

Paszkowski, J., and Saul, M. W. 1986. Direct gene transfer to plants, *Methods Enzymol.* 118:668–684.

Paszkowski, J., Shillito, R., Saul, M., Mandak, V., Hohn, T., Hohn, B., and Potrykus, I. 1984. Direct gene transfer to plants, *EMBO J.* 3:2712–2722.

Paszkowski, J., Baur, M., Bogucki, A., and Potrykus, I. 1988. Gene targeting in plants, *EMBO J.* 7:4021–4026.

Peralta, E. G., and Ream, L. W. 1985. T-DNA border sequences required for crown gall tumorigenesis, *Proc. Natl. Acad. Sci. USA* 82:5112–5116.

Peralta, E. G., Hellmiss, R., and Ream, R. 1986. *Overdrive*, a T-DNA transmission enhancer on the *A. tumefaciens* tumor-inducing plasmid, *EMBO J.* 5:1137–1142.

Potrykus, I., Saul, M. W., Petruska, J., Paskowski, J., and Shillito, R. D. 1985. Direct gene transfer to cells of a graminaceous monocot, *Mol. Gen. Genet.* 199:183–188.

Powell Abel, P., Nelson, R. S., De, B., Hoffman, H., Rogers, S. G., Fraley, R. T., and Beachy, R. N. 1986. Delay of disease development in transgenic plants that express the tobacco mosaic virus coat protein, *Science* 232:738–743.

Prols, M., Schell, J., and Steinbiss, H. -H. 1989. Critical evaluation of electro-mediated gene transfer and transient expression in plant cells. In *Electroporation and Electrofusion in Cell Biology*, E. Neumann, A. E. Sowers, and C. A. Jordon (eds.). Plenum, New York, in press.

Prosen, D. E., and Simpson, R. B. 1987. Transfer of a ten-member genomic library to plants using *Agrobacterium tumefaciens*, *Bio/Technology* 5:966–971.

Pryor, A. 1987a. The origin and structure of fungal disease resistance genes in plants, *Trends Genet.* 3:157–161.

Pryor, A. 1987b. Stability of alleles of *Rp* (resistance to *Puccinia sorghi*), *Maize Newslett.* 61:37–38.

Rech, E. L., Ochatt, S. J., Chand, P. K., Power, J. B., and Davey, M. R. 1987. Electro-enhancement of division of plant protoplast-derived cells, *Protoplasma* 141:169–176.

Rech, E. L., Ochatt, S. J., Chand, P. K., Davey, M. R., Mulligan, B. J., and Power, J. B. 1988. Electroporation increases DNA synthesis in cultured plant protoplasts, *Bio/Technology* 6:1091–1093.

Reich, T. J., Iyer, V. N., and Miki, B. L. 1986. Efficient transformation of alfalfa protoplasts by the intranuclear microinjection of Ti-plasmids, *Bio/Technology* 4:1001–1004.

Reiss, B., Sprengel, R., Will, H., and Schaller, H. 1984. A new sensitive method for qualitative and quantitative analysis of neomycin phosphotransferase in crude cell extracts, *Gene* 30:211–218.

Rezaian, M. A., Skene, K. G. M., and Ellis, J. G. 1988. Anti-sense RNAs of cucumber mosaic virus in transgenic plants assessed for control of the virus, *Plant Mol. Biol.* 11:463–471.

Rhodes, C. A., Pierce, D. A., Mettler, I. J., Mascarenhas, D., and Detmer, J. T. 1988. Genetically transformed maize plants from protoplasts, *Science* 240:204–210.

Richardson, M., Valdes-Rodriguez, S., and Blanco-Labra, A. 1987. A possible function for thaumatin and a TMV-induced protein suggested by homology to a maize inhibitor, *Nature* 327:432–434.

Robertson, J. D., Lyttleton, P., Bullivant, S., and Grayston, G. F. 1978. Membranes in lupine root nodules. I. The role of Golgi bodies in the biogenesis of infection threads and peribacteroid membranes, *J. Cell Sci.* 30:129–149.

Robertson, J., Wells, B., Brewin, N. J., Wood, E., Knight, C. D., and Downie, J. A. 1985. The legume-*Rhizobium* symbiosis: a cell surface interaction, *J. Cell Sci. Suppl.* 2:317–331.

Rodermel, S. R., Abbot, M. S., and Bogorad, L. 1988. Nuclear-organelle interactions: nuclear antisense gene inhibits ribulose bisphosphate carboxylase enzyme levels in transformed tobacco plants, *Cell* 55:673–681.

Rogers, S. G., Klee, H., Horsch, R. B., and Fraley, R. T. 1987. Improved vectors for plant transformation: expression cassette vectors and new selectable markers, *Methods Enzymol.* 153:253–277.

Rolfe, B., and Gresshoff, P. 1988. Genetic analysis of legume nodule initiation, *Annu. Rev. Plant Physiol. Plant Mol. Biol.* 39:297–319.

Ryan, C. A. 1987. Oligosaccharide signalling in plants, *Annu. Rev. Cell Biol.* 3:295–317.

Ryan, C. A., and An, G. 1988. Molecular biology of wound-inducible proteinase inhibitors in plants, *Plant Cell Env.* 11:345–350.

Ryder, T. B., Cramer, C. L., Bell, J. N., Robbins, M. P., Dixon, R. A., and Lamb, C. J. 1984. Elicitor rapidly induces chalcone synthase mRNA in *Phaseolus vulgaris* cells at the onset of the phytoalexin defense response, *Proc. Natl. Acad. Sci. USA* 81:5724–5728.

Saiki, R. K., Gelfand, D. H., Stoffel, S., Scharf, S. J., Higuchi, R., Horn, G. T., Mullis, K. B., and Erlich, H. A. 1988. Primer-directed enzymatic amplification of DNA with a thermostable DNA polymerase, *Science* 239:487–491.

Sanchez-Serrano, J. J., Deil, M., O'Conner, A., Schell, J., and Willmitzer, L. 1987. Wound-induced expression of a potato proteinase inhibitor gene in transgenic tobacco plants, *EMBO J.* 6:303–306.

Sanders, P. R., Winter, J. A., Barnason, A. R., Rogers, S. G., and Fraley, R. T. 1987. Comparisons of cauliflower mosaic virus 35S and nopaline synthase promoters in transgenic plants, *Nucleic Acids Res.* 15:1543–1558.

Schell, J. 1987. Transgenic plants as tool to study the molecular organization of plant genes, *Science* 237:1176–1183.

Schmidt, R. J., Burr, F. A., and Burr, B. 1987. Transposon tagging and molecular analysis of the maize regulatory locus *opaque-2*, *Science* 238:960–963.

Schmidt, W. E., and Ebel, J. 1987. Specific binding of a fungal glucan phytoalexin elicitor to membrane fractions from soybean *Glycine max*, *Proc. Natl. Acad. Sci. USA* 84:4117–4121.

Schroder, G., Brown, J. W. S., and Schroder, J. 1988. Molecular analysis of resveratrol synthase, *Eur. J. Biochem.* 172:161–169.

Scott, R. J., and Drager, J. 1987. Transformation of carrot tissues derived from proembryogenic suspension cells: a useful model system for gene expression studies in plants, *Plant Mol. Biol.* 8:265–274.

Sequeira, L. 1983. Mechanisms of induced resistance in plants, *Annu. Rev. Microbiol.* 37:51–79.

Sharp, J. K., Albersheim, P., Ossowski, P., Pilotti, A., Garegg, P., and Lindberg, P. 1984. Comparison of the structures and elicitor activities of a synthetic and a mycelial-wall-derived hexa-(beta-D-glucopyranosyl)-D-glucitol isolated from the mycelial walls of *Phytophthora megasperma* f. sp. *glycinea*, *J. Biol. Chem.* 259:11341–11345.

Shaw, C. H., Carter, G. H., Watson, M. D., and Shaw, C. H. 1984. A functional map of the nopaline synthase promoter, *Nucleic Acids Res.* 12:7831–7842.

Simpson, R. B., Spielmann, A., Margossien, L., and McKnight, T. D. 1986. A disarmed binary vector from *Agrobacterium tumefaciens* functions in *Agrobacterium rhizogenes*. Frequent cotransformation of two distinct T-DNAs, *Plant Mol. Biol.* 6:403–415.

Simoens, C., Alliotte, T., Mendel, R., Muller, A., Schiemann, J., Van Lijsabettens, M., Schell, J., Van Montagu, M., and Inze, D. 1986. A binary vector for transforming genomic libraries to plants, *Nucleic Acids Res.* 14:8073–8090.

Smith, C. J. S., Watson, C. F., Ray, J., Bird, C. R., Morris, P. C., Schuch, W., and Grierson, D. 1988. Antisense RNA inhibition of polygalacturonase gene expression in transgenic tomatos, *Nature* 334:724–726.

Somssich, I. E., Schmelzer, E., Bollmann, J., and Hahlbrock, K. 1986. Rapid activation by fungal elicitor of genes encoding "pathogenesis-related" proteins in cultured parsley cells, *Proc. Natl. Acad. Sci. USA* 83:2427–2430.

Somssich, I. E., Schmelzer, E., Kawalleck, P., and Hahlbrock, K. 1988. Gene structure and in situ transcript localization of pathogenesis-related protein 1 in parsley, *Mol. Gen. Genet.* 213:93–98.

Stab, M. R., and Ebel, J. 1987. Effects of Ca^{2+} on phytoalexin induction by fungal elicitor in soybean cells, *Arch. Biochem. Biophys.* 257:416–423.

Stachel, S. E., and Zambryski, P. C. 1986*a*. VirA and virG control the plant-induced activation of the T-DNA transfer process of *Agrobacterium tumefaciens*, *Cell* 46:325–333.

Stachel, S. E., and Zambryski, P. C. 1986b. *Agrobacterium tumefaciens* and the susceptible plant cell: a novel adaptation of extracellular recognition and DNA conjugation, *Cell* 47:155–157.

Stachel, S. E., Messens, E., Van Montagu, M., and Zambryski, P. 1985. Identification of signal molecules produced by wounded plant cells that activate T-DNA transfer in *Agrobacterium tumefaciens*, *Nature* 318:624–630.

Stachel, S. E., Timmerman, B., and Zambryski, P. 1987. Activation of *Agrobacterium tumefaciens vir* gene expression generates multiple single-stranded T-strand molecules from the pTIA6 T-region: requirement for 5′ *vir*D gene products, *EMBO J.* 4:857–863.

Staskawicz, B. J., Dahlbeck, D., and Keen, N. T. 1984. Cloned avirulence gene of *Pseudomonas syringae* pv. *glycinea* determines race-specific incompatibility of *Glycinae max* (L.) Merr, *Proc. Natl. Acad. Sci. USA* 81:6024–6028.

Stermer, B. A., and Bostock, R. M. 1987. Involvement of 3-hydroxy-3-methylglutaryl coenzyme A reductase in the regulation of sesquiterpenoid phytoalexin synthesis in potato, *Plant Physiol.* 84:404–408.

Stockhaus, J., Eckes, P., Blau, A., Schell, J., and Willmitzer, L. 1987. Organ-specific and dosage-dependent expression of a leaf/stem specific gene from potato after tagging and transfer into potato and tobacco plants, *Nucleic Acids Res.* 15:3479–3491.

Strasser, H., Hoffman, C., Grisebach, H., and Matern, U. 1986. Are polyphosphoinositides involved in signal transduction of elicitor-induced phytoalexin synthesis in cultured plant cells? *Z. Naturforsch.* 41C:717–724.

Tanksley, S. D., Miller, J., Paterson, A., and Bernatzky, R. 1988. Molecular mapping of plant chromosomes, in *Chromosome Structure and Function*, J. P. Gustafson and R. Appels (eds.). Plenum, New York, pp. 157–173.

Teeri, T. H., Herrera-Estrella, L., Depicker, A., Van Montagu, M., and Palva, E. T. 1986. Identification of plant promoters in situ by T-DNA mediated transcriptional fusions to the *nptII* gene, *EMBO J.* 5:1755–1760.

Templeton, M. D., and Lamb, C. J. 1988. Elicitors and defence gene activation, *Plant Cell Env.* 11:395–402.

Tepper, C. S., and Anderson, A. J. 1986. Two cultivars of bean display a differential response to extracellular components from *Colletotrichum lindemuthianum, Physiol. Mol. Plant Pathol.* 29:411–420.

Thornburg, R. W., Cleveland, T. E., and Ryan, C. A. 1987. Wound-inducible expression of potato Inhibitor II gene in transgenic tobacco plants, *Proc. Natl. Acad. Sci. USA* 84:744–748.

Topfer, R., Matzeit, V., Gronenborn, B., Schell, J., and Steinbiss, H. -H. 1987. A set of plant expression vectors for transcriptional and translational fusions, *Nucleic Acids Res.* 15:5890.

Topfer, R., Gronenborn, B., Schell, J., and Steinbiss, H. -H. 1989. Uptake and transient expression of chimeric genes in seed-derived embryos, *Plant Cell* 1:133–139.

Toriyama, K., Arimoto, Y., Uchimiya, H., and Hinata, K. 1988. Transgenic rice plants after direct gene transfer into protoplasts, *Bio/Technology* 6:1072–1074.

Tran Thanh Van, K., Toubart, P., Cousson, A., Darvill, A. G., Gollin, A. J., Celf, P., and Albersheim, P. 1985. Manipulation of the morphogenetic pathways of tobacco explants by oligosaccharins, *Nature* 314:615–617.

Tran Thanh Van, M. 1973. *In vitro* control of *de novo* flower, bud, root and callus differentiation from excised epidermal tissues, *Nature* 246:44–45.

Tran Thanh Van, M. 1981. Control of morphogenesis in *in vitro* cultures, *Annu. Rev. Plant Physiol.* 32:291–311.

Trinh, T. H., Mante, S., Pua, E. -C., and Chua, N. -H. 1987. Rapid production of transgenic flowering shoots and F1 progeny from *Nicotiana plumbaginifolia* epidermal peels, *Bio/Technology* 5:1081–1084.

Tumer, N. E., O'Connell, K. M. O., Nelson, R. S., Sanders, P. R., Beachy, R. N., Fraley, R. T., and Shah, D. M. 1987. Expression of alfalfa mosaic virus coat protein gene confers cross-protection in transgenic tobacco and tomato plants, *EMBO J.* 6:1181–1188.

Uchimiya, H., Fushimi, T., Hashimoto, H., Harada, H., and Syono, Y. 1986. Expression of a foreign gene in callus derived from DNA treated protoplasts of rice (*Oryza sativa* L.), *Mol. Gen. Genet.* 204:204–207.

Umbeck, P., Johnson, G., Barton, K., and Swain, W. 1987. Genetically transformed cotton (*Gossypium hirsutum* L.) plants, *Bio/Technology* 5:263–266.

Ursic, D., Slightom, J. L., and Kemp, J. D. 1983. *Agrobacterium tumefaciens* T-DNA integrates into multiple sites of the sunflower crown gall genome, *Mol. Gen. Genet.* 190:494–503.

Vance, C. P. 1983. *Rhizobium* infection and nodulation: a beneficial plant disease? *Annv. Rev. Microbiol.* 37:399–424.

Vance, C. P., Egli, M. A., Griffith, S. M., and Miller, S. S. 1988. Plant regulated aspects of nodulation and N_2 fixation, *Plant Cell Env.* 11:413–427.

Vaeck, M., Reynaerts, A., Hofte, H., Jansens, S., De Beuckeleer, M., Dean, C., Zabeau, M., Van Montagu, M., and Leemans, J. 1987. Transgenic plants protected from insect attack, *Nature* 328:33–37.

Van den Elzen, P. J. M., Townsend, J., Lee, K. Y., and Bedbrook, J. 1985. A chimeric hygromycin resistance gene as a selectable marker in plant cells, *Plant Mol. Biol.* 5:299–302.

van der Krol, A. R., Lenting, P. E., Veenstra, J., van der Meer, I. M., Koes, R. E., Gerats, A. G. M., Mol, J. N. M., and Stuitje, A. R. 1988a. An anti-sense chalcone synthase gene in transgenic plants inhibits flower pigmentation, *Nature* 333:866–869.

van der Krol, A. R., Mol, J. N. M., and Stuitje, A. R. 1988b. Modulation of eukaryotic gene expression by complementary RNA or DNA sequences, *Biotechniques* 6:958–976.

van Dun, C. M. P., Overduin, B., van Vloten-Doting, L., and Bol, J. F. 1988. Transgenic tobacco expressing tobacco streak virus or mutated alfalfa mosaic virus coat protein does not cross-protect against alfalfa mosaic virus infection, *Virology* 164:383–389.

Van Kammen, A. 1984. Suggested nomenclature for plant genes involved in nodulation and symbiosis, *Plant Mol. Biol. Rep.* 2:43–45.

Van Sluys, M. A., Tempe, J., and Fedoroff, N. 1987. Studies on the introduction and mobility of the maize *Activator* element in *Arabidopsis thaliana* and *Daucus carota*, *EMBO J.* 6:3881–3889.

Velten, J., and Schell, J. 1985. Selection-expression plasmid vectors for use in genetic transformation of higher plants, *Nucleic Acids Res.* 13:6981–6998.

Verma, D. P. S., Fortin, M. G., Stanley, J., Mauro, V. P., Purohit, S., and Morrison, N. 1986. Nodulins and nodulin genes of *Glycine max.*, a perspective, *Plant Mol. Biol.* 7:51–61.

Verma, D. P. S., and Nadler, K. 1984. Legume-*Rhizobium* symbiosis: host's point of view. In *Genes Involved in Microbe-Plant Interactions*, D. P. S. Verma and T. Hohn (eds.). Springer Verlag, Vienna, Austria.

Vornam, B., Schon, H., and Kindl, H. 1988. Control of gene expression during induction of cultured peanut cells: mRNA levels, protein synthesis and enzyme activity of stilbene synthase, *Plant Mol. Biol.* 10:235–243.

Walbot, V., and Warren, C. 1988. Regulation of *Mu* element copy number in maize lines with an active or inactive Mutator transposable element, *Mol. Gen. Genet.* 211:27–34.

Waldron, C., Murphy, E. B., Roberts, J. L., Gustafson, G. D., Armour, S. L., and Malcolm, S. K. 1985. Resistance to hygromycin B. A new marker for plant transformation, *Plant Mol. Biol.* 5:103–108.

Walter, M. H., Grima-Pettenatic, J., Grand, C., Boudet, A. M., and Lamb, C. J. 1988. Cinnamyl alcohol dehydrogenase, a molecular marker specific for lignin biosynthesis: cDNA cloning and mRNA induction by fungal elicitor, *Proc. Natl. Acad. Sci. USA* 85:5546–5550.

Wang, K., Herrera-Estrella, L., Van Montagu, M., and Zambryski, P. 1984. Right 25 bp terminus sequences of the nopaline T-DNA is essential for and determines direction of DNA transfer from *Agrobacterium* to the plant genome, *Cell* 38:455–462.

Wang, K., Genetello, C., Van Montagu, M., and Zambryski, P. 1987. Sequence context of the T-DNA border repeat element determines its relative activity during T-DNA transfer to plant cell, *Mol. Gen. Genet.* 210:338–346.

Wang, Y. C., Klein, T. M., Fromm, M., Cao, J., Sanford, J. C., and Wu, R. 1988. Transient expression of foreign genes in rice, wheat and soybean cells following particle bombardment, *Plant Mol. Biol.* 11:433–439.

Weinand, U., and Saedler, H. 1987. Plant transposable elements: unique structures for gene/transposable tagging and gene cloning, *Plant Gene Res.* 4:205–227.

Weising, K., Schell, J., and Kahl, G. 1988. Foreign genes in plants: transfer, structure, expression, and applications, *Annu. Rev. Genet.* 22:421–477.

White, R., and Lalouel, J. -M. 1988. Chromosome mapping with DNA markers, *Sci. Am.* 258:40–48.

Willmitzer, L., Dhaese, P., Schreier, P., Schmalenbach, W., Van Montagu, M., and Schell, J. 1983. Size, location and polarity of TDNA-encoded transcripts in nopaline crown gall tumors; common transcripts in octopine and nopaline tumors, *Cell* 32:1045–1056.

Wingate, V. P. M., Lawton, M. A., and Lamb, C. J. 1988. Glutathione causes a massive and selective induction of plant defense genes, *Plant Physiol.* 87:206–210.

Yao, P. Y., and Vincent, J. M. 1969. Host specificity in the root hair "curling factor" of *Rhizobium* sp., *Aust. J. Biol. Sci.* 22:413–423.

Zambryski, P. 1988. Basic processes underlying *Agrobacterium*-mediated DNA transfer to plant cells, *Annu. Rev. Genet.* 22:1–30.

Zambryski, P., Joos, H., Genetello, C., Leemans, J., Van Montagu, M., and Schell, J. 1983. Ti plasmid vector for the introduction of DNA into plant cells without alteration of their normal regenerative capacity, *EMBO J.* 2:2143–2150.

Zambryski, P., Tempe, J., and Schell, J. 1989. Transfer and function of T-DNA genes from Agrobacterium Ti and Ri plasmids in plants, *Cell* 56:193–201.

Zhou, J. C., Tchan, Y. T., and Vincent, J. M. 1985. Reproductive capacity of bacteroids in nodules of *Trifolium repens* L. and *Glycine max.* (L.) Merr, *Planta* 163:473–482.

Zook, M. N., Rush, J. S., and Kuc, J. A. 1987. A role for Ca^{2+} in the elicitation of rishitin and lubumin accumulation in potato tuber tissue, *Plant Physiol.* 84:520–525.

Recent Advances in Molecular Biology Techniques for Studying Phytosymbiotic Microbes

David Gerhold*

Department of Microbiology
University of Tennessee
Knoxville, Tennessee 37996

Gary Stacey

Department of Microbiology and
Graduate Program of Ecology
University of Tennessee
Knoxville, Tennessee 37996

Present address: Department of Plant Pathology, University of California, Riverside, California 92521

Introduction

Plant-microbe interactions have been the subject of an explosive research effort in recent years. This explosion is largely the result of the development of methods to manipulate microbial genes and gene products, in vitro as well as in vivo. Our understanding of plant-microbe interactions is thus primarily due to our newfound ability to alter and study symbiosis-related genes of the Gram-negative bacteria, filamentous fungi (molds), and viruses that associate with plants. These techniques, which arose largely from studies of *Escherichia coli* and its phages, are, in turn, giving rise to a similar revolution in plant molecular biology. This is occurring via adaptation of *Agrobacterium tumefaciens* and plant viruses for plant transformation.

From a plant's point of view (or a farmer's), microbial symbionts are either beneficial or harmful (i.e., pathogenic), with a "gray area" in between, populated by saprophytes and microbes that interact "loosely" with plants. Genetic study of the beneficial and pathogenic bacteria is relatively straightforward because each evolves a particular set of genes to modulate its interactions with the plant host. Thus, rhizobia carry *nodulation* (*nod*) genes, which are induced by host-plant signals (Downie and Johnston 1986), and agrobacteria carry host-induced virulence (*vir*) genes (Rogowsky et al. 1987), which act to transfer tumorigenic DNA (T-DNA) into plant cells. Relatively sophisticated genetic methods have been developed for these intensively studied Rhizobiaceae. Most of these methods are applicable to other Gram-negative bacteria such as pathogenic *Pseudomonas*, *Xanthomonas*, and *Erwinia* species, as well as the potentially important rhizosphere bacteria: nitrogen-fixing *Azospirillum* and growth-promoting *Pseudomonas* species.

Progress is being made on the study of plant DNA viruses [i.e., cauliflower mosaic virus (CaMV)] and RNA viruses [e.g., tobacco mosaic virus (TMV)], both for use as plant vectors and for diagnostic purposes. Difficulties in culturing the ubiquitous mycorrhizal fungi, however, still preclude development of suitable genetic systems for these beneficial symbionts.

In contrast, transformation systems have recently been developed for a variety of plant-pathogenic fungi including some ascomycetes: *Cochliobolus heterostrophus*, *Glomerella cingulata*, *Magnaporthe grisea* (Parsons et al. 1987); and the basidiomycete *Ustilago maydis* (Leong et al. 1988). Much of the molecular biology of the ascomycetes is being pioneered in *Neurospora crassa* (a nonpathogen) and *Aspergillus nidulans* (a common "pathogen" of bread).

By focusing on the above mentioned organisms, for which molecular genetics systems have been developed, we hope to provide a brief update covering techniques currently in use. Suggestions will be made concerning developments that may be extended to other genera when appropriate, and mention will be made of some general new techniques applicable to nucleic acid manipulations in bacteria, viruses, and fungi. Finally, this is not intended to be an exhaustive summary of methods, but rather an update on recent developments and future trends in molecular biology techniques for studies of plant-microbe interactions.

Bacterial Molecular Genetics

The majority of new molecular genetics methods were developed for use with bacterial genes. Many of these methods can be applied to studies of fungi and viruses as well. Thus, the bacterial genetics section that follows will also include more general methods for manipulations of nucleic acids.

Mutagenesis

Mutagenesis of symbiotic bacteria has traditionally been performed by UV irradiation or chemical (e.g., nitrosoguanidine) treatment of wild-type cells, followed by screening of survivors for alterations in the ability to interact with host plants. Such mutants can often be characterized by complementing mutations from a wild-type DNA library (Gutterson et al. 1986). The use of Tn5 and other transposons has become the method of choice for both random and site-directed mutagenesis, because mutant genes can be readily localized and cloned via the transposon's antibiotic markers. Transposon mutagenesis has also been adapted for most bacteria, both by utilizing a variety of transposons, and by the invention of several methods for introducing them into the bacterial genome. The Gram-negative transposons listed in Table 1 appear to function in any Gram-negative bacterium (Berg and Berg 1983). Transposons used for mutagenesis have been selected on the basis of low insertion-sequence specificity, and variety in native or engineered marker genes.

Several special applications of transposons have been developed, including use of in vivo transposon replacement (i.e., substituting one drug resistance marker for another by homologous recombination) to allow another cycle of mutagenesis (DeVos et al. 1986), inclusion of the mob site in Tn5 to allow conjugal transfer of nonconjugal plasmids (Simon 1984), and deletion mutagenesis mediated by two transposons (Michiels and Cornelis 1986).

For random mutagenesis (of an entire genome), transposons are usually introduced via conjugation of "suicide plasmids," which can-

TABLE 1   Transposons Useful for Mutagenesis

Transposon	Marker genes	Reference
$Tn5$	Km^{Ra}	deBruijn and Lupski 1984
$Tn5\text{-}mob^{b}$	Km^{R}	Simon 1984
$Tn5\text{-}132$	Km^{R}, Tc^{R}	Berg and Berg 1983
$Tn5\text{-}751$	Km^{R}, Tmp^{R}	Rella et al. 1985
$Tn5\text{-}233$	$Km^{R}, Gm^{R}, Str^{R}, Spc^{R}$	deVos et al. 1986
$Tn5\text{-}235$	$Km^{R}, Str^{R}, lacZ^{C}$	deVos et al. 1986
$Tn3$	Ap^{R}	Heffron 1983
$Tn3503$	Km^{R}	Gutterson et al. 1986
$Tn1721$	Tc^{R}	Ubben and Schmitt 1986
$Tn1725$	Cm^{R}	Ubben and Schmitt 1986
$Tn1732$	Km^{R*}	Ubben and Schmitt 1986

[a]The $Tn5$ Km^{R} gene also confers neomycin resistance and, in many nonenteric hosts, a second gene confers streptomycin resistance (Selvaraj and Iyer 1984, Putnoky et al. 1983).
[b]Allows transfer of resident replicons via triparental matings (Ditta et al. 1980).
Key: Km = kanamycin, Str = streptomycin, Spc = spectinomycin, Tc = tetracycline, Tmp = trimethoprim, Ap = ampicillin, Cm = chloramphenicol, Gm = gentamicin, and $lacZ^{C}$ indicates a constitutive gene for β-galactosidase.

not replicate in the target organism because they contain lethal genes (Johnston et al. 1978), narrow-host-range replication functions (Simon et al. 1983), or temperature-sensitive replication functions (Rella et al. 1985). For site-directed mutagenesis (of plasmid-cloned DNA), transposons have also been introduced via coliphages (lambda, Ruvkun and Ausubel 1981, or P1, Quinto and Bender 1984), or from a transposon resident in the chromosome (Ma et al. 1982). $Tn5$ insertions occurring in the target plasmid can then be selected by conjugation or transformation into a second host bacterium.

Despite its broad usefulness, $Tn5$ has several significant limitations. Weak promoters read out from the left and/or right ends of $Tn5$, (Berg et al. 1980), rendering some $Tn5$ mutations incompletely polar. Also, the insertion sequence 50 (IS50) elements bordering $Tn5$ can occasionally transpose to sites distant from the "parent" $Tn5$ (Berg and Berg 1983), causing secondary mutations. $Tn5$ insertions also appear to be not entirely random, favoring sites that are being actively transcribed (McKinnon et al. 1985).

Mutagenesis of cloned DNA can be performed by many methods. Deletions can be constructed by deleting restriction fragments from a plasmid, then ligating in a selectable marker such as kanamycin resistance (Km^{R}) from $Tn5$ or specially designed cassettes called "interposons" (Fellay et al. 1987). Alternatively, genomic residents (e.g., phage lambda prophages) carrying a selectable marker can be induced by heat, and excision/deletion events can be detected among survivors (Berg and Berg 1983). Synthetic oligonucleotide linkers can be inserted in sites scattered within a clone using "linker scanning" (Zhang et al. 1986).

Selectable mutations in cloned DNA are especially useful for locating bacterial (or fungal) genes following replacement of the wild-type gene with the mutant allele (Figure 1). This gene replacement event (also called "marker exchange," "transplacement," or "homogenotization") can often be selected by mobilizing the mutant allele from *E. coli* to the bacterium of interest on a narrow-host-range vector and selecting sequentially for maintenance of the mutation marker ("cointegration," Figure 2), followed by loss of the vector marker (Nieuwkoop et al. 1987, Jagadish et al. 1985). Alternatively, a wide-host-range vector carrying the mutant gene will be lost following homologous recombination events on both sides of the marker if a second incompatible plasmid is introduced by conjugation (Ruvkun and Ausubel 1981, Figure 1). It is imperative that putative "homogenotes" (organisms with wild-type genes replaced by mutant alleles) be confirmed by DNA hybridization analysis prior to analyzing their symbiotic phenotypes. This is done by probing electrophoretically separated restriction digests of genomic DNA with a probe containing the wild-type gene or DNA region in question, to ensure it has been interrupted in the homogenote.

Once a particular gene has been cloned, identified, and (usually) sequenced, a variety of new techniques can be used to perform fine-structure analysis of the gene. Single or multiple mutations can be designed and can be introduced at a precise location, by synthesizing an oligonucleotide that is complementary to the antisense strand of a bacteriophage M13 clone save for one or several base pairs (Geider 1986). By annealing this oligonucleotide to the noncoding strand of the gene in an M13 clone, then synthesizing the remainder of the com-

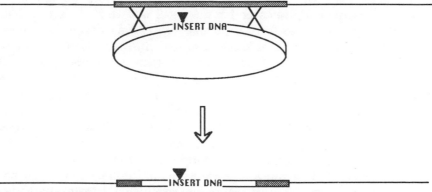

Figure 1   Transplacement. The wild-type gene is replaced by a mutant allele containing a transposon (black triangle) by homologous recombination (X's).

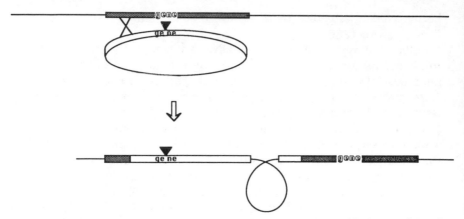

Figure 2 Cointegration. The mutant gene containing a transposon (black triangle) and the plasmid are incorporated into the chromosome by homologous recombination (X).

plementary strand using DNA polymerase, ligase, and a second primer oligonucleotide, the mutation can be incorporated into the M13 sense strand. An M13 clone carrying the desired oligo-directed mutation is then identified by colony hybridization using the synthetic oligonucleotide as probe. A similar, but simpler, procedure requires an oligonucleotide that bridges a restriction site in the gene and carries a noncomplementary mutation (e.g., extra deoxynucleotides) between the complementary sequences (Mandecki 1986, Figure 3). This oligonucleotide is annealed to restricted, denatured DNA, and the oligonucleotide–complementary-strand hybrid is used to directly transform *E. coli*. The screening of putative mutant plasmids is facilitated by the loss of the original restriction site. As in all the fine-structure mutagenesis schemes, DNA sequencing is necessary (these procedures generally are performed in M13 vectors amenable to "dideoxy" sequencing) to establish the exact nature of the mutations. Procedures designed to form a population of gene derivatives, each bearing a mutation at a single codon, employ incorporation of guanine 8-amino fluorene (Mitchell and Stoehrer 1986) or α-mercaptodeoxynucleotide (Abarzua and Marians 1984) analogues. Similar results can be obtained by "misincorporation" of 1 to 2 bp using Avian myeloblastosis-virus (AMV) reverse transcriptase (Champoux 1984) or terminal transferase (Sirotkin 1986). While these are powerful techniques for dissecting gene function, they tend to be relatively complex and difficult, and are most useful if a strong selection can be devised for mutant genes determining a particular phenotype. Another fine-structure technique, useful for determining functions of particular protein domains is construction of "chimeric" genes. In general, two homologous genes that confer different phenotypes can

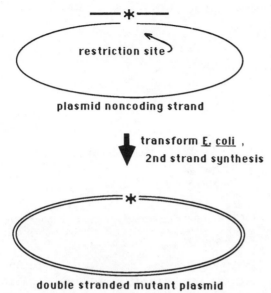

plasmid noncoding strand

transform E. coli ,
2nd strand synthesis

double stranded mutant plasmid

Figure 3 Oligonucleotide mutagenesis. A mutation (*) is introduced into a plasmid via a synthetic oligonucleotide (heavy line), which spans a restriction site. *E. coli* synthesizes the second strand and fills the gap opposite the mutation following transformation. (*Mandecki 1986.*)

be spliced together at a restriction site that is common to both (Horvath et al. 1987). The fusion protein from such a chimera will thus result from the 5′ portion of one gene and the 3′ portion of the second, homologous gene. Such chimeras have also been formed by selecting for recombination between two highly homologous genes (Lugtenberg 1988). A third procedure involves annealing lambda-exonuclease-generated protruding 3′ ends of two gene homologues followed by repair synthesis using DNA polymerase I and DNA ligase to generate the chimera (Matsui et al. 1987, Figure 4). This method is useful for making hybrids from genes that lack a common restriction site.

Cloning

Since numerous techniques for cloning DNA fragments are in common use, only new cloning strategies and particularly original or useful cloning vectors will be discussed. Also, since wide-host-range vectors are particularly appropriate for cloning and transfer of genes from bacterial symbionts of plants, these will be stressed. A thorough, well-organized catalogue of published vectors is sold by Elsevier

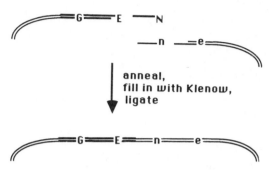

Figure 4 Hybrid gene formation. Two partially homologous genes (GENE and gene) are truncated at opposite ends by restriction endonucleases. The 5′ strands are then partially removed using lambda exonuclease, and the remaining strands are annealed. Single-stranded gaps are filled using Klenow, and the double-stranded hybrid plasmid is ligated and transformed into *E. coli*.

(*Cloning Vectors*) with updates added annually. (See under Pouwels et al. (eds.) in "References.")

Bacterial genes are usually cloned from libraries of DNA inserts in "cosmids," plasmids carrying the cohesive (*cos*) ends from lambda. In vitro lambda packaging of cosmids allows efficient transfection of DNA into *E. coli* and sets convenient lower and upper size limits on the DNA insert sizes that are cloned. Narrow-host-range cosmid pJB8, for example, allows cloning of *BamHI, Sau3A*, or *MboI* partial digest fragments of approximately 31 to 47 kb in size (Ish-Horowicz and Burke 1981). This cosmid also facilitates efficient ligations and cloning by introducing a method that prevents complex concatamer formation (Ish-Horowicz and Burke 1981, Figure 5). This scheme is broadly applicable, including broad-host-range cosmids (e.g., pLAFR3, Staskawicz et al. 1987).

The plasmids pLAFR3 and pRK311 (Ditta et al. 1985) have the additional advantage of containing a polylinker positioned between the promoter and structural parts of the *lacZ* gene, allowing cloning of a variety of restriction enzyme fragments, and differentiation between cloned DNA and religated vector sequences. The closely related pVK100-102 cosmids (Knauf and Nester 1982) are convenient because they allow insertional inactivation of the tetracycline resistance (Tc^R) or Km^R gene by *HindIII, SalI*, or *XhoI* fragments. Screening of cosmid libraries may be facilitated in some cases by amplification in *E. coli* strains adapted for repackaging cosmids (Jacobs et al 1986).

Genes of interest are usually selected from a cosmid library by using either a selectable marker such as Km^R from a *Tn5* insert, hybridization to select clones complementary to a probe from a similar gene,

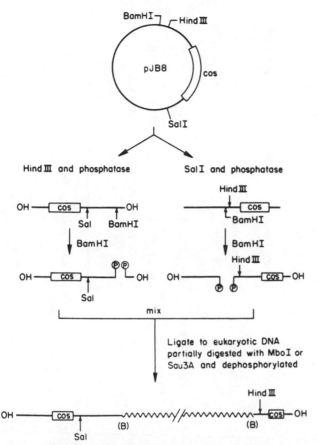

Figure 5 Cosmid cloning method of Ish-Horowicz and Burke (*1981; from Maniatis et al. 1982, with permission*). This method prevents self-ligation of vector DNA and cloning of multiple noncontiguous insert DNA fragments. Dephosphorylated insert fragments can be ligated only between phosphorylated *BamHI* ends to generate molecules (bottom of figure) that can be packaged into lambda phage in vitro.

or complementation to select clones that correct a mutation or negative phenotype. Broad-host-range mobilizable cosmids are particularly useful for selecting genes via complementation. A few of these versatile vectors are listed in Table 2. Conjugation of vectors containing the *mob* region from *E. coli* to other Gram-negative bacteria is generally accomplished via triparental matings (Ditta et al. 1980) using helper plasmids pRK2073 (trimethoprim resistant) or pRK2013 (kanamycin resistant).

Cloning of large DNA fragments from a few Gram-negative phytosymbiotic bacteria has been facilitated by the use of "R-primes."

TABLE 2 Broad-Host-Range Mobilizable Cosmids[a]

Cosmid	Size (kb)	Incompatibility group	Genetic markers[b]	Cloning sites[c]	Reference
pLAFR1	21.6	IncP	Tc^R	EcoRI, BstEII	Friedman et al. 1982
pRK311	22	IncP	Tc^R, lacZ	BamHI, HindIII, PstI	Ditta et al. 1985
pLAFR3 1987	22	IncP	Tc^R, lacZ	EcoRI, BstEII, BamHI	Staskawicz et al., 1987
pVK100	23	IncP	Tc^R, Km^R	EcoRI, HindIII, SalI, XhoI	Knauf and Nester 1982
pJRD215	10.2	IncQ	Km^R, Str^R	23 restriction sites	Davison et al. 1987
pSF6	11.1	IncW	Str^R/Spc^R	EcoRI, SalI, BamHI, HindIII, XbaI, PstI	Selvaraj et al. 1984
pKT247	11.5	IncQ	Ap^R, Sul^R Str^R	BamHI, EcoRI, SacI	Bagdasarian et al. 1981
pMMB33	13.8	IncQ	Km^R	BamHI, HpaI, SacI, EcoRI	Frey et al. 1983
pUCD200110	0.4	See reference	Ap^R, Km^R	SacI, ClaI, KpnI, PstI, SalI	Gallie et al. 1985
pGS72	13	IncP	Tc^R, Km^R	EcoRI, HindIII, SalI, XhoI	Selvaraj and Iyer 1985

[a]Abridged and updated from *Cloning Vectors* (Elsevier Publ.).
[b]Tc = tetracycline, Km = kanamycin, Str = streptomycin, Spc = spectinomycin, Sul = sulfonamide, Ap = ampicillin.
[c]Note that *BamHI* sites allow cloning of *BglII, BclI, XhoII, MboI,* and *Sau3A* fragments.

These plasmids, derived from drug resistance plasmid RP4, can spontaneously incorporate 100- to 330-kb DNA fragments and transfer them into another host via their inherent transfer (*tra*) functions (Moore et al. 1983, Banfalvi et al. 1983, Nayudu and Rolfe 1987). Such large R-prime plasmids may not be obtainable in all bacteria, and their large sizes can make them difficult to manipulate (N. Deshmane, unpublished).

Isolation of *E. coli* genes has been greatly simplified by application of transducing coliphages. Effective transducing phages have not been described for most nonenteric Gram-negative bacteria. The cloning and transfer of the *lamB* lambda receptor, however, appears to extend this phage's host range (deVries et al. 1984) to include nonenteric bacteria. Further adaptation of this principle may, in the future, allow gene transfer to and from *Rhizobium, Agrobacterium*, and *Pseudomonas* by simple transduction.

Several narrow-host-range plasmids are noteworthy for specialized applications. The M13-derived vectors offer advantages for ease of cloning small fragments, oligonucleotide-directed mutagenesis, high yield of insert DNA, and usefulness in "dideoxy" sequencing. The *lacZ* inactivation scheme, which allows one to distinguish clones from religated vector, is finding widespread application beyond the original M13- and pUC-series vectors (Geider 1986). Vectors that allow positive selection of cloned DNA are also becoming quite numerous (see Elsevier's *Cloning Vectors*). Such vectors are conditionally unable to replicate, or can kill particular *E. coli* hosts, in the absence of an insert (e.g., pHSG66, Hashimoto-Gotoh et al. 1986). For maximal amplification of cloned DNA, Frey and Timmis (1985) have developed a family of ColD-derived narrow-host-range plasmids that autoamplify to extremely high copy number in stationary-phase cells. A similar amplification of cloned DNA is possible in plasmid pRBG156, which is deleted for the site encoding RNAI, a transcript that normally represses the plasmid's replication (Gayle et al. 1986).

The construction of vectors with unusual restriction enzyme sites may be facilitated using restriction site banks carrying 13 sites (Davison et al. 1987). Finally, cocloning of the *lacZo* operator and insert DNA allows phenotypic selection of cloned inserts to simplify subcloning and deletion mutagenesis (Danner 1986). This selection relies on the ability of *lacZo* to bind to the *lac*-repressor protein, relieving repression of easily detectable *lacZ* gene expression.

Hybridization

Nucleic acid hybridizations have become an indispensable tool for screening DNA libraries, investigating the occurrence of homologous genes among different species or strains, and evaluating the transcription of selected genes. An important development in hybridization technology is the ability to detect shorter regions of homology by incorporation (into probes) of nucleotides with reduced base-pairing specificity. Decreasing specificity by incorporating, for example, 5-fluorodeoxyuridine (Habener et al. 1988), deoxyinosine (Miura et al. 1985), or triazolodeoxynucleotides (Huynh et al. 1985) in probes allows one to detect shorter regions of homology than the present limit of approximately 15 nucleotides. The ability to detect shorter regions of homology is especially important for attempts to identify genes using probes derived from short protein sequences (Fukuda et al. 1985). A mixed 25-mer was recently used to identify conserved (≈ 47 bp) promoter sequences, and thus, coregulated genes, by simple hybridization (Rostas et al. 1986, Gerhold et al. 1988). This scheme may become generally applicable to cloning of entire gene regulons if shorter promoters and/or operators can be detected by hybridization.

Several new procedures are available to increase the specific activities or simplify preparation of hybridization probes. Very short (6 bp) mixed oligonucleotides annealed to target DNA serve as "random primers" for [^{32}P]-dNTP Klenow (DNA polymerase I large fragment) incorporation and complementary strand synthesis. This allows Klenow synthesis of probes with $>10^9$ dpm/µg specific activities (Feinberg and Vogelstein 1983). Similar specific activities can be obtained by [^{32}P]-dAMP "tailing" of DNA fragments using nucleotide terminal transferase (Collins and Hunsaker 1985). Probes can also be made from M13 clones by extending sequencing primers with Klenow enzyme. Fixed-length probes can be made from M13 clones by blocking primer extension using a second dideoxy AMP oligonucleotide (Liu et al. 1986). Gamper *et al.* (1986) have devised an extremely sensitive method for solution hybridization. Hybridization is followed by covalent cross-linking of the labeled probe to complementary DNA and analysis by denaturing electrophoresis.

Several modifications of the original colony blot procedure of Grunstein and Hogness (1975) have increased its sensitivity and adapted it for detecting RNA transcripts. Steaming *E. coli* cells was found to enhance plasmid-filter binding (Maas 1983), and a protease IV digestion step aided lysis of *Rhizobium* cells (Hodgson and Roberts 1983). Development of mRNA-amplifying vectors (lambda gt11, Huynh et al. 1985, and lambda SWAJ, Palazzolo and Meyerowitz 1987) allows overproduction of RNA complementary to either the coding or noncoding DNA strand. This RNA can be labeled and used as a hybridization probe to screen cDNA libraries, quantify cellular message levels (Yu and Gorovsky 1986), and identify the DNA coding strand by hybridization to cellular messenger RNA (Palazzolo and Meyerowitz 1987).

DNA mapping and sequencing

Development of techniques for mapping and sequencing DNA is receiving much attention lately, due to the proposed sequencing of the human genome. In preliminary work, researchers have recently mapped the *E. coli* genome using overlapping large (20 to 1,000 kb) restriction fragments generated by restriction enzymes with 8-bp recognition sites (Smith et al. 1987). This genome was mapped at higher resolution independently, by making partial restriction digests of lambda library clones linearized at the *cos* site, followed by electrophoresis and blotting the digests to nitrocellulose, and probing with a fragment complementary to the right end of the lambda vector (Kohara et al. 1987). Cosmids were ordered using a computer program that aligned overlapping ends by their common restriction fragments (generated by five restriction enzymes).

The separation of large DNA fragments is facilitated by two new electrophoretic methods: orthagonal-field-alternation gel electrophoresis (OFAGE) and field-inversion gel electrophoresis (FIGE, Schwartz and Cantor 1984, Carle et al. 1986). These techniques are capable of separating DNA of up to 2,000 kb (Collins et al. 1987). Thus, these techniques facilitate cosmid library construction, mapping of large restriction fragments, and "chromosomal jumping" (see below), and may allow construction of "subgenomic libraries" from individual fungal chromosomes or organelles, or from bacterial megaplasmids.

A gene can often be cloned by "chromosomal walking" (Carlock, 1986) if it can be identified by genetic linkage with a second, cloned locus. In this instance, adjacent, overlapping cosmid clones are selected by hybridization, and used sequentially to hybrid-select further clones. "Walking" allows one to traverse approximately 20 kb per step, whereas "chromosomal jumping" allows one to traverse 100 kb or more per step. "Jumping" is performed by ligating the two ends of very large (e.g., 80 to 130 kb, Collins et al. 1986) contiguous DNA restriction fragments to a selectable gene such as *supF*, digesting completely with a second restriction endonuclease, and ligating the resulting left-end–*supF*–right-end fragment into cosmids. By probing such a jumping library with a defined clone one can "jump" along the chromosome. Finally, DNA containing the left or right ends of a jumping clone can be selected by probing a contiguous (walking) cosmid library with the jumping clone (Figure 6).

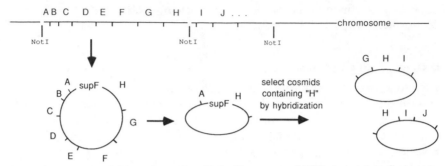

Figure 6 Chromosomal jumping using NotI. Chromosomal DNA is cut into fragments averaging 50 to 100 kb using 8-bp restriction enzyme NotI. Resulting fragments are ligated to the selectable marker *supF*, and digested with, e.g., *EcoRI*, generating fragments "B," "C," "D," etc. Fragments flanking *supF* are cloned into a cosmid vector to make a "jumping library." Clones containing, for example, fragment "A" are selected by hybridization, and used to select distant DNA fragment H by hybridization to a conventional cosmid library.

Several new methods exist for physical mapping of cosmid clones in addition to the hybridization technique described above (Kohara et al. 1987). In one method, a cosmid is partially digested with a restriction enzyme that cuts only within insert DNA, recircularized by ligase under dilute conditions, and transformed into *E. coli*. The restriction fragments within the cosmid can then be ordered by comparing the restriction patterns of approximately 10 to 20 such deleted derivatives (Buikema et al. 1983). A simpler technique is to do a partial restriction enzyme digest of a DNA fragment that is radioactively labeled at one end, and separate the resulting ladder of labeled fragments by electrophoresis. Single-end-labeled fragments can be generated using DNA kinase to label both ends of a fragment (with ^{32}P), cutting it in half using an appropriate restriction enzyme, and isolating the pieces electrophoretically (Smith and Bernstiel 1976).

The chemical sequencing procedure of Maxam and Gilbert (1980) has been largely supplanted by the dideoxy-chain-termination, method (Carlson and Messing 1984); however, new strategies are reviving the original chemical method. Versatile vectors pSP64CS and pSP65CS facilitate subcloning, single-strand labeling, and transcription of clones for chemical sequencing (Eckert 1987). Rosenthal et al. (1986) have increased the efficiency of the chemical degradation reactions by adopting them for DNA fragments immobilized on anion-exchange paper. The most exciting advance in chemical sequencing is the demonstration of "multiplex sequencing." Up to 20 restriction fragment clones are pooled, then chemically degraded and resolved on sequencing gels. The DNA is blotted to a nylon membrane and probed sequentially with oligonucleotides complementary to sequences unique to each of the 20 original vectors (Church and Kiefer-Higgins 1988).

Dideoxy-chain-termination DNA sequencing is experiencing a revolution based on the use of automated sequencers, but is still seriously limited by the time-consuming task of subcloning overlapping fragments of DNA templates. Since only 300 to 600 nucleotides can be read from one strand of a typical M13-vector clone, longer fragments must either be subcloned (Carlson and Messing 1984), or primed internally with a synthetic oligonucleotide dictated by the sequence. Plasmid pKUN9 allows sequencing of both strands of cloned DNA, since superinfection with M13/FF causes replication of the "F" strand, and superinfection with phage Ike allows "I" strand replication (Peeters et al. 1986).

Cloned restriction fragments can be progressively deleted from one end to form "nested sets" by relatively time-consuming procedures employing exonuclease III (Henikoff 1984, Yanisch-perron et al. 1985), or phage T_4 DNA polymerase (Dale et al. 1985). A faster procedure allows selection of deletions caused by a $Tn9$ fragment in pAA-pZ vectors (Peng and Wu 1986). However, this system may be fatally flawed by the occurrence of deletion endpoint "hot spots" in insert DNA (Peng and Wu 1986).

Sequencing of single-stranded M13 DNA has been extended to allow dideoxy sequencing of supercoiled dsDNA (Chen and Seeburg 1985) and of isolated restriction fragments. Restriction fragments can be primed for dideoxy sequencing by addition of poly (dA) tails followed by poly (dT) oligonucleotide primers (Stallard et al. 1987). Alternatively, a "sequencing primer linker" ("splinker") can be ligated to one end of a restriction fragment, allowing its self-complementary "hairpin" structure to prime the sequencing reaction (Kalisch et al. 1986).

Two automated systems have recently been developed to sequence DNA without radioactivity. The ABI system (Applied Biosystems, Inc., Foster City, California) uses dideoxynucleotides to terminate enzymatic extension of a primer, which is coupled to one of four fluorescent dyes. Four reactions are performed, using a different color dye for each dideoxynucleotide reaction. The DNA "ladders" resulting from the four reactions are mixed, resolved by electrophoresis, and the DNA base sequence is read by a laser fluorometer as the colored DNA fragments exit the bottom of the electrophoresis gel (Roberts 1988). The Du Pont system is similar except that the dyes are linked to the dideoxynucleotides, and are chemically very similar. These dyes necessitate use of a sophisticated laser fluorometer to read sequences, but minimize sequence errors by exerting a uniform effect on fragment mobility (Prober et al. 1988). These systems will probably be steadily refined as their use becomes widespread.

Once a locus of interest has been sequenced, it is useful to employ specialized computer programs to align overlapping fragments (Waterman 1986) and evaluate the sequences in various ways. In the absence of precise mapping of transcribed regions ("S1 nuclease mapping" or "primer extension," Fisher et al. 1987), it is useful to identify putative genes by delimiting and evaluating open reading frames (ORFs) (Doolittle 1986). Location of ORFs and promoter regions can be performed by searches for *E. coli*-type promoter homology (PROMSEARCH program, Mulligan and McClure 1986). Unfortunately, some symbiotic genes do not have typical *E. coli* $-10/-35$ consensus promoters. Thus, although *E. coli* promoters generally function in other Gram-negative bacteria, *Rhizobium* promoters for symbiotic genes, for example, often do not function well in *E. coli* (Kahn and Timblin 1984). Statistical analyses based on 6-bp helix-twist angle motifs (Tung and Harvey 1987), or pattern recognition (Galas et al. 1985) can also identify putative promoters. The latter two methods are potentially applicable to non-*E. coli*-type $(-10, -35)$ promoters, including eukaryotic fungal and viral promoters. Another versatile statistical tool can predict functionality of ORFs (Fickett 1982). This algorithm searches for 3-bp periodicity of codons, resulting from preferred codon usages.

In general, the software necessary to align sequences, identify ORFs, compare sequences, predict translated amino acids, and even predict protein folding, is supplied in packaged sets (reviewed by Cannon 1988). Finally, comparison of a newly sequenced gene to the collection of published gene sequences (GenBank, Los Alamos, New Mexico 87545) can provide clues to the identity and biochemical functions of the gene. Such clues must, of course, be substantiated by biochemical assays to determine gene functions.

Gene fusion

Investigation of gene regulation has long been performed by probing northern (RNA) blots with the gene sequence of interest and by probing western (protein) blots with labeled antibodies. Now many more assays can be performed and precisely quantitated by fusing a reporter gene to the regulatory sequences of the gene of interest. If the reporter gene retains its translational signals (ATG start codon and ribosome binding site in bacteria), expression of the reporter-gene product is regulated by the transcriptional signals (i.e., promoter and regulatory sequences) of the gene of interest. In translational fusions, reporter-gene expression is dependent on transcriptional *and* translational sequences from the gene of interest (Figure 7). Translational fusions must be formed in the correct reading frame, producing a chimeric protein product. The *E. coli lacZ* gene is the favorite reporter gene for use in bacteria; however, many other genes have also been adapted for this purpose: e.g., *uidA*, encoding β-glucuronidase (Jefferson et al. 1986); *galk*, encoding galactokinase (Schneider and Beck 1986); *phoA*, encoding alkaline phosphatase (Schneider and Beck 1986); *cat*, encoding chloramphenicol acetyl transferase (Osbourn et al. 1987); and bacterial or firefly *lux* genes (Engebrecht et al. 1985), which cause luminescence. The *lacZ* gene is also a convenient phenotypic marker for monitoring recombinant bacteria following their release into the environment (Drahos et al. 1986).

Transposon fusion vehicles are highly versatile, allowing fusions to cloned genes, or identification of symbiotically regulated genes in vivo. Symbiotic-gene–reporter-gene fusions can be constructed in vitro by insertion of promoter-probe cassettes or by cloning the 5′ portion of the symbiotic gene into a promoter-probe vector. Alternatively,

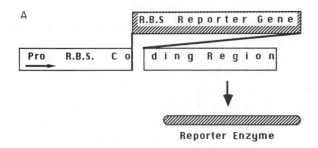

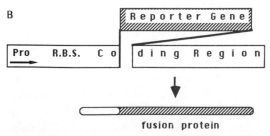

Figure 7 Transcriptional (*A*) and translational (*B*) reporter-gene fusions. The gene under investigation is drawn with respect to promoter (Pro), ribosome binding site (R.B.S.), and coding regions.

transposons bearing a reporter gene can transpose into the gene of interest to form fusions in vivo. By using random transposon–reporter-gene mutagenesis, followed by assay of replicate fusion strains for increased reporter-gene activity in the presence of plant extracts, Sadowski et al. (1988) were able to directly identify plant-regulated genes in *Rhizobium fredii*. More commonly, gene fusion vehicles are employed in site-directed mutagenesis of cloned sequences.

Probably the most versatile, widely used fusion vehicles are the "mini-mu" derivatives. These are deleted phage mu coliphage-transposon constructs containing the *lacZYA* operon (see Table 3). Some of these mini-mu vehicles also carry *mob* and *ori* segments allowing mobilization and maintenance of narrow-host-range plasmid fusion constructs in diverse Gram-negative bacteria. It should be noted that transcriptional mini-mu–*lac* and *Tn5–lac* vehicles bear the *trp* translational start sequence, which apparently acts as a promoter in various Gram-negative bacteria, causing a high background of β-galactosidase activity (Groisman and Casadaban 1986, and Gerhold, unpublished results).

While plasmidborne gene fusions seem to reflect the regulation of the genomic gene copies, the level of expression is probably altered in many cases by its presence on a multicopy plasmid, level of supercoiling, allosteric effects within the fusion protein, and so forth. Most of these inaccuracies can be overcome by recombining the gene-fusion allele into the gene's native site in the genome (Mulligan and Long 1985) as described earlier for transposon "gene replacement."

Some useful transposon-*lacZ* vehicles are listed in Table 3. *TnphoA*, a transposon containing the promoterless alkaline phosphatase gene, is useful for detecting envelope (inner membrane, outer membrane, and periplasmic) proteins from Gram-negative bacteria (Gutierrez et al. 1987).

As an alternative to transposition, genes can be fused to reporter genes in vitro using promoter-probe vectors. Again, translational fusions must be spliced together in frame if the vector lacks a ribosome binding site and initiation codon. Such fusions cannot be used to replace the wild-type gene in vivo, but cointegrate merodiploids can sometimes be constructed (Figure 2). If vector fusion constructs are to be maintained as plasmids, they must be cloned into a wide-host-range replicon. Some of the newly developed wide-host-range cloning vectors are listed in Table 4. Note that β-galactosidase, glucuronidase, alkaline phosphatase, and galactokinase are easily assayed using chromogenic substrates, whereas precise *lux* quantitation requires special instrumentation (Shaw and Kado 1986), and quantitation of antibiotic resistance is not very precise [except for chloramphenicol acetyl transferase (Osbourn et al. 1987].

TABLE 3 Transposon-*lac* Fusion Vehicles

Transposon-*lac* fusion vehicle	Size (kb)	Transcriptional or translational?	Genetic markers	Replicon incompatibility group	Reference
Mu dI1734	9.7	XS	Km^R	—	Castilho et al. 1984
Mu dII1734	9.7	XL	Km^R	—	Castilho et al. 1984
Mu dI1678	22.4	XS	Ap^R	—	Castilho et al. 1984
Mu dII1678	22.4	XL	Ap^R	—	Castilho et al. 1984
Mu dIIPR13	9.2	XL	Cm^R	—	Ratet et al. 1988
Mu dIIPR46	13.7	XL	Cm^R, Gm^R	NHR	Ratet et al. 1988
Mu dIIPR48	21.5	XL	Cm^R, Gm^R	IncRiHRI	Ratet et al. 1988
Mu dII4042	16.7	XL	Cm^R	NHR	Groisman and Casadaban 1986
Mu DI5155	15.6	XS	Km^R	NHR	Groisman and Casadaban 1986
Mu dI5166	15.8	XS	Cm^R	NHR	Groisman and Casadaban 1986
Mu dII5117	21.7	XL	Km^R, Spc^R/Str^R	IncW	Groisman and Casadaban 1986
Tn3-HoHo1	14.25	XL	Ap^R		Stachel et al. 1985
Tn3-HoHo2	14.25	XS	Ap^R		Stachel et al. 1985
Tn5-lac	12.2	XS	Km^R		Kroos and Kaiser 1984
Tn1737 Cm	10.75	XS	Cm^R		Ubben and Schmitt 1987
Tn1737 Km	9.45	XS	Km^R		Ubben and Schmitt 1987
Tn1737 Sm	10.5	XS	Str^R		Ubben and Schmitt 1987

Key: XS = transcriptional, XL = translational, Km = Kanamycin, Ap = ampicillin, Cm = chloramphenicol, Gm = gentamicin, Spc = spectinomycin, Str = streptomycin, NHR = narrow host range, Inc = incompatibility group.

TABLE 4 Wide-Host-Range Plasmids Containing Reporter Genes

Wide-host-range fusion vehicle	Size (kb)	Fusion gene	Genetic markers	Fusion cloning sites[a]	Reference
pUCD607	20.3	b-lux	Spc^R, Km^R, Ap^R	SalI	Shaw and Kado 1986
pIJ3100	13.2	cat	Str^R	SalI, HindIII, SmaI, BamHI	Osbourn et al. 1987
pMK341	15.4	npt	Tc^R, Cm^R	BglII	Kahn and Timblin 1984
pMK330	23.0	lacZ	Tc^R, Km^R	BamHI	Kahn and Timblin 1984
pVS2434	24.7	lacZ	Ap^R, Km^R	EcoRI, BamHI, SmaI	Simon and Schumann 1987
pUI108	14.9	lacZ	Str^R, Sul^R	BamHI, SmaI	Nano et al. 1984

[a]Note that a *BamHI* or *BglII* site allows cloning of *BglII, BclI, XhoII, MboI,* and *Sau3A* fragments.
Key: Spc = spectinomycin, Km = kanamycin, Ap = ampicillin, Str = streptomycin, Tc = tetracycline, Cm = chloramphenicol, Sul = sulfonamide. *b-lux* = luminescence genes from *Vibrio* bacteria, *cat* = chloramphenicol acetyl transferase, *npt* = neomycin (kanamycin) phosphotransferase, *lacZ* = E. coli β-galactosidase.

By constructing "reversed" gene fusions, that is, fusing the promoters of a well-characterized gene to the coding region of a symbiotic gene, one can regulate expression of the symbiotic gene product. This is usually performed by cloning the gene's coding sequence downstream from a strong and easily regulated promoter in an expression vector to allow overproduction of the protein, or to determine whether a given restriction fragment contains a functional ORF (gene). Commonly used controllable promoters are the phage lambda P_L promoter (Seth et al. 1986), *E. coli trp* promoter (Tacon et al. 1983), *lacUV5* promoter (Bittner and Vapnek 1981), or *trp-lac* hybrid "*tac*" promoter (Stark 1987). The regulation of such genes is particularly important for cloning of genes whose products may be lethal in large amounts, such as certain membrane proteins (Windle 1986). By fusing the 5' region of the symbiotic gene of interest to the 3' coding region of *lacZ* in an expression vector, one can easily purify the fusion protein on an affinity chromatography column that employs antibodies specific for β-galactosidase (5 Prime–3 Prime, Inc.). This purified fusion protein may then be used to elicit antibodies against the symbiotic gene product. The expression vector lambda gt11 can be used to make 5'*lacZ*–3' symbiotic-gene fusions, or cDNA expression libraries as described below for fungi. Since the available expression vectors are numerous, no attempt will be made to list them here (see *Cloning Vectors*, Elsevier).

Fungal Molecular Genetics

Molecular genetic analyses of phytosymbiotic fungi are in their infancy since transformation of molds pathogenic to plants has been accomplished only recently (Leong 1988). In general, successful transformation of molds has been performed by enzymatic cell wall digestion to form protoplasts, adding DNA in the presence of $CaCl_2$ and polyethylene glycol (PEG), and regenerating mycelia on selective media. Despite this successful adaptation of yeast transformation techniques, the relatively advanced yeast shuttle vectors are not maintained in filamentous fungi. Thus, stable transformation of molds has resulted from chromosomal integration of marker genes, and a stable replicating vector for mold nuclei is not available (Leong 1988). Since phytopathogenic fungi do not maintain yeast shuttle vectors, and seem to recognize different transcriptional signals (Timberlake 1986), molecular genetics techniques are being pioneered in other filamentous fungi. In particular, the filamentous ascomycetes *Neurospora crassa* and *Aspergillus nidulans* are model systems for study of phytopathogenic fungi. Since techniques for transformation, complementation, and studying gene expression are relatively sophisticated in these two species, and seem to be applicable in large part to other filamentous fungi, they will be discussed in this chapter.

Gene identification and cloning

Fungal genes can be initially identified by classical genetic analysis of spontaneous mutants, by hybridization to analogous genes from other organisms, or by complementation of phenotypic mutants (Leong 1988, Yelton et al. 1983). By making a random DNA library in a special shuttle vector, Yelton et al. (1985) cloned a gene from *A. nidulans* by complementing a mutant in the same organism. Such shuttle vectors incorporate an origin of replication and selectable antibiotic resistance markers for *E. coli* with a selectable marker for fungi. The marker for fungi should express a selectable phenotype to mark integration of cloned DNA into the fungal genome. Selectable fungal genes employed include nutritional markers (e.g., *trpC* to complement a tryptophan-auxotrophic host, Yelton et al. 1985), and antibiotic resistance markers for hygromycin B (Holden et al. 1988), benomyl (Orbach et al. 1985), or chloramphenicol (Hamer and Timberlake 1987). Timberlake (1986) has developed a "depletion hybridization" method for selecting cDNA clones from genes that are expressed during differentiation. This technique may be applicable to identify genes that are expressed during host invasion by plant pathogens. Such a scheme would entail copying cDNAs from an mRNA population representing fungi cultured with plant hosts. This cDNA would be depleted of "housekeeping" messages by hybridization to "poly A" mRNA from a free-living culture. The remaining free cDNA is then cloned and screened for symbiosis-specific expression by probing dot blots of each clone with the free-living and symbiotic mRNA populations. cDNA libraries in expression vectors such as lambda gt11 may be screened using either labeled DNA fragments or antibodies (Young and Davis 1983).

Several methods have been developed to mutagenize cloned fungal DNA (see under "Mutagenesis" above) and replace it into the fungal genome. Any of the site-directed mutagenesis methods described earlier, if they provide a selectable marker for filamentous fungi, can be applied. Mutagenized genes as linear DNA are then transformed into the host fungus, and homologous recombination into the genome is selected. As with bacteria, gene replacement by recombination may be performed in one step (Figure 1); or in two steps, by selecting for maintenance, and then loss of the plasmid marker (Figure 2, then Figure 1, Leong 1988). Alternatively, a cloned gene can be truncated at its 5′ and 3′ ends, and recombined into the genome as shown in Figure 2, to generate one gene copy lacking its 5′ terminus, and one copy lacking the 3' terminus (Shortle et al. 1982). Again, it is necessary to verify the nature of the recombination events by DNA hybridization (Miller et al. 1985).

Fungal gene regulation

The fundamental techniques for use of gene fusions are being successfully adapted for use in phytosymbiotic fungi. Although many filamentous fungi contain endogenous β-galactosidase activity, *trpC-lacZ* fusions have been used successfully in *A. nidulans* by repressing endogenous activity with glucose (Hamer and Timberlake 1987). By deleting successively longer portions of the 5'-flanking region of the *trpC* gene, then recombining these 5' *trpC-lacZ* fusions into the *argB* locus of *A. nidulans*, the authors were able to assess how much 5' DNA is necessary for promoter activity (Hamer and Timberlake 1987). Preliminary results from this study and sequence comparisons of *A. nidulans* and *N. crassa* genes suggest that the CCAAT and TATAAA canonical sequences common to other eukaryotic promoters, or the TATA elements upstream of yeast genes, are not present. Other results suggest that fungal gene regulation is position-dependent, and so gene fusions should be recombined into their natural loci when possible (Miller et al. 1987).

Plant Viruses and Mycoplasmas

Plant viruses and related infectious agents (viroids, satellites, and mycoplasmas) are drawing the interest of molecular biologists for several reasons. Molecular techniques have revolutionized diagnosis of plant diseases caused by mycoplasmas and viruses, and hold promise for protecting plants from these parasites. DNA and RNA viruses also hold some promise for use as plant vectors, despite the widespread use of *A. tumefaciens* vectors for plant transformation. Since plant transformation is reviewed in the plant molecular genetics chapter of this book, we will focus here on techniques for genetic manipulation of plant viruses.

Nucleic acid hybridization provides a rapid, sensitive, and specific test for the presence of particular plant viruses and mycoplasmas. By extracting nucleic acids from diseased plant tissue and immobilizing them on nitrocellulose or nylon filters, viral sequences can be detected using labeled DNA hybridization probes (Symons 1984). Cauliflower mosaic virus (CaMV) or related viruses with DNA genomes can be detected by probing with nick-translated plasmid clones of DNA virus sequences. RNA virus probes may be made by isolation of viral RNA, reverse transcriptase copying into complementary DNA (cDNA), and cloning of the double-stranded form of cDNA; cDNA clones that hybridize to authentic infected plant samples, but not to axenic plants, may be labeled for use as diagnostic probes. Recently cDNA has been cloned from a nonculturable plant mycoplasma using RNA extracted from its insect host (Kirkpatrick et al. 1988). Antibody techniques for detecting plant viruses also deserve mention. These include the sensitive enzyme-linked immunosorbent assay (ELISA) and radioimmunoassay (RIA), as well as the immunosorbent electron microscopy assay (ISEM), which allows one to localize virus antigens within tissues (ELISA, RIA, and ISEM techniques; Symons 1984).

Integration of viral nucleic acid sequences into plant genomes can sometimes provide protection from related viruses, a strategy made possible by viral DNA–cDNA cloning techniques. The related "cross-protection" phenomenon was defined earlier as protection of plants from a virulent viral strain by preexposing plants to a related strain (McKinney 1929). Subgenomic virus RNAs that can replicate without the parent virus ("viroids"), or only in the presence of a virus ("satellites"), may also protect plants by interfering with virus replication (Palukaitis and Zaitlin 1984). Study of viroids using viroid cDNA may be useful for determining how viruses replicate (i.e., for possible use in replicating plant vectors) and how they protect plants from related viruses. Transgenic plants containing virus-derived DNA have been constructed by fusing plant promoters to viral gene cDNAs in vitro. These constructs were then integrated into plant genomes using *A. tumefaciens* vectors (French et al. 1986, Cees et al. 1987). It is not yet known whether this "genetically engineered cross-protection" is manifested by interference in viral uncoating, replication, or another step in viral infection. cDNA cloning techniques have been successfully applied to cloning of viral sequences in recent years. Since most plant viruses have RNA genomes (even CaMV replicates via a genomic RNA template), and many of these consist of large RNA molecules, or multiple ("mosaic") RNAs, it is often necessary to clone several cDNAs in order to obtain the entire genome (see Ahlquist and Janda 1984 and Maniatis et al. 1982 for cDNA cloning methods). Plasmid PM1 was designed to allow in vitro transcription of cloned cDNA by *E. coli*. RNA polymerase, and capping of the transcripts (Ahlquist and Janda 1984). For a number of phytopathogenic viruses, in vitro translation using reticulocyte or wheat germ systems has allowed identification of virus-encoded proteins (Shih and Kaesburg 1976). cDNA clones of plant viruses have also been used for infection of plant cells using in vitro transcripts (Ahlquist and Janda 1984), or by introducing cloned CaMV DNA into plants (Robertson et al. 1983). These infection strategies are important for the development of virus-derived vectors, as well as for studying viral replication and the cross-protection phenomenon.

The ongoing development of virus-derived plant vectors holds several potential benefits over the *A. tumefaciens* vectors now in use. Viral vectors may allow extrachromosomal maintenance and intercellular transfer of introduced genes as replicating RNA, and may facilitate gene transfer into monocots that are refractory to *Agrobacterium*-mediated transfer (Siegel 1985). Reporter genes have been incorporated into several viruses as a first step in vector development, including BMV::*cat*, TMV::*cat*, and CaMV::*lux* insertions (French et al. 1986, Shaw et al. 1988, Ow et al. 1986). The reporter genes are generally fused to a viral promoter sequence to mediate expression in plants. The *E. coli* β-glucuronidase (*gus*) gene promises to be a particularly useful reporter gene in plants due to the lack of endogenous activity in plants, and the ease and sensitivity of Gus assays (Jefferson 1987).

Examination of reporter genes after multiple passages or long-term maintenance *in planta* may facilitate measures of mutations caused by RNA virus replication errors. If this mutation rate is too high, it may limit the usefulness of virus-based cloning vectors containing genes that are not under selection pressure (Vloten-Doting et al. 1985). It is also noteworthy that the ssRNA viruses are gaining most attention as potential vectors due to the stringent packaging limits of the dsDNA caulimoviruses and the restriction of ssDNA geminiviruses predominantly to the plant phloem (Siegel 1985).

It has been suggested that "antisense" RNA expression may be a useful tool for studying plant gene expression (Marx 1984). By cloning a viral gene cDNA "backwards" into an RNA-expression vector such as PMI, the strand complementary to the viral coding or "sense" strand can be expressed. Transfer of such a construct into a plant allowed inactivation of the viral gene message via RNA–RNA annealing in vivo (Hemenway et al. 1988).

Prospects

The current emphasis on model systems and application of new techniques promises to provide a basic understanding of how microbial genes and plant genes interact to shape symbiotic interactions. *Agrobacterium* T-DNA vectors have arrived as the most efficient method for plant transformation, and further increases in efficiency may allow screening of immense plant genome libraries by complementation *in planta*. Such advances may hinge on the rapid ongoing elucidation of the molecular mechanism of plant transformation by *Agrobacterium*, and the functions of the virulence genes. Progress in understanding *Rhizobium* nodulation is slower, and may require characterization of plant nodulin genes as well as bacterial nodulation genes. Progress in understanding genes involved in competition between various *Rhizobium* and *Bradyrhizobium* strains, and establishment of suitable federal guidelines for release of recombinant organisms may eventually allow engineering and release of improved inoculum strains. Work with readily manipulated bacterial plant pathogens such as *Pseudomonas syringae* and *Xanthomonas campestris* is on the verge of yielding the function(s) of avirulence genes (Staskawicz et al. 1987, Gabriel et al. 1986). This work promises to elucidate the molecular mechanisms underlying gene-for-gene pathogen–host-plant interactions, and clarify how plants defend against pathogens.

Sudden successes in transforming filamentous fungal phytopathogens (Leong et al. 1988) pave the way for elucidation of how economically important fungal pathogens attack plants and mutate to overcome resistant plant varieties. Adaptation of techniques developed in *Neurospora* and *Aspergillus* should lead to rapid advances in this area. Such work will also rely on the pioneering research to characterize mechanisms of gene regulation and expression in phytopathogenic molds (Leong 1988).

Plant virus research is experiencing a revolution in diagnostic methodology, and current work on cross-protection is suggesting ways to interfere with virus replication at the molecular level (Kerr 1987). Although the value of viruses as plant vectors is debatable, the abilities of viruses to replicate, move intercellularly, and transfer genes into monocots are potentially valuable for plant transformation (Siegel 1985).

Molecular genetics techniques have not yet had a significant impact on agriculture, perhaps because techniques for studying plant symbionts have been developed only in the last ten years. These ten years have, however, seen a tremendous increase in our understanding of selected phytosymbionts. Thus, we can expect engineered *Rhizobium, Agrobacterium*, and perhaps virus-derived plant vectors, and pathogen-resistant plants, to revolutionize agriculture in the near future.

Acknowledgments

The authors thank Dr. Brad Reddick, Dr. Sally Leong, Dr. William Timberlake, Dr. Desh Pal Verma, Dr. Joachim Messing, Dr. Frans de Bruijn, and Dr. Olin Yoder for communicating published and unpublished results. Work cited from the author's laboratory was supported by grants from the U.S. Department of Agriculture (86-CRCR-1-2120 and 87-CRCR-1-2512) and the National Institutes of Health (#2-R01 GM33494) and GM 40183.

References

Abarzua, P., and Marians, K. J. 1984. Enzymatic techniques for the isolation of random single base substitutions *in vitro* at high frequency.

Ahlquist, P., and Janda, M. 1984. cDNA cloning and *in vitro* transcription of the complete Brome Mosaic Virus genome. *Mol. Cell. Biol.* 4(12):2876–2880.

Bagdasarian, M., Lurz, R., Ruckert, B., Franklin, F. C. H., Bagdasarian, M. M., Frey, J., and Timmis, K. N. 1981. Specific purpose plasmid cloning vectors. II. Broad host range, high copy number, RSF1010-derived vectors, and a host system for gene cloning in *Pseudomonas*. *Gene* 16:237–247.

Bagdasarian, M. M., Amann, E., Lurz, R., Ruckert, B., and Bagdasarian, M. 1983. Activity of the hybrid *trp-lac* (*tac*) promoter of *E. coli* in *Pseudomonas putida*. Construction of broad host range, controlled-expression vectors. *Gene* 26:273–282.

Banfalvi, Z., Gursharan, S. R., Kondoros, E., Kiss, A., and Kondorosi, A. 1983. Construction and characterization of R-prime plasmids carrying symbiotic genes of *R. meliloti*. *Mol. Gen. Genet.* 189:129–135.

Berg, D. E., and Berg, C. M. 1983. The procaryotic transposon Tn5. *Biol. Technol.* 1:417–435.

Berg, D. W., Weiss, A., and Crossland, L. 1980. Polarity of Tn5 insertion mutations in *E. coli*. *J. Bacteriol.* 142:439–446.

Bittner, M., and Vapnek, D. 1981. Versatile cloning vectors derived from the runaway-replication plasmid pKN402. *Gene* 15:319–329.

Bruijn, F. J. de, and Lupski, J. R. 1984. Use of transposon Tn5 mutagenesis in rapid generation of correlated genetic and physical maps of DNA segments cloned into multicopy plasmids—a review. *Gene* 27:131–149.

Buikema, W. J., Long, S. R., Brown, S. R., van den Bos, R. C., Earl, C. D., and Ausubel, F. M. 1983. Cosmid cloning of a large region of symbiotic genes from the megaplasmid of *Rhizobium meliloti*. *J. Mol. Appl. Genet.* 2:249–260.

Cannon, G. 1988. Sequence analysis on microcomputers. *Science* 238:97–103.

Carle, G. F., Frank, M., and Olson, M. V. 1986. Electrophoretic separations of large DNA molecules by periodic inversion of the electric field. *Science* 232:65–68.

Carlock, L. R. 1986. Analyzing lambda libraries. *FOCUS* 8(2):6–8.

Carlson, J., and Messing, J. 1984. Efficiency in cloning and sequencing using the single-stranded bacteriophage M-13. *J. Biotechnol.* 1(5–6):253–264.

Castilho, B. A., Olfson, P., and Casadaban, M. J. 1984. Plasmid insertion mutagenesis and *lac* gene fusion with mini-Mu bacteriophage transposons. *J. Bacteriol.* 158:488–495.

Cees, M. P., Bol. J. F., and Vloten-Doting, L. V. 1987. Expression of alfalfa mosaic virus and tobacco rattle virus coat protein genes in transgenic tobacco plants. *Virology* 159:299–305.

Champoux, J. J. Efficient misincorporation by avian myeloblastosis virus reverse transcriptase in the presence of a single deoxyribonucleotide triphosphate. 1984. *J. Mol. Appl. Genet.* 2(5):454–464.

Chen, E. Y., and Seeburg, P. H. 1985. Supercoil sequencing: a fast and simple method for sequencing plasmid DNA. *DNA* 4:165–170.

Church, G. M., and Kieffer-Higgins, S. 1988. Multiplex DNA Sequencing. *Science* 240:117–256.

Collins, F. S., Drumm, M. L., Cole, J. L., Lockwood, W. K., Vande Woude, G. F., and Ianuzzi, M. C. 1986. Construction of a general human chromosome jumping library, with application to cystic fibrosis. *Science* 235:1046–1048.

Collins, F. S., and Weissman, S. M. 1984. Directional cloning of DNA fragments at a large distance from an initial probe: a circularization method. *Proc. Natl. Acad. Sci. USA 81*:6812–6816.

Collins, M. L., and Hunsaker, W. R. 1985. Improved hybridization assays employing tailed oligonucleotide probes: a direct comparison with 5′ end-labeled oligonucleotide probes and nick-translated plasmid probes. *Anal. Biochem. 151*:211–224.

Dale, R. M. K., McClure, B. A., and Houchins, J. P. 1985. A rapid single-stranded cloning strategy for producing a sequential series of overlapping clones for use in DNA sequencing: application to sequencing the corn mitochondrial 18S ribosomal DNA. *Plasmid 13*(1):31–40.

Danner, D. B. 1986. The *lac* operator as a phenotypic label for DNA fragments cloned in *Eschericia coli. Gene 44*:193–199.

Davison, J., Heusterspreute, M., Chevalier, N., Ha-Thi, V., and Brunel, F. 1987. Vectors with restriction site banks. V. pJRD215, a wide host range cosmid vector with multiple cloning sites. *Gene 51*:272–280.

DeVos G., Walker, G. L., and Signer, E. R. 1986. Genetic manipulations in *Rhizobium meliloti* utilizing two new transposon Tn5 derivatives. *Mol. Gen. Genet. 204*:485–491.

deVries, G. E., Raymond, C. K., and Ludwig, R. A. 1984. Extension of bacteriophage lambda host range: selection, cloning, and characterization of a constitutive lambda receptor gene. *Proc. Natl. Acad. Sci. USA 81*:6080–6084.

Ditta, G., Schmidhauser, T., Yakobson, E., Lu, P., Liang, X., Finlay, D. R., Guiney, D., and Helinski, D. R. 1985. Plasmids related to the broad host range vector pRK290, useful for gene cloning and monitoring gene expression. *Plasmid 13*:149–153.

Ditta, G., Stanfield, S., Corbin, D., and Helinski, D. R. 1980. Broad host range DNA cloning system for Gram-negative bacteria: construction of a gene bank of *Rhizobium meliloti. Proc. Natl. Acad. Sci. USA 77*:7347–7351.

Doolittle, R. F. 1986. *Of URFs and ORFs: A Primer On How to Analyze Derived Amino Acid Sequences.* University Science Books, Mill Valley, California.

Downie, J. A., and Johnston, A. W. B. 1986. Nodulation of legumes by *Rhizobium. Cell 47*:153–154.

Drahos, D. J., Hemming, B. C., and McPherson S. 1986. Tracking recombinant organisms in the environment—beta-galactosidase as a selectable nonantibiotic marker for fluorescent *Pseudomonas. Bio/Technology 4*:439–444.

Eckert, R. L. 1987. New vectors for rapid sequencing of DNA fragments by chemical degradation. *Gene 51*:247–254.

Engebrecht, J. A., Simon, M., and Silverman, M. 1985. Measuring gene expression with light. *Science 227*:1345.

5 Prime–3 Prime, Inc., 19 East Central Avenue, Paoli, Pennsylvania.

Feinberg, A. P., and Vogelstein, B. 1983. A technique for radio labeling DNA restriction endonuclease fragments to high specific activity. *Anal. Biochem. 132*(1):6–13.

Fellay, R., Frey, J., and Krisch, H. 1987. Interposon mutagenesis of soil and water bacteria: a family of DNA fragments designed for *in vitro* insertional mutagenesis of Gram-negative bacteria. *Gene 52*:147–154.

Fickett, J. W. 1982. Recognition of protein-coding regions in DNA sequences. *Nucleic Acids Res. 10*:5303–5318.

Fisher, R. F., Brierley, H. L., Mulligan, J. T., and Long, S. R. 1987. Transcription of *Rhizobium meliloti* nodulation genes. *J. Biol. Chem. 262*(14):6849–6855.

French, R., Janda, M., and Ahlquist, P. 1986. Bacterial gene inserted in an engineered RNA virus: efficient expression in monocotyledonous plant cells. *Science 231*:1294–1297.

Frey, J., Bagdasarian, M., Feiss, D., Franklin, C. H., and Deshusses, J. 1983. Stable cosmid vectors that enable the introduction of cloned fragments into a wide range of Gram negative bacteria. *Gene 24*:299–308.

Friedman, A. M., Long, S. R., Brown, S. E., Buikema, W. J., and Ausubel, F. 1982. Construction of a broad host range cosmid cloning vector and its use in the genetic analysis of *Rhizobium meliloti. Gene 18*:289–296.

Fukuda, R., Yano, R., Fukui, T., Hase, T., Ishihama, A., and Matsubara, H. 1985. Cloning of the *Escherichia coli* gene for the stringent starvation protein. *Mol. Gen. Genet.* 201(2):151–157.

Gabriel, D. W., Burges, A., and Lazo, G. R. 1986. Gene-for-gene interactions of five cloned avirulence genes from *Xanthomonas campestris* pv. *malvacearum* with specific resistance genes in cotton. *Proc. Natl. Acad. Sci. USA* 83:6415–6418.

Galas, D. J., Eggert, M., and Waterman, M. S. 1985. Rigorous pattern-recognition methods for DNA sequences. *J. Mol. Biol.* 186:117–128.

Gallie, D. R., Novak, S., and Kado, C. I. 1985. Novel high- and low-copy cosmids for use in *Agrobacterium* and *Rhizobium*. *Plasmid* 14:171–175.

Gamper, H. B., Cimino, G. D., Isaacs, S. T., Ferguson, M., and Hearst, T. E. 1986. Reverse Southern hybridization. *Nucleic Acids Res.* 14(24):9943–9954.

Gayle, R. B., Vermersch, P. S., and Bennett, G. N. 1986. Construction and characterization of pBR322-derived plasmids with deletions of the RNA I region. *Gene* 41:281–288.

Geider, K. 1986. DNA cloning vectors utilizing replication functions of the filamentous phages of *Escherichia coli*. *J. Gen. Virol.* 67:2287–2303.

Gerhold, D., Stacey, G., and Kondorosi, A. 1989. Use of a synthetic "nod box" probe to identify two nodulation operons in *Rhizobium meliloti*. *Plant Mol. Biol.*, 12:181–188.

Groisman, E. A., and Casadaban, M. J. 1986. Mini-Mu bacteriophage with plasmid replicons for *in vivo* cloning and *lac* gene fusing. *J. Bacteriol.* 168:357–364.

Grunstein, M., and Hogness, D. S. 1975. Colony hybridization: a method for the isolation of cloned DNA's that contain a specific gene. *Proc. Natl. Acad. Sci. USA* 72:3961–3965.

Gutierrez, C., Barondess, J., Manoil, C., and Beckwith, J. 1987. The use of transposon Tn*phoA* to detect genes for cell envelope proteins subject to a common regulatory stimulus. *J. Mol. Biol.* 195:289–297.

Gutterson, N. I., Layton, T. J., Ziegle, T. S., and Warren, G. J. 1986. Molecular cloning of genetic determinants for inhibition of fungal growth by a fluorescent Pseudomonad. *J. Bacteriol.* 165:696–703.

Habener, J. F., Vo, C. D., Le, D. B., Gryan, G. P., Ercolani, L., and Wang, A. H. J. 1988. 5-Fluorodeoxyuridine as an alternative to the synthesis of mixed hybridization probes for the detection of specific gene sequences. *Proc. Natl. Acad. Sci. USA* 85:1735–1739.

Hamer, J. E., and Timberlake, W. E. 1987. Functional organization of the *Aspergillus nidulans trpC* promoter. *Mol. Cell. Biol.* 7:2352–2359.

Hashimoto-Gotoh, T., Kume, A., Wakako, M., Takeshita, S., and Fukuda, A. 1986. Improved vector, pHSG664, for direct streptomycin-resistance selection: cDNA cloning with G:C tailing procedure and subcloning of double-digest DNA fragments. *Gene* 41:125–128.

Heffron, F. 1983. Tn*3* and its relatives. In J. Shapiro (ed.), *Mobile Genetic Elements*. Academic, New York, pp. 223–260.

Hemenway, C., Fang, R. X., Kaniewski, W. K., Chua, N. H., and Tumer, N. E. 1988. Analysis of the mechanism of protection in transgenic plants expressing the potato virus X coat protein or its antisense RNA. *EMBO J.* 7(5):1273–1280.

Henikoff, S. 1984. Unidirectional digestion with exonuclease III creates targeted breakpoints for DNA sequencing. *Gene* 28:351–358.

Hodgson, A. L. M., and Roberts, W. P. 1983. DNA colony hybridization to identify *Rhizobium* strains. *J. Gen. Microbiol.* 129:207–212.

Holden, D. W., Wang, J., and Leong, S. A. 1988. DNA-mediated transformation of *Ustilago bordei* and *Ustilago nigra*. *Physiol. Mol. Plant Pathol.* (in press).

Horvath B., Bachem C. W. B., Schell, J., and Kondorosi, A. 1987. Host-specific regulation of nodulation genes in *Rhizobium* is mediated by a plant-signal, interacting with the *nod* D gene product. *EMBO J.* 6(4):841–848.

Huynh, T. V., Young, R. A., and Davis, R. W. 1985. Constructing and screening cDNA libraries in lambda gt10 and lambda gt11. In *DNA Cloning—A Practical Approach*, D. Glover (ed.) IRL Press, Washington, D.C., pp. 49–78.

Ish-Horowicz, D., and Burke, J. F. 1981. Rapid and efficient cosmid vector cloning. *Nucleic Acids Res.* 9:2989–2996.

Jacobs, W. R., Barrett, J. F., Clark-Curtiss, J. E., and Curtiss, R., III. 1986. *In vivo* repackaging of recombinant cosmid molecules for analyses of *Salmonella typhimurium, Streptococcus mutans* and mycobacterial genomic libraries. *Infect. Immun.* 52(1):101–109.

Jagadish, M. N., Bookner, S. D., and Szalay, A. A. 1985. A method for site-directed transplacement of *in vitro* altered DNA sequences in *Rhizobium. Mol. Gen. Genet.* 199:249–255.

Jefferson, R. A. 1987. Assaying chimeric genes in plants: the GUS gene fusion system. *Plant Mol. Biol. Rep.* 5(4):387–405.

Jefferson, R. A., Burgess, S. M., and Hirsh, D. 1986. Beta-glucuronidase from *Escherichia coli* as a gene-fusion marker. *Proc. Natl. Acad. Sci. USA* 83:8447–8451.

Johnston, A. W. B., Beynon, J. L., Buchanan-Wollaston, A. V., Setchell, S. M., Hirsch, P. R., and Beringer, J. E. 1978. Transfer of the drug-resistance transposon Tn5 to *Rhizobium. Nature* 276:633–636.

Kahn, M. L., and Timblin, C. R. 1984. Gene fusion vehicles for the analysis of gene expression in *Rhizobium meliloti. J. Bacteriol.* 158:1070–1077.

Kalisch, B. W., Krawetz, S. A., Schoenwalder, K. H., and vande Sande, J. H. 1986. Covalently linked sequencing primer linkers (splinkers) for sequence analysis of restriction fragments. *Gene* 44:263–270.

Kerr, A. 1987. The impact of molecular genetics on plant pathology. *Annu. Rev. Phytopathol.* 25:87–110.

Kinscherf, T. G., and Leong, S. A. 1988. Molecular analysis of the karyotype of *Ustilago maydis. Chromosoma* (in press).

Kirkpatrick, B. C., Stenger, D. C., Morris, T. J., and Purcell, A. H. 1988. Cloning and detection of DNA from a nonculturable plant pathogenic mycoplasma-like organism. *Science* 238:197–200.

Knauf, V. C., and Nester, E. W. 1982. Wide host range cloning vectors: a cosmid clone bank of an *Agrobacterium* Ti plasmid. *Plasmid* 8:45–54.

Kohara, Y., Akiyama, K., and Isono, K. 1987. The physical map of the whole *E. coli* chromosome: application of a new strategy for rapid analysis and sorting of a large genomic library. *Cell* 50(3):485–508.

Kondorosi, A., Kondorosi, E., Pankhurst, C. E., Broughton, W. J., and Banfalvi, Z. 1982. Mobilization of an *R. meliloti* megaplasmid carrying nodulation and nitrogen fixation genes into other *Rhizobia* and *Agrobacterium. Mol. Gen. Genet.* 188:433.

Kroos, L., and Kaiser, D. 1984. Construction of Tn5 *lac*, a transposon that fuses *lacZ* expression to exogenous promoters, and its introduction into *Myxococcus xanthus. Proc. Natl. Acad. Sci. USA* 81:5816–5820.

Leong, S. A. 1988. Recombinant DNA research in phytopathogenic fungi. In *Advances in Plant Pathology*, vol. 6.. Academic, New York (in press).

Liu, J., Lanclos, K. D., and Huisman, T. 1986. Synthesis of a fixed-length single-stranded DNA probe by blocking primer extension in bacteriophage M13. *Gene* 42:113–117.

Lugtenberg. 1988. Regulation of *nod* gene expression and nodulation. Presented at the 7th Intl. Congress on Nitrogen Fixation, Koln, West Germany.

Ma, Q. S., Johnston, A. W. B., Hombrecher, G., and Downie, H. 1982. *Mol. Gen. Genet.* 187:166.

Maas, R. 1983. An improved colony hybridization method with significantly increased sensitivity for detection of single genes. *Plasmid* 10:296–298.

Mandecki, W. 1986. Oligonucleotide-directed double strand break repair in plasmids of *Escherichia coli*: A method for site-specific mutagenesis. *Proc. Natl. Acad. Sci. USA* 83:7177–7181.

Maniatis, T., Fritsch, E. F., and Sambrook, J. 1982. *Molecular cloning*: A laboratory manual. Cold Spring Harbor Publ., Cold Spring Harbor, NY.

Marx, J. L. 1984. New ways to "mutate" genes. *Science* 225:819.

Matsui, T., Hirano, M., Naoe, T., Yamada, K., and Kurosawa, Y. 1987. Production of chimeric protein coded by the fused viral H-*ras* and human N-*ras* genes in *Eschericia coli*. *Gene* 52:215–224.

Maxam, A. M., and Gilbert, W. 1980. Sequencing end-labeled DNA with base specific chemical cleavages. *Methods Enzymol.* 65:499–560.

McKinney, H. H. 1929. Mosaic diseases in the Canary Islands, West Africa, and Gibraltar. *J. Agric. Res.* 39:557–578.

McKinnon, R. D., Waye, J. S., Bautista, D. S., and Graham, F. L. 1985. Non-random insertion of Tn5 into cloned human adenovirus DNA. *Gene* 40(1):31–38.

Michiels, T., and Cornelis G. 1986. Tn951 derivatives designed for high-frequency plasmid-specific transposition and deletion mutagenesis. *Gene* 43:175–181.

Miller, B. L., Miller, K. Y., Robert, K. A., and Timberlake, W. E. 1987. Position-dependent and -independent mechanisms regulate cell-specific expression of the SpoCl gene cluster of *Aspergillus nidulans*. *Mol. Cell. Biol.* 7:427–434.

Miller, B. L., Miller, K. Y., and Timberlake, W. E. 1985. Direct and indirect gene re-placements in *Aspergillus nidulans*. *Mol. Cell. Biol.* 5(7):1714–1721.

Mitchell, N., and Stoehrer, G. 1986. Mutagenesis originating in site-specific DNA damage. *J. Mol. Biol.* 191(2):177–180.

Miura, N., Ohtsuka, E., Yamaberi, N., Ikehara, M., Uchida, T., and Ikada, Y. 1985. Use of the deoxyinosine-containing probe to isolate and sequence complementary DNA encoding the fusion glycoprotein of sendai virus. *Gene* 38(1–3):271–274.

Moore, A. T., Nayudu, M., and Holloway, B. W. 1983. Genetic mapping in *Methylophilus methylotrophus* AS1. *J. Gen. Microbiol.* 129:785–799.

Mulligan, J. T., and Long, S. R. 1985. Induction of *Rhizobium meliloti nodC* expression by plant exudate requires *nodD*. *Proc. Natl. Acad. Sci. USA* 82:6609–6613.

Mulligan, M. E., and McClure, W. R. 1986. Analysis of the occurrence of promoter-sites in DNA. *Nucleic Acids Res.* 14:109–126.

Nano, F. E., Shepert, W. D., Watkins, M. M., Kuhl, S. A., and Kaplan, S. 1984. Broad host range plasmid vector for the *in vitro* construction of *lac* fusions. *Gene* 34:219–226.

Nayudu, M., and Rolfe, B. G. 1987. Analysis of R-primes demonstrates that genes for broad host range nodulation of *Rhizobium* strain NGR234 are dispersed on the Sym plasmid. *Mol. Gen. Genet.* 206:326–337.

Nieuwkoop, A. J., Banfalvi, Z., Deshmane, N., Gerhold, D., Schell, M. G., Sirotkin, K. M., and Stacey, G. 1987. A locus encoding host range is linked to the common nodulation genes of *Bradyrhizobium japonicum*. *J. Bacteriol.* 169:2631–2638.

Orbach, M., Vollmer, S., and Yanofsky, C. 1985. *J. Cell. Biochem.* Suppl. 9C Abstr. 1567.

Osbourn, A. E., Barber, C. E., and Daniels, M. J. 1987. Identification of plant-induced genes of the bacteria pathogen *Xanthomonas campestris* using a promoter-probe plasmid. *EMBO J.* 6:23–28.

Ow, D. W., Wood, K. V., DeLuca, M., de Wet J. R., Helinski, D. R., and Howell, S. H. 1986. Transient and stable expression of the firefly luciferase gene in plant cells and transgenic plants. *Science* 234:856–859.

Palazzolo, M. J., and Meyerowitz, E. M. 1987. A family of lambda phage cDNA cloning vectors, lambda SWAJ, allowing the amplification of RNA sequences. *Gene* 52:197–206.

Palukaitis, P., and Zaitlin, M. 1984. A model to explain the "cross protection" phenom-enon shown by plant viruses and viroids. In *Plant-Microbe Interaction: Molecular and Genetic Perspectives*, T. Kosuge and E. Nester (eds.). Macmillan, New York, pp. 420–429.

Parsons, K. A., Chumley, F. G., and Valent, B. Genetic transformation of the fungal pathogen responsible for vice blast disease. *Proc. Natl. Acad. Sci. USA* (in press).

Peeters, B. P., Schoenmakers, J. G., and Konings, R. N. 1986. Plasmid pKUN9, a ver-satile vector for the selective packaging of both DNA strands into single-stranded DNA-containing phage-like particles. *Gene* 41:39–46.

Peng, Z., and Wu, R. 1986. A simple and rapid nucleotide sequencing strategy and its application in analyzing a rice histone 3 gene. *Gene* 45:247–252.

Pouwels, P. H., Enger-Valk, B. E., and Brammer, W. J. (eds.). 1987. *Cloning Vectors*. Elsevier, Amsterdam.

Prober, J. M., Trainer, G. L., Dam, R. J., Hobbs, F. W., Robertson, C. W., Zagursky, R. J., Cocuzza, A. J., Jensen, M. A., and Baumeister, K. 1988. A system for rapid DNA sequencing with fluorescent chain-terminating dideoxynucleotides. *Science* 238:336–341.

Putnoky, P., Kiss, G. B., Ott, I., and Kondorosi, A. 1983. Tn5 carries a streptomycin resistance determinant downstream from the kanamycin resistance gene. *Mol. Gen. Genet.* 191:288–294.

Quinto, M., and Bender, R. A. 1984. Use of bacteriophage P1 as a vector for Tn5 insertion mutagenesis. *Appl. Environ. Microbiol.* 47:436–438.

Ratet, P., Schell, J., and de Bruijn, F. J. 1988. Mini-Mu*lac* transposons with broad host range origins of conjugal transfer and replication designed for gene regulation studies in Rhizobiaceae. *Gene* 63:41–52.

Rella, M., Mercenier, A., and Haas, D. 1985. Transposon insertion mutagenesis of *Pseudomonas aeruginosa* with a transposon Tn5 derivative: application to physical mapping of the arc gene cluster. *Gene* 33(3):293–304.

Roberts, L. 1988. New sequencers to take on the genome. *Science* 238:271–273.

Robertson, H. D., Howell, S. H., Zaitlin, M., and Malmberg, R. L. (eds.). 1983. *Plant Infectious Agents*. Cold Spring Harbor Press, Cold Spring Harbor, New York.

Rogowsky, P. M., Close, T. J., Chimera, J. A., Shaw, T. J., and Kado, C. I. Regulation of the *vir* genes of *Agrobacterium* tumefaciens plasmid pTiC58. *J. Bacteriol.* 169:5101–5112.

Rosenthal, A., Jung, R., and Hunger, H. D. 1986. Optimized conditions for solid-phase sequencing: simultaneous chemical cleavage of a series of long DNA fragments immobilized on CCS anion-exchange paper. *Gene* 42:1–9.

Rostas, K., Kondorosi, E., Horvath, B., Simoncsits, A., and Kondorosi, A. 1986. Conservation of extended promoter regions of nodulation genes in *Rhizobium*. *Proc. Natl. Acad. Sci. USA* 83:1757–1761.

Ruvkun, G. B., and Ausubel, F. M. 1981. A general method for directed mutagenesis in prokaryotes. *Nature* 289:85–88.

Sadowski, M. J., Olson, E. R., Foster, V. E., Kosslak, R. M., and Verma, D. P. S. 1988. Two host-inducible genes of *Rhizobium fredii* and characterization of the inducing compound. *J. Bacteriol.* 170:171–178.

Schneider, K., and Beck, C. F. 1986. Promoter-probe vectors for the analysis of divergently arranged promoters. *Gene* 42:37–48.

Schwartz, D. C., and Cantor, C. R. 1984. Separation of yeast chromosome-sized DNAs by pulsed field gradient field electrophoresis. *Cell* 37:67.

Selvaraj, G., Fong, Y. C., and Iyer, V. N. 1984. A portable DNA sequence carrying the cohesive site (cos) of bacteriophage lambda and the mob (mobilization) region of the broad-host-range plasmid RK2—a module for the construction of new cosmids. *Gene* 32:235–241.

Selvaraj, G., and Iyer, V. N. 1985. A small mobilizable IncP group plasmid vector packageable into bacteriophage capsids *in vitro*. *Plasmid* 13:70–74.

Selvaraj, G., and Iyer, V. N. 1984. Transposon Tn5 specifies streptomycin resistance in *Rhizobium* spp. *J. Bacteriol.* 158:580–584.

Seth, A., Lapis, P., Vandewoude, G. F., and Papas, T. 1986. High level expression vectors to synthesize unfused proteins in *Escherichia coli*. *Gene* 42:49–57.

Shaw, J., Bubrick, P., Lewandowski, D., Culver, J., Hilf, M., Grantham, G., Desjardins, P., and Dawson, W. O. 1988. Expression of foreign genes by tobacco mosaic virus. Abstract from 11th Symposium on Plant Physiology, University of California at Riverside.

Shaw, J. J., and Kado, C. I. 1986. Development of a *Vibrio* bioluminescence gene-set to monitor phytopathogenic bacteria during the ongoing disease process in a nondisruptive manner. *Biotechnology* 4:560–564.

Shih, D. S., and Kaesberg, P. 1976. Translation of the RNA's of Brome Mosaic Virus: The monocistronic nature of RNA1 and RNA2. *J. Mol. Biol.* 103:77–88.

Shortle, D., Haber, J., and Botstein, D. 1982. *Science 217*:371–373.

Siegel, A. 1985. Plant virus-based vectors for gene transfer may be of considerable use despite a presumed high error frequency during RNA synthesis. *Plant Mol. Biol. 4*:327–328.

Simon, R. 1984. High frequency mobilization of Gram-negative bacterial replicas by the *in vitro* constructed Tn5-mob transposon. *Mol. Gen. Genet. 196*:413–420.

Simon, R., Priefer, U., and Puhler, A. 1983. A broad host range mobilization system for *in vitro* genetic engineering: transposon mutagenesis in Gram negative bacteria. *Bio/Technology 1*:784–791.

Simon, V., and Schumann W. 1987. *In vivo* formation of gene fusions in *Pseudomonas putida* and construction of versatile broad host range vectors for direct subcloning of Mud1 and Mud2 fusions. *Appl. Environ. Microbiol. 53*:1649–1654.

Sirotkin, K. 1986. Advantages to mutagenesis techniques generating populations containing the complete spectrum of single codon changes. *J. Theoret. Biol. 123*:261–279.

Smith, C. L., Econome, J. G., Schutt, A., Klco, S., and Cantor, C. R. 1987. A physical map of the *Escherichia coli* K12 genome. *Science 236*:1448–1453.

Smith, H. O., and Bernstiel, M. L. 1976. A simple method for DNA restriction site mapping. *Nucleic Acids Res. 3*:2387–2390.

Stachel, S. E., An, G., Flores, L., and Nester, E. W. 1985. A Tn3*lacZ* transposon for the random generation of beta-galactosidase gene fusions: application to the analysis of gene expression in *Agrobacterium*. *EMBO J. 4*(4):891–898.

Stallard, R. L., Certa, U., and Bannwarth, W. 1987. Dideoxy sequencing of double-stranded DNA from poly(A)-extended 3' ends. *Anal. Biochem. 162*:197–201.

Stark, M. J. R. 1987. Multicopy expression vectors carrying the *lac* repressor gene for regulated high-level expression of genes in *Escherichia coli*. *Gene 51*:255–267.

Staskawicz, B., Dahlbeck, D., Keen, N., and Napoli, C. 1987. Molecular characterization of cloned avirulence genes from Race 0 and Race 1 of *Pseudomonas syringae* pv. *glycinae*. *J. Bacteriol. 168*(12):5789–5794.

Symons, R. H. 1984. Diagnostic approaches for the rapid and specific detection of plant viruses and viroids. In *Plant-microbe interactions*, T. Kosuge and E. Nester (eds.), vol. I. Macmillan, New York, pp. 93–124.

Tacon, W. C. A., Bonass, W. A., Jenkins, B., and Emtage, J. S. 1983. Expression plasmid vectors containing *E. coli* tryptophan promoter transcriptional units lacking the attenuator. *Gene 23*:255–265.

Timberlake, W. E. 1986. Isolation of stage- and cell-specific genes from fungi. In *Biology and Molecular Biology of Plant Pathogen Interactions*, J. Bailey (ed.). Springer-Verlag, Heidelberg.

Tung, C. S., and Harvey, S. C. 1987. A common structural feature of promoter sequences of *E. coli*. *Nucleic Acids Res. 15*:4973–4985.

Ubben, D., and Schmitt R. 1986. Tn*1721* derivatives for transposon mutagenesis, restriction mapping, and nucleotide sequence analysis. *Gene 41*:145–152.

Van Larebeke, N., Engler, G., Holsters, M., Van den Elsackers, Zaenen I., Schilperoort, R. A., and Schell, J. 1974. Large plasmid in *Agrobacterium tumefaciens* essential for crown gall-inducing ability. *Nature 252*:169–170.

Vloten-Doting L., van, Bol., J. F., and Cornelissen, B. 1985. Plant-virus-based vectors for gene transfer will be of limited use because of the high error frequency during viral RNA synthesis. *Plant Mol. Biol. 4*:323–326.

Vries, G. E., de-, Raymond, C. K., and Ludwig, R. A. 1984. Extension of bacteriophage lambda host range. I. Selection, cloning, and characterization of a constitutive lambda receptor gene. *Proc. Natl. Acad. Sci. USA 81*:6080–6084.

Waterman, M. S. 1986. Multiple sequence alignment by consensus. *Nucleic Acids Res. 14*(22):9095–9102.

Windle, B. E. 1986. Phage lambda and plasmid expression vectors with multiple cloning sites and *lacZ* alpha-complementation. *Gene 45*:95–99.

Yanisch-perron, C., Vieira, J., and Messing, J. 1985. Improving M13 phage cloning vectors and host strains: nucleotide sequences of the M13 mp18 and pUC19 vectors. *Gene 33*:103–119.

Yelton, M. M., Hamer, J. E., de Sousa, E. R., Mullaney, E. J., and Timberlake, W. E. 1983. Developmental regulation of the *Aspergillus nidulans trpC* gene. *Proc. Natl. Acad. Sci. USA 80*:7576–7580.

Yelton, M. M., Timberlake, W. E., and Handel, C. A. 1985. A cosmid for selecting genes by complementation in *Aspergillus nidulans*: selection of the developmentally regulated yA locus. *Proc. Natl. Acad. Sci. USA 82*:834–838.

Young, R. A., and Davis, R. W. 1983. Efficient isolation of genes by using antibody probes. *Proc. Natl. Acad. Sci. USA 80*:1194–1198.

Yu, S. M., and Gorovsky, M. A. 1986. *In situ* dot blots: quantitation of mRNA in intact cells. *Nucleic Acids Res. 14*(19):7597–7616.

Zhang, F., Denome, R. M., and Cole, C. N. 1986. Fine-structure analysis of the processing and polyadenylation region of the herpes simplex virus type 1 thymidine kinase gene by using linker scanning, internal deletion, and insertion mutations. *Mol. Cell. Biol. 6*(12):4611–4623.

Use of Genetically Altered Bacteria to Achieve Plant Frost Control

Steven E. Lindow

Department of Plant Pathology
Hilgard Hall
University of California at Berkeley
Berkeley, California 94720

Introduction

Information and concepts from several different disciplines, including plant pathology, plant physiology, ecology, and molecular biology will be addressed in this chapter to define a new approach to the biological control of plant frost injury. I will first summarize information related to the ecology of bacteria important in inciting plant frost injury and pursue recent studies of their interactions with the plants on which they live, as well as with each other. This will be followed by a discussion of current research in molecular genetic studies pertaining to the biological control of frost injury to plants and the releases of recombinant bacteria for this purpose. Ice-nucleation-active bacteria on plants are the focus of this review, and the parameters influencing the structure of microbial communities including these species on leaves will be emphasized. Strategies for manipulating these communities will then be addressed. It should be noted, however, that much of the information presented here should be applicable to modifications of analogous communities of plant-pathogenic microorganisms to achieve a consequent reduction in disease severity. However, to retain a focus, emphasis will be placed on the ecology of the epiphytic ice-nucleation-active bacterium *Pseudomonas syringae* and its role in causing frost injury to the plants on which it lives. It is recognized that many other phytopathogenic bacteria have epiphytic habitats and behaviors similar to those of this species, and therefore information presented here should easily be transferrable.

The Nature of Plant Frost Injury

Low-temperature injury to plants is one of the major factors limiting the cultivation of plants worldwide (Chandler 1958, Cary and Mayland 1970, Burke et al. 1976, White and Haas 1975). Frost injury to plants can occur at all temperatures lower than 0°C. Many researchers, however, distinguish "warm-temperature" frost injury, occurring at temperatures above about −5°C, from frost injury that occurs to plants at colder temperatures (Chandler 1958, Cary and Mayland 1970, Burke et al. 1976). The plant frost injury to be discussed here will be that which occurs solely at temperatures above about −5°C.

Plants differ in their ability to tolerate the formation of ice within their tissues. A large group of frost-tolerant plants have adaptations which permit them to tolerate the formation of ice in their tissues (Chandler 1958, Cary and Mayland 1970, Burke et al. 1976, Levitt 1972, Mazur 1969, Olien 1967). A smaller group of frost-sensitive plants, but one that includes most important agricultural plants, have no ability to tolerate ice formation. Frost-tolerant plants generally survive ice formation inside their tissues by restricting it to the intercellular spaces. Once initiated, ice spreads rapidly in the intercellular spaces of such plants. When grown under the proper environmental conditions (such as with cold nights and appropriate short-day-length conditions), alterations in the cell membrane, cell wall constituents, and cytoplasmic constituents occur (Ketchie 1985, Li 1984). The water potential of the cytoplasm of plant cells rapidly comes into equilibrium with the ice that has formed intercellularly (Chandler 1958, Cary and Mayland 1970, Burke et al. 1976, Ketchie

1985). As the temperature of plants containing ice decreases, more intracellular water freezes intercellularly (Quamme 1983). Plant adaptations induced under hardening conditions apparently provide a barrier for ice movement into plant cells and permit the survival of plant cells at decreasing water potentials during the dehydration that occurs during the freezing process (Fennell and Li 1986). Some cells, particularly those of woody perennial plant species, are apparently physically isolated from the ice that forms in the remainder of the plant due to morphological adaptations such as the formation of small cells in meristematic tissues (Andrews and Proebsting 1986, 1987, Andrews et al. 1987, Ashworth 1984, Ashworth and Abeles 1984, Quamme 1985, Anderson and Ashworth 1986). Such localized and physically isolated cells do not freeze and therefore do not have to cope with the presence of ice (Quamme 1985, Anderson and Ashworth 1986, Wisniewski and Ashworth 1986a). In both cases, however, plant frost injury results when the rate of ice formation in the tissue is sufficiently large or cold-hardiness adaptations are inadequate to prevent the breach of the cell membrane by ice crystals. Intracellular ice formation is invariably lethal to plants (Chandler 1958, Cary and Mayland 1970, Burke et al. 1976). Death can occur by mechanical disruption by penetration of ice through cells or due to dehydration stresses occurring during ice formation or possibly during the rehydration of dehydrated cells upon the melting of ice (Chandler 1958, Cary and Mayland 1970, Burke et al. 1976). Since frost-hardy plants have endogenous ice-tolerance mechanisms, they will not be considered in most of this chapter.

Frost-sensitive plants are distinguished from frost-hardy plants by their inability to tolerate intercellular ice formation. Ice crystals are not restricted to the intercellular spaces in such plants, even at temperatures only slightly below 0°C (Chandler 1958, Cary and Mayland 1970, Burke et al. 1976). Plants such as corn, beans, tomatoes, the flowers of deciduous fruits such as pear, and subtropical plant species, such as the fruit of citrus, apparently possess none of the adaptations that allow the membranes of other plants to be differentially permeable to ice. Frost injury results when ice crystals enter the cells of such plants. Ice propagation and subsequent damage can occur at temperatures only slightly lower than 0°C (Chandler 1958, Cary and Mayland 1970, Burke et al. 1976). This chapter will focus on frost-sensitive plants that have no ability to tolerate ice formation and must avoid ice formation to prevent frost injury.

Plant Supercooling and Plant Frost Injury

Many frost-sensitive plants have the capacity to supercool and avoid damaging ice formation (Chandler 1958, Kaku 1964, 1966, 1971a,b,

1973, 1975, Hendershott 1962, Marcellos and Single 1979, 1976, Modlibowska 1962, Proebsting et al. 1982, Lucas 1954). Small volumes of pure water can supercool to temperatures approaching $-40°C$ before homogeneous nucleation occurs (Bigg 1953). Ice formation at temperatures above $-40°C$ generally occurs by a process of heterogeneous ice nucleation, whereby ice embryos are initiated by the orientation of water molecules by organic or inorganic substances (Bigg 1953, Hobbs 1974, Mason 1960, Kozloff et al. 1984, Mason and Maybank 1958, Power and Power 1962, Rosinski and Parungo 1966, Schnell and Vali 1972, 1973, 1976). Ice-nucleating agents associated with herbaceous plant tissues that are active at temperatures above $-5°C$ are rare (Kaku 1964, 1966, 1971a,b, 1973, 1975, Hendershott 1962, Marcellos and Single 1976, 1979, Modlibowska 1962, Proebsting et al. 1982, Lucas 1954). Several reports indicate that ice-nucleating events active at temperatures of about $-2°C$ occur in woody tissues of some plant species such as peach when collected from the field (Anderson et al. 1986, Ashworth et al. 1985a,b, Cody et al. 1987, Gross et al. 1984b). The difficulty of producing axenically grown perennial woody tissues has precluded a definite determination as to whether ice-nucleating agents of nonplant origin are associated with such tissues. However, most herbaceous plants, especially those grown under axenic conditions, exhibit little or no ice-nucleation activity at temperatures above $-5°C$. Ice can propagate from localized ice-nucleation sites in a plant to adjacent parts of the plant and therefore cause frost injury (Chandler 1958, Cary and Mayland 1970, Burke et al. 1976). Therefore the probability of frost injury to plants that cannot tolerate ice formation would be expected to increase with increasing size of the plant, assuming that the numbers of ice nuclei per unit mass of the plant remain constant. If the number of ice-nucleation events associated with a plant or plant part is sufficiently low, however, plant parts or entire plants can avoid damaging ice formation altogether.

Plant-Associated Ice-Nucleating Bacteria

Several plant-associated bacteria are highly active in ice nucleation and may be responsible for frost injury to the plants on which they live. Approximately half of all pathovars (Dye et al. 1980) of *P. syringae* (Paulin and Luisetti 1978, Arny et al. 1976, Maki et al. 1974, Hirano et al. 1978), most strains of *Xanthomonas campestris* pathovar *translucens* (Lim et al. 1987), *Pseudomonas viridiflava* (Paulin and Luisetti 1978), and some strains of *Erwinia herbicola* (Yankofsky et al. 1981, Lindow et al. 1978b, Makino 1982) and *Pseudomonas fluorescens* (Maki and Willoughby 1978, Lindemann and Suslow 1987) are active in ice nucleation at temperatures above $-5°C$. These bacteria nucleate ice at temperatures between -1.5 and $-10°C$. On average, 1 *Pseudomonas syringae* cell out of 1,000 is active in ice nucleation at $-5°C$ (Hirano et al. 1978). The fraction of cells of different strains that are active at a given temperature (for example, $-2°C$) varies greatly (Hirano et al. 1978). Similarly, the efficiency with which ice nucleation is expressed by a given strain on several different host plants varies greatly (O'Brien and Lindow 1988). No example has been found in which every cell of a given bacterial strain has been active in ice nucleation on a plant surface.

The population size of ice-nucleation-active bacteria varies greatly among plants and temporally on a given plant species under field conditions (Hirano and Upper 1983, 1986, Gross et al. 1983, 1984*a*, Lindow 1982*a*, 1983*b,c* 1985*a,b*, 1987, Lindow and Connell 1984, Lindow et al. 1978*a,c*, 1983*a*, Lindow and Panopoulos 1988, Hirano et al. 1982). The population size of ice-nucleation-active bacteria is on average approximately the same on frost-tolerant as on frost-sensitive plants. However, it varies greatly among plant species exposed to similar environmental conditions. For example, the average population size of ice-nucleation-active bacteria on navel orange leaves in California is often as low as 10^2 to 10^3 cells per gram fresh weight, but population sizes of over 10^7 cells per gram fresh weight have been observed on nearby almond and pear trees (Lindow 1983*c*, Lindow and Connell 1984). Large differences in population size of ice-nucleation-active bacteria on the same plant grown in a small geographical region can occur. These differences in population size may be associated with the use of different agrochemicals used for pest management in different fields (Andrews and Kenerley 1978). However, ice-nucleation-active bacteria have been detected in the lower atmosphere, and a substantial rate of deposition of *P. syringae* occurs in the vicinity of plants that are a good source of this bacterial species (Lindemann et al. 1984, Lindemann and Upper 1985, Lindemann et al. 1982, Andersen and Lindow 1985). Immigration and short-distance movement of bacteria active in ice nucleation may therefore be important in determining the initial inoculum on plants. Aerosols may provide a substantial and continual source of immigrant cells to plant surfaces. The population size of ice-nucleation-active bacteria generally increases with the chronological age of the plant tissues on which they reside (Hirano and Upper 1983, Gross et al. 1983, Lindow et al. 1978*a*). Young vegetative tissues of most plants have a small population size of total bacteria, as well as ice-nucleation-active bacterial species such as *P. syringae*. When environmental conditions, such as periods of rainfall, permit the increase in population size of bacteria on plants, the population of *P. syringae* may increase rapidly (Hirano and Upper 1986). Generation times of approximately 4 hours have been observed for *P. syringae* on plant surfaces under field conditions. Large daily changes in population size of ice-nucleation-active bacteria can sometimes occur, but population size may appear relatively invariant when considered over time periods of several days or more (Hirano and Upper 1986).

Most ice-nucleation-active bacteria occur as epiphytes on the surface of leaves. Generally, as with other epiphytic bacteria, over 90% of the cells of species such as *P. syringae* can be killed by the application of topical disinfectants such as cupric sulfate or hydrogen peroxide, or after exposure to nonionizing radiation such as ultraviolet light (Haefele and Lindow 1982, Young 1978, Spurr 1979, Miedema et al. 1980, Sztejnberg and Blakeman 1973, Pennycook and Newbook 1982). The leaf-surface habitats colonized by *P. syringae* are unknown. It is thought that most epiphytic bacteria preferentially inhabit depressions between epidermal cells and at the bases of hairs, and crevices formed along veins (Leben 1988, Schneider and Grogan 1977, Leben 1961). Such sites may have a physical environment that is more conducive to cell survival than more exposed regions on the surface of the plant. Approximately 0.1 to 1% of the total surface area of a plant is covered by bacterial cells. Since ice-nucleation-active bacteria including *P. syringae* generally comprise no more than 0.1 to 10% of the total bacteria on a leaf, the total leaf surface that is colonized by such bacteria is rather low. However, since the total population size of epiphytic bacteria generally never exceeds 10^7 cells per square centimeter, some biological or physical factors must limit the population size of bacteria that can be supported on leaves. This feature will be addressed in more detail later in conjunction with a discussion of interactions of microorganisms on leaf surfaces.

The population size of ice-nucleation-active bacteria is generally sufficiently high to account for the ice formation that is necessary for frost damage to most frost-sensitive plant species (Gross et al. 1983, Lindow et al. 1978*a*, Lindow 1982*a,b,c*, 1983*a*, Hirano et al. 1982, Lester et al. 1977). Because most plants have a population size of *P. syringae* or other ice-nucleation-active bacteria of at least 100 cells per gram fresh weight, significant numbers of bacterial ice nuclei would be formed on plants even if on average only 1 cell in 1,000 was active in ice nucleation at a given time. For example, approximately one ice nucleus would be found per approximately 10 g of citrus leaf tissue under California growing conditions. Since a citrus leaf averages approximately 3 g in mass, one ice-nucleation event could be expected at $-5°C$ within a collection of every three leaves chosen at random. Even at population sizes of *P. syringae* that are difficult to detect (10^2 cells per gram fresh weight), a large number of ice-nucleation events would be initiated within a canopy of leaves in a citrus tree. If subsequent ice propagation from one nucleated leaf to an adjacent leaf on which ice nucleation has not occurred is allowed to proceed, it is easy to see that most leaves in a tree would likely freeze given sufficient time for ice propagation to occur. A direct relationship exists between frost injury to plants and the logarithm of population size of $P.$ $syringae$ on those plants (Lindow et al. 1978c, Lindow 1982a). Similarly, the supercooling point of a leaf decreases proportionally to the increase in the logarithm of the population size of $P.$ $syringae$ (Hirano et al. 1985, Lindow 1985b). For example, a plant with an average $P.$ $syringae$ leaf-surface population size of 10^8 cells per gram will be unlikely to supercool to lower than $-2°C$ before ice formation, and therefore frost damage, occurs. In contrast, a tree such as citrus with an average population size of $P.$ $syringae$ of 10^2 to 10^3 cells per gram fresh weight or lower can supercool to $-4°C$ or colder before ice formation is likely to occur. However, a large change in the population size of ice-nucleation-active bacteria is necessary to substantially reduce the average supercooling point of leaves (and therefore the temperature at which frost damage is likely to occur). For example, decreasing the population size of ice-nucleation-active bacteria by 100-fold decreases the average supercooling point of plant tissue by approximately $1°C$ (Hirano et al. 1985, Lindow 1985b). It is important to note that most plants contain substantial numbers of ice nuclei active at temperatures lower than $-5°C$ (Lindow 1982a, Kaku 1964, 1966, 1971a,b, 1973, 1975, Marcellos and Single 1979, 1976). The concentration of ice nuclei for most plants increases logarithmically with decreasing temperature (Lindow 1982a). While herbaceous plants contain no more than about 10^{-3} ice nuclei per gram at temperatures of -3 to $-5°C$, at -6 to $-7°C$ more than 1 nucleus per gram is found in many plant tissues (Lindow et al. 1978c, 1982a,c). Therefore most plants, particularly herbaceous annual plants, cannot be expected to supercool to lower than about $-5°C$, even in the absence of ice-nucleation-active bacteria.

Several species of lichens have recently been reported to contain large numbers of ice nuclei active at temperatures of -2 to $-5°C$ (Kieft 1988). Some of these lichen species occur on plant surfaces in California (Lindow, unpublished observation). Their abundance on plants would be expected to increase only slowly with time because of their slow growth rates. Therefore annual plants may contain few or no detectable ice-nucleating lichens, while perennial plants may have substantial numbers of lichen cells associated with them. The substantial ice-nucleation activity associated with dormant perennial woody plants may be due, at least partially, to the presence of ice-nucleation-active lichen species.

Existing methods of plant frost control generally attempt to maintain the temperature of plant tissues to above 0°C to prevent ice nucleation and/or ice propagation in plants. Common methods used for the heating of plant tissues under field conditions include the release of warm water onto the soil surface, where it can release heat directly by conduction, or where it can freeze and release its latent heat of fusion (Parsons and Tucker 1984, Parsons et al. 1985a,b). Application of water directly to freezing plant surfaces also will maintain plant surfaces at 0°C because of the release of latent heat of fusion (Blanc 1963). If sufficient amounts of water are applied to plants during the freezing process, so that free water remains available to freeze at all times, plant temperature will not drop below 0°C; ice will not penetrate into such ice-encrusted plants because of the slightly lower freezing point of solute-containing plant cells than that of pure water on the leaves. Large fans are also used for some crops to mix warmer air found 3 to 10 m above the soil with cold air that accumulates near the soil surface on calm, clear nights during radiative frosts (Renquist 1985). These methods of frost protection are all quite labor- or capital-intensive and generally increase the temperature of plants less than 2 to 3°C. If the ice-nucleation temperature of the plant tissue is reached in spite of such heating attempts, frost injury will result.

Molecular Determinants of Bacterial Ice Nucleation

The mechanisms of bacterial ice nucleation have recently been further elucidated using molecular genetic approaches as well as through classical biochemical studies. The genes conferring ice nucleation have been cloned and at least partially characterized from *P. syringae, E. herbicola*, and *P. fluorescens* (Orser et al. 1983, 1985, Green and Warren 1985, Corotto et al. 1986, Warren et al. 1986). In all cases, a single, contiguous DNA region of approximately 4,000 bp is necessary and sufficient for the expression of ice-nucleation activity in both the DNA source strains and heterologous recipients such as *E. coli*. Sequence analyses of these cloned regions have revealed the presence of a single, large, open reading frame of variable length (from 3.4 to 3.8 kb) (Green and Warren 1985, Warren et al. 1986, Corotto et al. 1986). The translation products predicted from the cloned *inaZ* and *inaW* genes of *P. syringae* and *P. fluorescens*, respectively, are approximately 150,000 Da in mass (Green and Warren 1985, Warren et al. 1986). A unique, tandemly repeating amino acid structure is predicted for the majority of the internal region of the mature protein encoded by both of these genes. A 16-amino-acid sequence that is tandemly repeated approximately 120 times comprises between 80 and 90% of the gene product of the *inaZ* and *inaW* genes (Warren et al. 1986). A very high extent of conservation of this repeated motif is apparent in both genes. Nonrepeating carboxyl and amino terminal sequences do occur, however. It is thought that, if properly folded, a protein with such a repetitious amino acid composition that is high in polar amino acids such as serine and threonine might directly orient water molecules into an icelike lattice. Individual hydrophilic residues on the protein might contribute to the overall ice-nucleation activity exhibited by the molecule as a whole.

The *iceC* and *inaW* gene products and ice-nucleation activity are localized primarily in the outer membrane of the respective source strain and in *E. coli* (Wolber et al. 1986, Deininger et al. 1988, Sprang and Lindow 1981, Govindarajan and Lindow 1985). Considerable amounts of such proteins accumulate in the cytoplasm of *E. coli* when overexpressed on appropriate expression vectors (Wolber et al. 1986). The natural abundance of the ice proteins is estimated to be less than 300 to 1,000 molecules per cell (Lindow and Govindarajan, unpublished data). The ice protein expresses ice-nucleation activity only when incorporated into the outer membrane of bacterial cells, when reconstituted into lipid-containing vesicles, or in certain detergent micelles (Wolber et al. 1986, Govindarajan and Lindow 1985, 1988). Thus the hydrophobic environment contributed by a membrane or a membranelike micelle may be required for the stabilization of or correct conformation of an ice-nucleation protein. Phosphatidyl inositol may be associated with ice-nucleation proteins (Kozloff et al. 1984). Such lipids have been implicated in the anchoring and proper association of other membrane-associated proteins.

Ice-nucleation sites on bacterial cells are generally larger than the size of individual ice-nucleation proteins (Govindarajan and Lindow 1988). Such sites have been hypothesized to be comprised of aggregates of the ice-nucleation protein into different sized multilinkers. Ice-nucleation sites active at temperatures of $-3°C$ are approximately 8 million Da in size, which would correspond to approximately 60 ice-nucleating proteins. Ice-nucleation sites decrease in size logarithmically with decreasing temperature at which ice-nucleation activity is expressed (Govindarajan and Lindow 1988). Thus at temperatures of $-14°C$ or colder (the lowest temperature at which ice-nucleation activity is detected in a population of bacterial cells), catalytic sites approximate the size of a single ice-nucleation protein. As expected, the concentration of ice nuclei expressed in a population of cells increases logarithmically with increasing concentration of ice-nucleation protein in the membranes (Southworth et al. 1988, Lindgren et al. 1988). Thus, macromolecular aggregates comprised of ice-nucleation proteins may be much more likely to form as the concentration of the ice protein, usually in low abundance in the cell membrane, increases.

The unique structure of the gene product of cloned ice-nucleation genes makes it unlikely that they have enzymatic activities. Since the expected amino acid sequence of these gene products is largely repetitive, they probably function as structural proteins similar to those found on the surfaces of other cells. Their low abundance in the cell probably also precludes them from participating in major metabolic pathways in the cell. The low abundance of ice-nucleation proteins in cell membranes probably also accounts for the low probability with which a given cell expresses ice-nucleation activity. It is unknown whether the concentrations of ice-nucleation proteins in different cells of a homogeneous cell population are similar, but if large aggregates of ice proteins are required for expression of ice-nucleation activity at temperatures above $-5°C$, and if the molecular abundance of the ice-nucleation protein in an average cell approaches this size, it is unlikely that all such molecules will be found in such aggregates at a given time.

Modification of Microbial Populations and Plant Frost Injury

Treatments of leaves with chemical or biological agents that reduce the population size of ice-nucleation-active bacteria reduce the incidence of frost injury to plants under greenhouse and field conditions (Anderson and Ashworth 1986, Cody et al. 1987, Marshall 1988, Lindow et al. 1978c, 1983b, Lindow and Connell 1984, Lindow 1982a, 1983a,c, 1985a,b, 1987, 1988b). Bactericides, including streptomycin and oxytetracycline as well as copper hydroxide, reduce the incidence of frost injury to treated plants significantly compared to untreated plants when applied at least 7 days prior to freezing temperatures. Such bactericide treatments usually function as preventative agents to reduce subsequent increases in population size of P. syringae and other bacteria on plants that would otherwise have occurred in the absence of such bacteriostatic chemicals. While such bactericides can kill bacteria immediately after application to plants, even "dead" cells may still retain ice-nucleation activity and, because these cells are not removed from plants by bactericide treatment, they initiate damaging ice formation. Non-ice-nucleation-active bacteria also reduce the subsequent population size of ice-nucleating bacteria when applied to plants both in the greenhouse and in the field. In all cases, a direct relationship is observed between the incidence of freezing injury and the logarithm of the population size of ice-nucleation-active bacteria on plants at the time of freezing (Lindow et al. 1983a, Lindow 1982a).

Several factors determine the effectiveness of antagonistic non-ice-nucleation-active bacteria as biological control agents of frost injury to plants. Such bacteria have only been observed to prevent increases of *P. syringae* subsequent to treatment of plants (compared with untreated plants), rather than to displace populations of *P. syringae* already established on plants (Lindow et al. 1983*a,b*, Lindow 1983*c*, 1982*a*). For example, in greenhouse studies, the population size of an ice-nucleation-active *E. herbicola* strain that was attained on plants was decreased compared to the population on control plants only when non-ice-nucleation-active bacteria were inoculated onto plants prior to or at the same time as inoculation of an ice-nucleation-active strain (Lindow et al. 1983*b*). Similarly, non-ice-nucleation-active bacteria applied to pear trees subsequently decreased the population size of *P. syringae* only when the population size of *P. syringae* was low at the time of inoculation and not on trees where high population sizes of *P. syringae* existed at the time of inoculation (Lindow 1982*a*). It appears that non-ice-nucleation-active bacteria can only prevent the growth of *P. syringae* on plants and not kill or displace this species once already established on plants. Non-ice-nucleation-active bacteria differ greatly in their ability to prevent the growth of *P. syringae* on leaves. For example, the population size of *P. syringae* increased substantially even in the presence of established populations of about half of all non-ice-nucleation-active bacterial strains previously applied to corn leaves, but increased only slightly in the presence of established populations of a few other strains (Lindow 1985*a*). Thus considerable differences in the ability of epiphytic bacteria to coexist on leaf surfaces are evident.

Several different mechanisms of interaction among epiphytic bacteria on leaves have been hypothesized. Direct parasitism of one bacterial strain by another (such as parasitism of *Pseudomonas syringae* pv. *glycinea* by *Bdellovibrio* species) is probably uncommon on leaves (Scherff 1973). Some bacteria, such as *E. herbicola*, may modify the chemical environment of their habitat during their growth (Riggle and Klos 1972, Chatterjee et al. 1969, Hsieh and Buddenhagen 1974). Some strains of this species produce acid by-products of their metabolism that may alter the pH of localized habitats sufficiently that certain other strains may not be able to coexist (Chatterjee et al. 1969). Antagonism among bacterial strains has also been hypothesized to be due to the production of inhibitory compounds by one strain that are active against another strain (Leben 1964, Blakeman 1981, 1982, Blakeman and Fokkema 1982, Crosse 1965, 1971, Goodman 1967, Leben and Daft 1965, Garrett and Crosse 1975, Gibbins 1972). Such inhibitory compounds include antibiotics, bacteriocins, siderophores, cyanide, and other types of toxic compounds. While direct evidence for the activity and presence of some of these compounds on roots has appeared, little direct evidence for their involvement in the interactions of bacteria on leaves has appeared. On the contrary, substantial evidence indicates that antibiosis is not a major mechanism determining the coexistence of bacteria on leaf surfaces (Lindow 1985a, 1988b). Only about half of a collection of 88 non-ice-nucleation-active bacteria that prevented the growth of *P. syringae* on plants in greenhouse and field tests produced antibiotics active against *P. syringae* on any culture media tested. Mutants of nearly all antibiotic-producing strains tested that were deficient in antibiotic production did not differ from their parental strains in their ability to prevent the growth of *P. syringae* on leaves when pretreated onto plant surfaces (Lindow 1985a, 1988b). It is unknown whether the antibiotics that such strains produce in culture were not produced on plants, whether *P. syringae* was insensitive to such compounds on leaves, or rather that they were produced in adequate amounts to account for inhibition of *P. syringae* growth on leaves. The production of a fluorescent siderophore by *P. syringae* on leaf surfaces has, however, recently been assessed by indirect methods (Loper and Lindow 1987). Results of these studies did not support the common assumption that siderophore are produced on leaf surfaces. Antibiosis therefore is probably not sufficient to account for the major interactions occurring between *P. syringae* and other bacteria on leaves. Competition for limiting environmental resources, on the other hand, may constitute the major mechanism by which such bacterial strains interact (Lindow 1985a, 1986a, 1987, Lindow et al. 1987). The nature of the leaf-surface resources for which bacterial strains may compete is unknown. Considerable quantities of nutrients are present on the surface of leaves (Tukey 1966, 1970, 1971, Fossard 1981, Tukey and Morgan 1963). It is unknown, however, what the spatial distribution, quantity, or composition of such nutrients in localized microhabitats such as those occupied by bacteria might be. I hypothesize that *P. syringae* occupies a hierarchy of preferred sites on the surface of leaves. It is further hypothesized that the concentration or quality of nutrients that supports the growth of *P. syringae* at such sites is distinct and discontinuous from those at other sites on the leaf surface. The physical environment of habitats occupied by *P. syringae* on leaves is probably also somewhat more favorable than the many parts of the leaf that *P. syringae* does not occupy (Lindow 1985b). Therefore, certain non-ice-nucleation-active bacteria may occupy the same or closely adjacent sites to those occupied by *P. syringae*, and thereby compete directly for limiting environmental resources such as nutrients, sites for growth and survival, and other unrecognized resources.

Studies utilizing mutants of *P. syringae* deficient in ice nucleation have given direct evidence for competition for limiting environmental resources as a major mechanism determining the coexistence of bacteria on leaves (Lindow 1985a, 1987). It was hypothesized that bacterial strains of a similar genotype (and therefore having similar ecological habitat requirements) would compete more directly for limiting environmental resources than dissimilar strains (Lindow 1985c, 1988a). Mutants of *P. syringae* that were deficient in ice nucleation were selected following mutagenesis with ethylmethane sulfonate. Such near-isogenic bacterial strains were utilized in pairwise competition studies to determine the magnitude of any competition that could be exhibited between such isogenic strains. All chemically induced non-ice-nucleation-active (Ice⁻) mutants of *P. syringae* reduced the population size of ice-nucleation-active parental strains of *P. syringae* that were challenge-inoculated onto plants pretreated with mutant strains (Lindow 1987). Thus, mechanisms other than competition for limiting resources appear unlikely to account for the major interactions among these near-isogenic bacterial strains on leaves. Ice⁻ mutants of *P. syringae* also decreased the population size of indigenous ice-nucleation-active *P. syringae* strains when applied to pear trees in the field prior to the colonization of these trees by ice-nucleation-active-bacteria (Lindow 1987). Competition therefore appeared to be the major interaction occurring among even presumably heterologous strains of this species under field conditions. However, the incomplete exclusion of *P. syringae* by Ice⁻ mutants of this species raises the question of the specificity with which different strains compete for limiting environmental resources. For example, do different strains of *P. syringae* have different ecological niche requirements and therefore differ in their magnitude of competition? It would be expected that bacterial strains with dissimilar ecological niche requirements could coexist on leaves. Any one Ice⁻ mutant of *P. syringae* might therefore not exclude all other ice-nucleation-active strains of this species.

Studies of the specificity of competition of bacteria have been greatly facilitated by molecular genetic approaches that have permitted the construction of truly isogenic strains of *P. syringae*. If strains of *P. syringae* differ in ecological habitat requirements, then different strains of this bacterium should coexist to different extents when coinoculated onto plants (Lindow 1988a). This phenomenon was most easily studied by examining the ability of different Ice⁺ parental *P. syringae* strains to coexist with a collection of isogenic Ice⁻ *P. syringae* strains and heterogeneous Ice⁻ *P. syringae* strains constructed in vitro. The cloned *iceC* gene of *P. syringae* permitted the construction of site-directed deletion mutants of the source strain as well as other *P. syringae* strains (Lindow 1985c). Deletions internal to the cloned *P. syringae* strain Cit7 *iceC* gene were constructed by sequential *sall* endonuclease digestion and ligation (Orser et al. 1984). A subset of hybrid plasmids obtained following religation lacked one or more internal *sall* restriction fragments. Many such internal deletion-containing genes were completely deficient in conferring ice-nucleation activity in *E. coli* (Orser et al. 1984). Since defined and genetically stable Ice⁻ *P. syringae* strains were desired in tests of competition specificity as defined above, site-directed mutagenesis of dissimilar ice-nucleation-active (Ice⁺) *P. syringae* strains was conducted utilizing this and similar deletion-containing *ice* genes constructed in vitro.

Site-directed mutagenesis was conducted by a process similar to that initially described by Ruvkun and Ausubel (1981). For example, the deletion-containing *iceC* gene was introduced into strains of *P. syringae* to be mutagenized on a nonreplicative plasmid (pBR325) by conjugation. Because of homology of the cloned ice-nucleation gene flanking the internal deletion with a chromosomal counterpart, homologous recombination causes the integration of the deletion-containing *iceC* gene and the plasmid vector into the chromosome of recipient *P. syringae* strains (Lindow 1985c, 1988a). When such *cis*-merodiploid strains were grown in the absence of selection for the plasmid vector, secondary recombinational events, in which a recombinational event distal to the first event causes the reciprocal exchange of the deletion-containing *iceC* gene into the chromosome for the indigenous functional gene, accumulate within the population of cells to a frequency of approximately 10^{-4}. No detectable DNA alterations outside the ice-nucleation region were detected utilizing this "homogenotization" procedure.

Tests of Competitive Exclusion Using Ice⁻ Recombinant Strains

Recombinant Ice⁻ deletion mutant strains of *P. syringae* do not appear to differ from their Ice⁺ parental strains in their behavior on leaves (Lindemann and Suslow 1985, 1987, Lindow 1985c, 1988a). No differences in competitive abilities of Ice⁺ and Ice⁻ strains were detected in laboratory experiments (Lindow 1985c, 1988a, Lindemann and Suslow 1985). No change in the relative frequency of Ice⁺ and Ice⁻ strains initially applied in mixtures to plants were observed after growth for approximately 10 generations on leaf surfaces. Ice⁻ recombinant strains also do not differ from parental strains in their survival in soil, their survival of freezing and thawing cycles in aqueous suspensions, or in their ability to grow and survive when inoculated individually onto any of 65 plant species (Lindow 1985c, Lindemann and Suslow 1985). Inactivation of the Ice phenotype by deletions constructed in vitro apparently has not interfered with important adaptations that *P. syringae* has for the colonization of leaf surfaces. Consequently, Ice⁻ deletion mutant strains of *P. syringae* should be useful in tests of a novel procedure by which biological control agents are constructed from deleterious plant-associated microorganisms through deletion of deleterious traits.

The competitive exclusion of bacteria from leaf surfaces by prior inoculation by Ice⁻ *P. syringae* strains appears to be a population-size-dependent, and not a genotype-dependent, phenomenon. Competitive exclusion of Ice⁺ *P. syringae* strains with nonrecombinant non-ice-nucleation-active bacteria increased with increasing log population size of the competitor strain at the time of inoculation with Ice⁺ bacteria (Lindow 1988*a*). The population size of both isogenic Ice⁺ parental strains and heterologous Ice⁺ strains was lower subsequent to inoculation onto plants previously colonized by Ice⁻ strains than on control plants (Lindemann and Suslow 1987, Lindow 1985*c*, 1988*a*). However, appreciable competitive exclusion of Ice⁺ *P. syringae* strains occurred on plants only when the population size of Ice⁻ strains was at least 10^5 cells per gram fresh weight.

Statistical analysis of the results of competitive exclusion of a number of Ice⁺ *P. syringae* strains with isogenic and heterologous Ice⁻ strains revealed some interaction between the competitor and target organism (Lindow 1986*b*). Thus some specificity appears to exist in the competitive interaction exhibited among different strains of *P. syringae*. Most noteworthy, isogenic Ice⁻ strains were not always superior to heterologous Ice⁻ *P. syringae* strains in preventing population increases of a given Ice⁺ strain (Lindow 1986*b*). This raises the possibility that some Ice⁺ *P. syringae* strains may escape competition against certain Ice⁻ *P. syringae* strains. Field studies would therefore be useful to evaluate the potential for some Ice⁺ *P. syringae* strains to coexist with Ice⁻ recombinant *P. syringae* strains on the same plants. Plants inoculated in the field with Ice⁻ *P. syringae* strains could therefore be used as "enrichment trap plants" to select for indigenous Ice⁺ *P. syringae* strains that were able to coexist with such Ice⁻ strains. Identification of Ice⁺ strains that escape competition would be useful in further studies of the nature of competitive interactions of bacteria on leaves.

Field studies in which Ice⁻ recombinant *P. syringae* strains were used to study competition within this species and to assess biological frost control have recently been completed (Lindow and Panopoulos 1988). Because in vitro techniques were used in the construction of Ice⁻ *P. syringae* strains by the homogenotization procedure discussed earlier, special authorization was required to test such strains without containment under field conditions in the United States. Petitions were first made to the National Institutes of Health (NIH) by researchers at the University of California, Berkeley, in 1982. Although permission to begin field tests was obtained from the NIH in 1983, a series of legal challenges and changes in federal regulations pertaining to the release of recombinant microorganisms into the environment delayed initial field experiments using Ice⁻ *P. syringae* strains until April 1987. Initial field trials of Ice⁻ *P. syringae* strains applied to strawberry and potato were carefully designed and intensively monitored to restrict and describe the distribution of recombinant strains in and around experimental plots and to determine accurately the ultimate fate of recombinant strains released at these field sites. Only limited dispersal of the recombinant Ice⁻ *P. syringae* strains that were released at these experimental sites has occurred (Lindow and Panopoulos 1988). Ice⁻ *P. syringae* strains inoculated onto potato plants comprised a significant part of the total epiphytic microflora for 4 to 6 weeks after inoculation. The population size of Ice⁺ *P. syringae* strains on plants colonized by Ice⁻ *P. syringae* strains was significantly decreased compared to that on uninoculated control plants (Lindow et al., unpublished data). The incidence of frost injury to potato plants inoculated with Ice⁻ *P. syringae* strains was significantly lower than to uninoculated control plants in several natural field frost events that occurred in the field in a northern California test site. Further reports on the specificity of competitive exclusion of indigenous *P. syringae* strains at these field sites by Ice⁻ *P. syringae* strains have not yet appeared.

Integrated Chemical and Biological Control of Frost Injury

Temporal and spatial variations in the population size of ice-nucleation-active bacteria require that chemical and biological agents be utilized in limited strategies to be effective. For example, competitive non-ice-nucleation-active bacteria are effective in preventing the increase in population size of *P. syringae* on leaves only when they are established in large population sizes in advance of target *P. syringae* strains (Lindow et al. 1983*a,b*, Lindow 1982*a*, 1983*c*). Because of this, they apparently can be applied to plants only when a "biological vacuum" exists on plants (Lindow 1985*b*). For example, competitive microorganisms used for frost control have been effective only when applied to young vegetative tissues under field conditions because of the relative lack of indigenous bacterial populations (particularly of target strains) on plants at these times. Not only can competitive microorganisms prevent the growth of Ice⁺ *P. syringae* strains by competition, but they themselves can be excluded from plants by the prior existence of other microorganisms. Chemical bactericides can also prevent the multiplication of *P. syringae* on leaves (Lindow 1983c). Bactericides are also most effective in preventing population increases if applied prior to the development of large population sizes of *P. syringae*. Unlike competitive microorganisms, however, chemical bactericides have a limited duration on leaf surfaces and are quickly removed from leaves by weathering processes such as wind and rain, and are not usually distributed across leaves subsequent to application, unlike competitive microorganisms. For this reason, chemical bactericides such as cupric hydroxide often must be applied frequently (every 3 to 5 days) to actively growing plant tissues to maintain a sufficient quantity of bactericide on young developing plant materials to prevent the growth of *P. syringae*. Repeated applications of bactericides during the spring on actively growing plant material limits the usefulness of bactericides for frost control because of economic considerations. The application of antagonistic microorganisms, on the other hand, is often limited to a narrow window of time shortly after vegetative growth of tissues. Subsequent applications of antagonists are unlikely to increase their numbers on plants and therefore do not improve biological control (Lindow et al. 1983a). Antagonistic microorganisms can be applied to plants successfully only when unoccupied habitats exist on leaves. Application of bactericides (such as cupric hydroxide) to leaves can kill indigenous microorganisms at habitats preferred by biological control strains (Lindow and Connell 1984). Applications of bactericides prior to the application of biological control organisms may therefore create an opportunity for such strains to grow on previously colonized plants. The microbial composition of leaf surfaces could be made simpler and the relative population size of inoculant Ice$^-$ *P. syringae* strains used for biological control could be increased on plants, possibly at any time of the year, by the integrated use of chemical bactericides and such organisms.

Many different strains of *P. syringae* have been recently discovered to be resistant to copper ions (Andersen and Lindow 1986). Copper bactericides have been used widely for the control of blast and canker, induced by *P. syringae*, on trees such as almond. Copper-tolerant strains of *P. syringae* have been detected from almond and citrus trees in all geographical areas of California (Andersen and Lindow 1986). Such strains exhibit no reduction of growth rate on plants treated with copper bactericides (such as cupric hydroxide). Similarly, the population size of copper-resistant strains of *P. syringae* is not reduced on plants when copper hydroxide is applied to plants colonized with these strains. While the genetic determinants for copper tolerance have not been fully characterized in *Pseudomonas syringae* pv.

syringae strains, such genes have been identified and partially characterized in *Pseudomonas syringae* pv. *tomato* (Bender and Cooksey 1986, Mellano and Cooksey 1988, Bender and Cooksey 1987).

The integrated use of cupric hydroxide and copper-tolerant Ice⁻ *P. syringae* strains is attractive for the biological control of plant frost injury. When cold temperatures occur in the spring shortly after the application of antagonistic microorganisms to young vegetative plants, frost injury can be mitigated by applications of the antagonist at a single time. However, should frost occur in the fall (several months after young vegetative tissues available for colonization have appeared), applications of antagonists at this time are unlikely to be successful. Successional changes in the composition of populations of epiphytic bacteria presumably will also occur subsequent to application of antagonistic Ice⁻ bacteria to young plants at a single time in the spring. Therefore, it would be desirable to create a niche by which Ice⁻ *P. syringae* strains could be applied to mature plants prior to frosts that occur in the fall. Application of cupric hydroxide would reduce greatly the population size of Ice⁺ *P. syringae* strains on plants and presumably other indigenous bacteria that may occupy habitats that could be subsequently colonized by coinoculant copper-tolerant Ice⁻ *P. syringae* strains. The integrated control of bacterial speck of tomato following application of nonpathogenic copper-tolerant strains of *P. syringae* pv. *tomato* and cupric hydroxide in this manner has recently been reported (Cooksey 1988). Copper-tolerant Ice⁻ *P. syringae* strains that could be used in the integrative control of frost injury could be constructed either by mobilization of copper resistance determinants from this species into existing Ice⁻ *P. syringae* strains, or by constructing Ice⁻ derivatives of copper-tolerant Ice⁺ *P. syringae* strains that have already been described. The latter procedure is preferred at this time because it does not require introduction of novel genetic material into bacterial strains destined for field release.

Research Needs to Improve Biological Frost Control

While an understanding of the ecology of Ice⁺ *P. syringae* strains and interactions of these strains with non-ice-nucleating bacteria has greatly increased our knowledge of their role in plant frost injury, there exists a critical need to enhance our understanding of the role of bacteria in the freezing processes of plants. For example, the relative importance of the nucleation of individual plant parts versus the propagation of ice from distant plant parts to cause damage in many plant species is poorly understood. Strategies of frost control based on the enhancement of supercooling of plants are likely to be more successful if numerous ice-nucleation events are required to initiate damaging ice formation in a plant. For example, if individual flowers are each nucleated independently, then reduction of the numbers of ice nuclei by procedures discussed earlier will more directly affect the probability of frost injury than if ice propagation from flower to flower or from more distant plant parts occurs. More information is also required on the indigenous ice-nucleating agents apparently associated with some woody plant species. It appears unlikely that endogenous ice-nucleating materials are associated with axenically grown herbaceous plants, but woody plant species appear to have many more ice nuclei than can be accounted for by either viable or nonviable ice-nucleating bacteria. While lichens may account for some or most of the ice nuclei associated with these plants, the source, localization, and activity of ice-nucleating sites on woody plant species needs to be much better understood.

Molecular genetic techniques should allow us to make great advances in our understanding of the interactions of bacteria on plants. Very little is known of the adaptive traits of ice-nucleation-active bacteria that permit them to colonize plants under field conditions. Similarly, the nature of the leaf-surface habitat colonized by these bacteria is unknown. The models of interaction of bacteria on leaves as discussed above are heavily dependent on our assumptions as to the nature of the leaf-surface environment that is inhabited by ice-nucleating bacteria as well as other bacteria that may compete with such strains for existence on leaves. For example, do different strains or species of bacteria occupy distinct habitats defined by physical or chemical factors on leaf surfaces? Does the distinct nature of habitats that might be occupied by different bacteria account for the ability of different strains to coexist on leaves? What are the limiting environmental resources for a given bacterial strain or species such as Ice$^+$ *P. syringae* strains in the habitats that they choose to occupy on leaves? Is competition for limiting environmental resources the sole determinant of the coexistence of bacterial species on leaves? What is the real activity of bacteria on leaves? Until now, many assumptions have been made concerning these very important questions. Molecular genetic techniques such as those discussed above now offer us the ability to develop simplified microbial communities to better address these important questions. The results of ongoing field studies in which Ice$^-$ recombinant *P. syringae* strains have been applied to plants in the field to test the specificity of competition on leaves should provide important information with regard to at least some of the questions posed above. It is a challenge to plant pathologists and microbial ecologists to define studies that can better address the remaining questions.

References

Andersen, G. L., and Lindow, S. E. 1985. Local differences in epiphytic bacterial population size and supercooling point of citrus correlated with type of surrounding vegetation and rate of bacterial immigration. *Phytopathology* 75:1321.

Andersen, G. L., and Lindow, S. E. 1986. Occurrence and control of copper-tolerant strains of *Pseudomonas syringae* on almond and citrus in California. *Phytopathology* 76:1118.

Anderson, J. A., Ashworth, E. N., and Davis, G. A. 1986. Non-bacterial ice nucleation in peach shoots. *J. Am. Soc. Hort. Sci.* 112:215–218.

Anderson, J. A., and Ashworth, E. N. 1986. The effects of streptomycin, desiccation, and UV radiation on ice nucleation by *Pseudomonas viridiflava*. *Plant Physiol.* 80:956–960.

Andrews, J. H., and Kenerley, C. M. 1978. The effects of a pesticide program on nontarget epiphytic microbial populations of apple leaves. *Can. J. Microbiol.* 24:1058–1072.

Andrews, P. K., and Proebsting, E. L., Jr. 1986. Development of deep supercooling in acclimating sweet cherry and peach flower buds. *HortScience* 21:99–100.

Andrews, P. K., and Proebsting, E. L., Jr. 1987. Effects of temperature on the deep supercooling characteristics of dormant and deacclimating sweet cherry flower buds. *J. Am. Soc. Hort. Sci.* 112:334–340.

Andrews, P. K., Proebsting, E. L., Jr., and Lee, G. S. 1987. A conceptual model of changes in deep supercooling of dormant sweet cherry flower buds. *J. Am. Soc. Hort. Sci.* 112:320–324.

Arny, D. C., Lindow, S. E., and Upper, C. D. 1976. Frost sensitivity of *Zea mays* increased by application of *Pseudomonas syringae*. *Nature* 262:282–284.

Ashworth, E. N. 1984. Xylem development in *Prunus* flower beds and the relationship to deep supercooling. *Plant Physiol.* 74:862–865.

Ashworth, E. N., and Abeles, F. B. 1984. The freezing behavior of water in small pores and the possible relationship to freezing of plant tissues. *Plant Physiol.* 76:201–204.

Ashworth, E. N., Anderson, J. A., and Davis, G. A. 1985a. Properties of ice nuclei associated with peach trees. *J. Am. Soc. Hort. Sci.* 110:287–291.

Ashworth, E. N., Davis, G. A., and Anderson, J. A. 1985b. Factors affecting ice nucleation in plant tissues. *Plant Physiol.* 79:1033–1037.

Bender, C. L., and Cooksey, D. A. 1986. Indigenous plasmids in *Pseudomonas syringae* pv. *tomato*: conjugative transfer and role in copper resistance. *J. Bacteriol.* 165:534–541.

Bender, C. L., and Cooksey, D. A. 1987. Molecular cloning of copper resistance genes from *Pseudomonas syringae* pv. *tomato*. *J. Bacteriol.* 169:470–474.

Bigg, E. K. 1953. The supercooling of water. *Proc. Phys. Soc.* B 66:688–694.

Blakeman, J. P. (ed.). 1981. *Microbial Ecology of the Phylloplane*. Academic, London, 502 pp.

Blakeman, J. P. 1982. Phylloplane interactions. In *Phytopathogenic Prokaryotes*, vol. 1, M. S. Mount and G. H. Lacy (eds.). Academic, New York, pp. 307–333.

Blakeman, J. P., and Fokkema, N. J. 1982. Potential for biological control of plant diseases on the phylloplane. *Annu. Rev. Phytopathol.* 20:167–192.

Blanc, M. L. 1963. Protection against Frost Damage. Technical Note no. 51, World Meteorological Organization, Geneva.

Burke, M. J., Gusta, L. A., Quamme, H. A., Weiser, C. J., and Li, P. H. 1976. Freezing and injury to plants. *Annu. Rev. Plant Physiol.* 27:507–528.

Cary, J. A., and Lindow, S. E. 1986. Ice nucleating *Pseudomonas syringae* and bean leaf water relations. *HortScience* 21:1417–1418.

Cary, J. W., and Mayland, H. F. 1970. Factors influencing freezing of supercooled water in tender plants. *Agron. J.* 62:715–719.

Chandler, W. H. 1958. Cold resistance in horticultural plants: a review. *Proc. Am. Soc. Hort. Sci.* 64:552–572.

Chatterjee, A. K., Gibbins, L. N., and Carpenter, J. A. 1969. Some observations on the physiology of *Erwinia herbicola* and its possible implication as a factor antagonistic to *Erwinia amylovora* in the "fireblight syndrome." *Can. J. Microbiol.* 15:640–642.

Cody, Y. S., Gross, D. C., Proebsting, E. L., Jr., and Spotts, R. A. 1987. Suppression of ice nucleation active *Pseudomonas syringae* by antagonistic bacteria in fruit tree orchards and evaluations of frost control. *Phytopathology* 77:1036–1044.

Cooksey, D. A. 1988. Reduction of infection by *Pseudomonas syringae* pv. *tomato* using a nonpathogenic, copper-resistant strain combined with a copper bactericide. *Phytopathology* 78:601–603.

Corotto, L. U., Wolber, P. K., and Warren, G. J. 1986. Ice nucleation activity of *Pseudomonas fluorescens*: mutagenesis, complementation analysis and identification of a gene product. *EMBO J.* 5:231–236.

Crosse, J. E. 1965. Bacterial canker of stone fruits. IV. Inhibition of leaf scar infection of cherry by a saprophytic bacterium from the leaf surface. *Ann. Appl. Biol.* 56:149–160.

Crosse, J. E. 1971. Interactions between saprophytic and pathogenic bacteria in plant disease. In *Ecology of Leaf Surface Microorganisms*, T. F. Preece and C. H. Dickinson (eds.). Academic, London, pp. 283–290.

Deininger, C. A., Mueller, G. M., and Wolber, P. K. 1988. Immunological characterization of ice nucleation proteins from *Pseudomonas syringae, Pseudomonas fluorescens*, and *Erwinia herbicola*. *J. Bacteriol.* 170:669–675.

Dye, D. W., Bradbury, J. F., Gato, M., Hayward, A. C., Lelliott, R. A., and Schroth, M. N. 1980. International standards for naming pathovars of phytopathogenic bacteria and a list of pathovar names and pathotype strains. *Annu. Rev. Plant Pathol.* 59:153–168.

Fennell, A., and Li, P. H. 1986. Temperature response of plasma membranes in tuber-bearing *Solanum* species. *Plant Physiol.* 80:470–472.

Fossard, R. 1981. Effect of guttation fluids on growth of microorganisms on leaves. In *Microbial Ecology of the Phylloplane*, J. P. Blakeman (ed.). Academic, London, pp. 213–216.

Garrett, C. M. E., and Crosse, J. E. 1975. Interaction between *Pseudomonas morsprunorum* and other pseudomonads in leaf-scar infection of cherry. *Physiol. Plant Pathol.* 5:89–94.

Gibbins, L. N. 1972. Relationship between pathogenic and nonpathogenic bacterial inhabitants of aerial plant surfaces. In *Proceedings of the 3rd International Conference on Plant Pathogenic Bacteria, 14–21 April 1971*, H. P. Maas Geesteranus (ed.). Wageningen, The Netherlands, pp. 15–24.

Goodman, R. N. 1967. Protection of apple stem tissue against *Erwinia amylovora* infection by avirulent strains and three other bacterial species. *Phytopathology* 57:22–24.

Govindarajan, A. G., and Lindow, S. E. 1985. Identification of *Pseudomonas syringae* ice gene translational product and reconstitution of ice nucleation activity *in vitro*. *Phytopathology* 75:1380.

Govindarajan, A. G., and Lindow, S. E. 1988. Phospholipid requirements for expression of ice nuclei in *Pseudomonas syringae* and *in vitro*. *J. Biol. Chem.* 263:9333–9338.

Green, R. L., and Warren, G. J. 1985. Physical and functional repetition in a bacterial ice nucleation gene. *Nature* 317:645–648.

Govindarajan, A. G., and Lindow, S. E. 1988. Size of bacterial ice nucleation sites measured *in situ* by gamma radiation inactivation analysis. *Proc. Natl. Acad. Sci. USA* 85:1334–1338.

Gross, D. C., Cody, Y. S., Proebsting, E. L., Radamaker, G. K., and Spotts, R. A. 1983. Distribution, population dynamics, and characteristics of ice nucleation active bacteria in deciduous fruit tree orchards. *Appl. Environ. Microbiol.* 46:1370–1379.

Gross, D. C., Cody, Y. S., Proebsting, E. L., Radamaker, G. L., and Spotts, R. A. 1984*a*. Ecotypes and pathogenicity of ice nucleation active *Pseudomonas syringae* isolated from deciduous fruit tree orchards. *Phytopathology* 74:241–248.

Gross, D. C., Proebsting, E. L., Jr., and Andrews, P. K. 1984*b*. The effects of ice nucleation-active bacteria on the temperatures of ice nucleation and low temperature susceptibilities of *Prunus* flower buds at various stages of development. *J. Am. Soc. Hort. Sci.* 109:375–380.

Haefele, D., and Lindow, S. E. 1982. Localization and quantification of ice nuclei and ice nucleation active bacteria associated with dormant and growing pear tissue. *Phytopathology* 72:946.

Hendershott, C. H. 1962. The response of orange trees and fruit to freezing temperatures. *Proc. Am. Soc. Hort. Sci.* 80:247–254.

Hirano, S. S., and Upper, C. D. 1983. Ecology and epidemiology of foliar plant pathogens. *Annu. Rev. Phytopathol.* 21:243–269.

Hirano, S. S., and Upper, C. D. 1986. Temporal, spatial and genetic variability of
leaf-associated bacterial populations. In *Microbiology of the Phyllosphere*, N. J.
Fokkema, and Van Den Heuvel (eds.). Cambridge University Press, London, pp.
235–251.

Hirano, S. S., Maher, E. A., Kelman, A. and Upper, C. D. 1978. Ice nucleation activity
of fluorescent plant pathogenic pseudomonads. In *Proceedings of the 4th International
Conference on Plant Pathogenic Bacteria, Angers, France*, Station de Pathologie
Vegetale et Phytobacteriologie, (ed.), vol 2. Institut National de la Recherche
Agronomique, Beaucouze, France, pp. 717–725.

Hirano, S. S., Nordheim, E. V., Arny, D. C., and Upper, C. D. 1982. Lognormal distribution of epiphytic bacterial populations on leaf surfaces. *Appl. Environ. Microbiol.*
44:695–700.

Hirano, S. S., Baker, L. S., and Upper, C. D. 1985. Ice nucleation temperature of individual leaves in relation to population sizes of ice nucleation active bacteria and frost
injury. *Plant Physiol.* 77:259–265.

Hobbs, P. V. 1974. *Ice Physics*. Clarendon, Oxford, U.K., pp. 461–523.

Hsieh, S. P. Y., and Buddenhagen, I. W. 1974. Suppressing effects of *Erwinia herbicola*
on infection by *Xanthomonas oryzae* and on symptom development in rice.
Phytopathology 64:1182–1185.

Kaku, S. 1964. Undercooling points and frost resistance in mature and immature leaf
tissues of some evergreen plants. *Bot. Mag.* 77:283–289.

Kaku, S. 1971a. Changes in supercooling in growing leaves of some evergreen plants
and their relation to intercellular space, osmotic value and water content. *Bot. Mag.*
79:98–104.

Kaku, S. 1971b. Changes in supercooling and freezing processes accompanying leaf
maturation in *Buxus*. *Plant Cell Physiol.* 12:147–155.

Kaku, S. 1973. High ice nucleating ability in plant leaves. *Plant Cell Physiol.*
14:1035–1038.

Kaku, S. 1975. Analysis of freezing temperature distribution in plants. *Cryobiol.*
12:154–159.

Ketchie, D. O. 1985. Cold resistance of apple trees through the year and its relationship
to the physiological stages. *Acta Horticulturae* 168:131–137.

Kieft, T. L. 1988. Ice nucleation activity in lichens. *Appl. Environ. Microbiol.*
54:1678–1681.

Kozloff, L. M., Lute, M., and Westaway, D. 1984. Phosphatidylinositol as a component of
the ice nucleation site of *Pseudomonas syringae* and *Erwinia herbicola*. *Science*
226:845–846.

Krezdorn, A. H., and Martsolf, J. D. 1984. Review of effects of cultural practices on frost
hazard. *Proc. Fla. State Hort. Soc.* 97:21–24.

Leben, C. 1961. Microorganisms on cucumber seedlings. *Phytopathology* 51:553–557.

Leben, C. 1964. Influence of bacteria isolated from healthy cucumber leaves on two leaf
diseases of cucumber. *Phytopathology* 54:405–408.

Leben, C. 1988. Relative humidity and the survival of epiphytic bacteria with buds and
leaves of cucumber plants. *Phytopathology*
78:179–185.

Leben, C., and Daft, G. C. 1965. Influence of an epiphytic bacterium on cucumber
anthracnose, early blight of tomato, and northern leaf blight of corn.
Phytopathology 55:760–762.

Lester, D. T., Lindow, S. E., and Upper, C. D. 1977. Freezing injury and shoot elongation in balsam fir. *Can. J. Forestry Res.* 7:584–588.

Levitt, J. 1972. *Responses of Plants to Environmental Stresses*. Academic, New York, pp.
306–398.

Li, P. H. 1984. Sub-zero temperature stress physiology of herbaceous plants. *Hort. Revs.*
6:373–417.

Lim, H. K., Orser, C. S., Lindow, S. E., and Sands, D. C. 1987. *Xanthomonas campestris* pv. *translucens* strains active in ice nucleation. *Plant Dis.* 71:994–997.

Lindemann, J., and Suslow, T. V. 1985. Characteristics relevant to the question of environmental fate of genetically engineered INA⁻ deletion mutant strains of *Pseudomonas*. In *Proceedings of the 6th International Conference on Plant Pathogenic Bacteria*, E. Civerolo, (ed.). U.S. Department of Agriculture, College Park, Maryland, pp. 1005–1012.

Lindemann, J., and Suslow, T. V. 1987. Competition between ice nucleation active wild-type and ice nucleation deficient deletion mutant strains of *Pseudomonas syringae* and *P. fluorescens* biovar I and biological control of frost injury on strawberry blossoms. *Phytopathology* 77:882–886.

Lindemann, J., and Upper, C. D. 1985. Aerial dispersal of epiphytic bacteria over bean plants. *Appl. Environ. Microbiol.* 50:1229–1232.

Lindemann, J., Constantinidou, H. A., Barchet, W. R., and Upper, C. D. 1982. Plants as sources of airborne bacteria, including ice nucleation-active bacteria. *Appl. Environ. Microbiol.* 44:1059–1063.

Lindemann, J., Arny, D. C., and Upper, C. D. 1984. Epiphytic populations of *Pseudomonas syringae* pv. *syringae* on snap bean and non-host plants and incidence of bacterial brown spot disease in relation to cropping patterns. *Phytopathology* 74:1329–1333.

Lindgren, P. B., Govindarajan, A. G., Frederick, R., Panopoulos, N. J., Staskawicz, B. J., and Lindow, S. E. 1989. An ice nucleation reporter gene system: identification of inducible pathogenicity genes in *Pseudomonas syringae* pv. *phaseolicola*. *EMBO J.* (in press).

Lindow, S. E. 1982a. Population dynamics of epiphytic ice nucleation active bacteria on frost sensitive plants and frost control by means of antagonistic bacteria. In *Plant Cold Hardiness*, P. H. Li and A. Sakai (eds.). Academic, New York, pp. 395–416.

Lindow, S. E. 1982b. Epiphytic ice nucleation active bacteria. In *Phytopathogenic Prokaryotes*, G. Lacy and M. Mount (eds.). Academic, New York, pp. 334–362.

Lindow, S. E. 1983a. The role of bacterial ice nucleation in frost injury to plants. *Annu. Rev. Phytopathol.* 21:363–384.

Lindow, S. E. 1983b. The importance of bacterial ice nuclei in plant frost injury. *Curr. Top. Plant Biochem. Physiol.* 2:119–128.

Lindow, S. E. 1983c. Methods of preventing frost injury caused by epiphytic ice nucleation active bacteria. *Plant Dis.* 67:327–333.

Lindow, S. E. 1985a. Integrated control and role of antibiosis in biological control of fireblight and frost injury. In *Biological Control on the Phylloplane*, C. Windels and S. E. Lindow (eds.). American Phytopathological Society Press, Minneapolis, pp. 83–115.

Lindow, S. E. 1985b. Strategies and practice of biological control of ice nucleation active bacteria on plants. In *Microbiology of the Phyllosphere*, N. Fokkema (ed.). Cambridge University Press, pp. 293–311.

Lindow, S. E. 1985c. Ecology of *Pseudomonas syringae* relevant to the field use of Ice⁻ deletion mutants constructed *in vitro* for plant frost control. In *Engineering Organisms in the Environment: Scientific Issues*. American Society for Microbiology, Washington, D.C., pp. 23–25.

Lindow, S. E. 1986a. *In vitro* construction of biological control agents. In *Biotechnology and Plant Improvement and Protection*, P. Day (ed.). BCPC Monograph no. 34, British Crop Protection Council, Cambridge, U.K., pp. 185–198.

Lindow, S. E. 1986b. Specificity of epiphytic interactions of *Pseudomonas syringae* strains on leaves. *Phytopathology* 76:1136.

Lindow, S. E. 1987. Competitive exclusion of epiphytic bacteria by Ice⁻ mutants of *Pseudomonas syringae*. *Appl. Environ. Microbiol.* 53:2520–2527.

Lindow, S. E. 1988a. Construction of isogenic Ice⁻ strains of *Pseudomonas syringae* for evaluation of specificity of competition on leaf surfaces. In *Microbial Ecology*, F. Megusar and M. Gantar (eds.). Slovene Society for Microbiology, Ljuvljana, Yugoslavia, pp. 509–515.

Lindow, S. E. 1988b. Lack of correlation of antibiosis in antagonism of ice nucleation active bacteria on leaf surfaces by non-ice nucleation active bacteria. *Phytopathology* 78:444–450.

Lindow, S. E., and Connell, J. E. 1984. Reduction of frost injury to almond by control of ice nucleation active bacteria. *J. Am. Soc. Hort. Sci.* 109:48–53.

Lindow, S. E., and Panopoulos, N. J. 1988. Field tests of recombinant Ice⁻ *Pseudomonas syringae* for biological frost control in potato. In *Proceedings of the First International Conference on Release of Genetically Engineered Microorganisms*, M. Sussman, C. H. Collins, and F. A. Skinner (eds.). Academic, London, pp. 121–138.

Lindow, S. E., Arny, D. C., and Upper, C. D. 1978a. Distribution of ice nucleation active bacteria on plants in nature. *Appl. Environ. Microbiol.* 36:831–838.

Lindow, S. E., Arny, D. C., and Upper, C. D. 1978b. *Erwinia herbicola*: a bacterial ice nucleus active in increasing frost injury to corn. *Phytopathology* 68:523–527.

Lindow, S. E., Arny, D. C., and Upper, C. D. 1982. Bacterial ice nucleation: a factor in frost injury to plants. *Plant Physiol.* 70:1084–1089.

Lindow, S. E., Arny, D. C., and Upper, C. D. 1983a. Biological control of frost injury II: establishment and effects of an antagonistic *Erwinia herbicola* isolate on corn in the field. *Phytopathology* 73:1102–1106.

Lindow, S. E., Arny, D. C., and Upper, C. D. 1983b. Biological control of frost injury I: an isolate of *Erwinia herbicola* antagonistic to ice nucleation-active bacteria. *Phytopathology* 73:1097–1102.

Lindow, S. E., Gies, D. R., Willis, D. K., and Panopoulos, N. J. 1987. Molecular analysis of the *Pseudomonas syringae* pv. *syringae* ice gene and construction and testing of Ice⁻ deletion mutants for biological frost control. In *Plant Pathogenic Bacteria*, E. Civerolo, A. Collmer, R. Davis, and Gillaspie (eds.). Martinus Nijhoff, Boston, p. 1030.

Loper, J. E., and Lindow, S. E. 1987. Lack of evidence for *in situ* fluorescent pigment production by *Pseudomonas syringae* pv. *syringae* on bean leaf surface. *Phytopathology* 77:1449–1454.

Lucas, J. W. 1954. Subcooling and ice nucleation in lemon. *Plant Physiol.* 29:245.

Maki, L. R., and Willoughby, K. J. 1978. Bacteria as biogenic sources of freezing nuclei. *J. Appl. Meteorol.* 17:1049–1053.

Maki, L. R., Galyon, E. L., Chang-Chien, M., and Caldwell, D. R. 1974. Ice nucleation induced by *Pseudomonas syringae*. *Appl. Microbiol.* 28:456–460.

Makino, T. 1982. Micropipette method: a new technique for detecting ice nucleation activity of bacteria and its application. *Ann. Phytopathol. Soc. J.* 48:452–457.

Marcellos, H. W., and Single, W. V. 1976. Ice nucleation on wheat. *Agric. Metereol.* 16:125–129.

Marcellos, H., and Single, W. V. 1979. Supercooling and heterogeneous nucleation of freezing in tissues of tender plants. *Cryobiology* 16:74–77.

Marshall, D. 1988. A relationship between ice nucleation active bacteria, freeze damage, and genotype in oats. *Phytopathology* 78:952–957.

Mason, B. J. 1960. Ice-nucleating properties of clay minerals and stony meteorites. *Q. J. R. Metereol. Soc.* 84:553–556.

Mason, B. J., and Hallett, J. 1957. Ice-forming nuclei. *Nature* 197:357–359.

Mason, B. J., and Maybank, J. 1958. Ice-nucleating properties of some natural mineral dusts. *Q. J. R. Metereol. Soc.* 84:235–241.

Mazur, P. 1969. Freezing injury to plants. *Annu. Rev. Plant Physiol.* 20:419–448.

Mellano, M. A., and Cooksey, D. A. 1988. Nucleotide sequence and organization of copper resistance genes from *Pseudomonas syringae* pv. *tomato*. *J. Bacteriol.* 170:2879–2883.

Miedema, P., Groot, P. J., and Zuidgeest, J. H. M. 1980. Vegetative propagation of *Beta vulgaris* by leaf cuttings. *Euphytica* 29:425–432.

Modlibowska, I. 1962. Some factors affecting supercooling of fruit blossoms. *J. Hort. Sci.* 37:249–261.

O'Brien, R. D., and Lindow, S. E. 1988. Effect of plant species and environmental conditions on ice nucleation activity of *Pseudomonas syringae* on leaves. *Appl. Environ. Microbiol.* 54:2281–2286.

Olien, C. R. 1967. Freezing stresses and survival. *Annu. Rev. Plant Physiol.* 18:387–408.

Orser, C. S., Staskawicz, B. J., Loper, J., Panopoulos, N. J., Dahlbeck, D., Lindow, S. E., and Schroth, M. N. Cloning of genes involved in bacterial ice nucleation and fluorescent pigment/siderophore production. In *Molecular Genetics of Bacterial-Plant Interactions*, A. Puhler (ed.). Springer-Verlag, Berlin, pp. 353–361.

Orser, C. S., Lotstein, R., Willis, D. K., Papp, J., Panopoulos, N. J., and Lindow, S. E. 1984. Analysis of the *Pseudomonas syringae* pv. *syringae* ice region and construction and testing of site-directed deletion mutants for biological frost control. Proc. Molecular Basis of Plant Disease Conference (abstr.).

Orser, C. S., Staskawicz, B. J., Panopoulos, N. J., Dahlbeck, D., and Lindow, S. E. 1985. Cloning and expression of bacterial ice nucleation genes in *Escherichia coli*. *J. Bacteriol.* 164:359–366.

Parsons, L. R., and Tucker, D. P. H. 1984. Sprinkler irrigation for cold protection in citrus groves and nurseries during an advective freeze. *Proc. Fla. State Hort. Soc.* 97:28–30.

Parsons, L. R., Combs, B. S., and Tucker, D. P. H. 1985a. Citrus freeze protection with microsprinkler irrigation during an advective freeze. *HortScience* 20:1078–1080.

Parsons, L. R., Wheaton, T. A., and Stewart, I. 1985b. Observations on the use of water and coverings for cold protection during an advective freeze. *Proc. Fla. State Hort. Soc.* 98:57–60.

Paulin, J. P., and Luisetti, J. 1978. Ice nucleation activity among phytopathogenic bacteria. In *Proceedings of the 4th International Conference on Plant Pathogenic Bacteria, Angers, France*, Station de Pathologie Vegetale et Phytobacteriologie (ed.), vol 2. Institut National de la Recherche (Agronomique, Beaucouze,) France, pp. 725–733.

Pennycook, S. R., and Newbook, F. J. 1982. Ultraviolet sterilization in phylloplane studies. *Trans. Br. Mycol. Soc.* 78:360–361.

Power, B. A., and Power, R. F. 1962. Some amino acids as ice nucleators. *Nature* 194:1170–1171.

Proebsting, E. L., Andrews, P. K., and Gross, D. 1982. Supercooling young developing fruit and floral buds in deciduous orchards. *Hortic. Sci.* 17:67–68.

Quamme, H. A. 1983. Relationship of air temperature to water content and supercooling of overwintering peach flower buds. *J. Am. Soc. Hort. Sci.* 108:697–701.

Quamme, H. A. 1985. Avoidance of freezing injury in woody plants by deep supercooling. *Acta Horticulturae* 168:11–30.

Renquist, A. R., 1985. The extent of fruit bud radiant cooling in relation to freeze protection with fans. *Agric. For. Meteorol.* 36:1–6.

Riggle, J. H., and Klos, E. J. 1972. Relationship of *Erwinia herbicola* to *Erwinia amylovora*. *Can. J. Bot.* 50:1077–1083.

Rosinski, J., and Parungo, F. 1966. Terpene-iodine compounds as ice nuclei. *J. Appl. Meteorol.* 5:119–123.

Ruvkun, G. B., and Ausubel, F. M. 1981. A general method for site-directed mutagenesis in prokaryotes. *Nature* 289:85–88.

Scherff, R. H. 1973. Control of bacterial blight of soybean by *Bdellovibrio bacteriovorus*. *Phytopathology* 63:400–402.

Schneider, R. W., and Grogan, R. G. 1977. Tomato leaf trichomes, a habitat for resident populations of *Pseudomonas tomato*. *Phytopathology* 67:898–902.

Schnell, R. C., and Vali, G. 1972. Atmospheric ice nuclei from decomposing vegetation. *Nature* 236:163–165.

Schnell, R. C., and Vali, G. 1973. World-wide sources of leaf-derived freezing nuclei. *Nature* 246:212–213.

Schnell, R. C., and Vali, G. 1976. Biogenic ice nuclei. Part 1. Terrestrial and marine sources. *J. Atmos. Sci.* 33:1554–1564.

Southworth, M. W., Wolber, P. C., and Warren, G. J. 1988. Nonlinear relationship between concentration and activity of a bacterial ice nucleation protein. *J. Biol. Chem.* 263:15211–15216.

Sprang, M. L., and Lindow, S. E. 1981. Subcellular localization and partial characterization of ice nucleation activity of *Pseudomonas syringae* and *Erwinia herbicola*. *Phytopathology* 71:256.

Spurr, H. W. 1979. Ethanol treatment—a valuable technique for foliar biocontrol studies of plant disease. *Phytopathology* 69:773–776.

Sztejnberg, A., and Blakeman, J. P. 1973. Ultraviolet-induced changes in populations of epiphytic bacteria on beetroot leaves and their effect on germination of *Botrytis cinerea* spores. *Physiol. Plant Pathol.* 3:443–451.

Thomson, S. V., Schroth, M. N., Moller, W. J., and Reil, W. O. 1976. Efficacy of bactericides and saprophytic bacteria in reducing colonization and infection of pear flowers by *Erwinia amylovora. Phytopathology* 66:1457–1459.

Tukey, H. B. 1966. The leaching of metabolites from above-ground plant parts and its implications. *Bull. Torrey Botan. Club* 93:385–401.

Tukey, H. B., Jr. 1970. The leaching of substances from plants. *Annu. Rev. Plant Physiol.* 21:305–324.

Tukey, H. B. 1971. Leaching of substances from plants. In *Ecology of Leaf Surface Microorganisms*, T. F. Preece and C. H. Dickinson (eds.). Academic, London, pp. 67–80.

Tukey, H. B., and Morgan, J. V. 1963. Injury to foliage and its effect upon the leaching of nutrients from above-ground plant parts. *Physiol. Plant* 16:557–564.

Warren, G., Corotto, L., and Wolber, P. 1986. Conserved repeats in diverged ice nucleation structural genes from two species of *Pseudomonas. Nucleic Acids Res.* 14:8047–8060.

Wolber, P. K., Deininger, C. A., Southworth, M. W., Van de Kerckhove, J., Van Montague, M., and Warren, G. J. 1986. Identification and purification of a bacterial ice nucleation protein. *Proc. Natl. Acad. Sci. USA* 83:7256–7260.

White, G. F., and Haas, J. E. 1975. *Assessment of Research on Natural Hazards*. MIT Press, Cambridge, Massachusetts, pp. 304–312.

Wisniewski, M., and Ashworth, E. N. 1984. Ultrastructural modification in response to a freezing stress and patterns of cell injury in xylem ray tissues of peach [*Prunus persica* (L.) Bartsch.]. *Am. J. Bot.* 71:55.

Wisniewski, M., and Ashworth, E. N. 1986a. A comparison of seasonal ultrastructural changes in stem tissues of peach [*Prunus persica* (L.) Bartsch.] that exhibit contrasting mechanisms of cold hardiness. *Bot. Gaz.* 147:407–417.

Wisniewski, M., and Ashworth, E. N. 1986b. Seasonal variation in deep supercooling and dehydrative resistance. *HortScience* 21:503–505.

Yankofsky, S. A., Levin, Z., and Moshe, A. 1981. Association with citrus of ice-nucleating bacteria and their possible role as causative agents of frost damage. *Curr. Microbiol.* 5:213–217.

Young, J. M. 1978. Survival of bacteria on *Prunus* leaves. In *Proceedings of the 4th International Conference on Plant Pathogenic Bacteria, Angers, France*, Station de Pathologie Vegetale et Phytobacteriologie (ed.), vol. 2. Institut National de la Recherche Agronomique, Beaucouze, France, pp. 779–786.

Young, R. H. 1966. Freezing points and lethal temperatures of citrus leaves. *Proc. Am. Soc. Hort. Sci.* 88:272–279.

Commercial Development of *Bacillus thuringiensis* Bioinsecticide Products*

Thomas C. Currier

Sterling Drug, Inc.
9 Great Valley Parkway
Malvern, Pennsylvania 19355

Cynthia Gawron-Burke

Ecogen Inc.
2005 Cabot Boulevard West
Langhorne, Pennsylvania 19047-1810

Introduction

The majority of sales of all biological pest control products worldwide are of the bacterial agent *Bacillus thuringiensis*, or "BT" as it is commonly known. BT is a Gram-positive sporulating bacterium that produces a crystalline inclusion (parasporal crystal) in sporulating cells. The crystal protein(s), sometimes referred to as "δ-endotoxins" (DET)*, contained within the parasporal body are toxic to certain insect pests. Commercially, the mixture of BT spores and crystals that results from a standard fermentation is concentrated following fermentation and then formulated for application with standard spray equipment. When a susceptible insect feeds on foliage that has been treated with BT, ingestion of a critical dose of toxin crystals is followed within a few minutes by gut paralysis, which leads to a cessation of feeding. This effect is usually followed by a generalized paralysis, and death typically occurs in 3 to 5 days (29).

A strain developed by the U.S. Department of Agriculture (HD-1, var. *kurstaki*), and derivatives of it, have been employed since the early 1970s for the commercial production of fermentation preparations used to control lepidopteran caterpillars (e.g., cabbage loopers, cabbageworms, spruce and tobacco budworms, and gypsy moth larvae)(26). A second BT strain, BT var. *israelensis* (BTI), is produced by several companies for control of the larvae of mosquito and blackfly species (36). Recently, several BT strains that produce crystals with activity against coleoptera (beetles) have been discovered and are either being sold or are under commercial development (24,49,67). A partial list of current BT products, their manufacturers, and bacterial active ingredients is shown in Table 1.

In this chapter, we will briefly review the molecular biology of BT crystal proteins and their mode of action as insect toxins. (See Refs. 4 and 11 for more detailed reviews.) The possibility of insect resistance will also be discussed (See Ref. 34 for recent review). Our main focus will be on the development and commercialization strategies for BT-based insect control products, with an emphasis on conventional

---

*These toxic proteins have been referred to as "δ-endotoxins" ("DET"). In this review, however, the terms "DET" or "toxin" will refer to proteolytically activated polypeptide of about 65 kDa.

TABLE 1   Commercial BT Products

Active ingredient (BT strain)	Product	Producer	Target insects
Bacillus thuringiensis var. *aizawai*	Certan[a]	Sandoz Crop Protection Corp.	Wax moth larvae
Bacillus thuringiensis var. *israelensis*	Vectobac-AS[a]	Abbott Laboratories	Mosquito and blackfly larvae
	Skeetal[a]	Novo Laboratories	
	Teknar[a]	Sandoz Crop Protection Corp.	
Bacillus thuringiensis var. *kurstaki*	Dipel[a]	Abbott Laboratories	Lepidopteran larvae
	Bactospeine[a]	Duphar B. V.	
	Thuricide[a]	Sandoz Crop Protection Corp.	
	Javelin[a]	Sandoz Crop Protection Corp.	
Bacillus thuringiensis var. *san diego*	M-One[a]	Mycogen Corporation	Colorado potato beetle larvae

[a]Trademark

spray-on BT products. We will compare the development and commercialization processes for spray-on BT products with those required for engineering BT toxin genes into plants and microbes.

Molecular Genetic Basis of Insecticidal Activity

BT is a complex bacterial group with more than 30 recognized subspecies. Distinctions among different strains of BT are based primarily on the antigenic determinants of the flagella (20) and on the immunological properties of the crystals (70). Different varieties, and even strains within a single variety, exhibit substantial diversity with regard to insecticidal activity (26).

The differential insecticidal activities or δ-endotoxin specificities exhibited by various BT strains are thought in part to reflect differences in the nature of the δ-endotoxin(s) that compose the parasporal crystal(s) (26). BT strains can exhibit a variety of crystal morphologies, and a single strain can harbor more than one type of crystal. For example, certain strains of BT var. *kurstaki*, including strain HD-1, produce a small cuboidal crystal, which may be attached to the larger bipyramidal crystal (59). Solubilization of these crystals in alkali followed by fractionation by polyacrylamide gel electrophoresis shows the bipyramidal crystal to contain proteins (P1-type) of 130 to 140 kDa in size; the cuboidal crystal contains a smaller protein (P2) of about 65 kDa (59,112). The terms "P1" and "P2" initially referred to column chromatography peaks obtained during crystal-protein fractionation (112). Subsequently, the terms "P1" and "P2" have also been used to describe crystal morphology, bipyramidal and cuboidal, respectively, in addition to referring to distinct crystal proteins. In addition to size differences, P1 and P2 proteins differ in immunological specificities, isoelectric properties, and insecticidal activities. The P1 proteins appear active against only lepidopteran insects, while P2 has activity against certain dipterans (specifically, mosquitoes) as well as lepidopterans (112).

P1 proteins exist in the bipyramidal crystal in an inactive ("protoxin") form of 130 to 140 kDa (55). Prolonged incubation of solubilized bipyramidal crystals in vitro generates a 65- to 70-kDa protein that is the presumed active toxin, with a concomitant decrease in the amount of 135- to 140-kDa protoxin. The activation process in vivo is believed to be due to proteolytic cleavage of the protoxin by proteases in the insect midgut (29), although it is possible that bacterial proteases associated with the crystal may contribute to this activation.

In some cases, the toxic activity of a BT spore-crystal preparation is due primarily to the crystalline endotoxin. In other cases, spore and crystal mixtures are more toxic than the purified crystal protein alone. In fact, lepidopterous larvae have been classified into three groups by Heimpel and Angus (44) on the basis of their susceptibility to crystalline endotoxin, spores, or mixtures of the two. Type I insects are killed only by the toxin; spores do not increase toxicity. Type II insects can be killed by endotoxin alone, but the presence of spores enhances toxicity. Type III insects are killed only when both spores and crystalline endotoxin are ingested. In cases in which spores enhance activity, the effect can be dramatic. In these cases the spores synergize the toxicity of the crystal protein, although some component of the spore activity is undoubtedly due to the presence of protoxin in the spore coat (98). An example of toxin-spore synergy was recently reported for *Galleria mellonella* (greater wax moth) with crystal protein from a BT *aizawai* strain. In this case toxicity was increased 36% when the crystal preparation contained as little as 0.001% spores (72).

Genetic analyses have shown that insecticidal crystal-protein production is correlated with the presence of specific extrachromosomal elements termed "plasmids" (39,98). This correlation of crystal production with the presence of certain plasmids has been confirmed by the use of a native BT conjugal transfer system (12,37). Crystal-producing (Cry$^+$) BT strains of various types can serve as donors of certain crystal-protein-encoding plasmids to cured derivatives of other BT types or to *Bacillus cereus* in broth matings (37). A recipient acquiring a crystal-protein-encoding plasmid from a donor gains the ability to produce insecticidal crystals characteristic of the donor (37).

Lepidopteran-active crystal-protein genes

Genes that encode δ-endotoxins with lepidopteran activity and that comprise a P1-type crystal (i.e., P1-type toxin genes) have been cloned from BT subspp. *kurstaki* (2,46,91,109), *thuringiensis* (52,64), *aizawai* (41,65,89), *subtoxicus* (65), and *sotto* (65,95). These genes have been cloned in either *Escherichia coli* or *Bacillus subtilis* host-vector systems, and their characterization has revealed important structural features of P1-type toxin genes.

The DNA sequence of a cloned P1-type toxin gene isolated from the active ingredient of Dipel®, BT subsp. *kurstaki* strain HD-1, contains an open reading frame coding for 1,186 amino acids (corresponding to a 133,500-Da polypeptide) (93), and deletion studies have shown that the N-terminal 645 codons of the structural gene encode a protein with insecticidal activity (92). Recently DNA sequences were determined for several other P1-type toxin genes originating from *kurstaki* strains HD-73, HD-1, and HD-244 and *thuringiensis* strain *berliner* 1715 (2,19,52,100). Significant sequence homology was present between the promoter regions, the N-terminal regions (base pairs 0 to 850) and the C-terminal regions (base pair 2,200 to end of gene) of these sequences and the original HD-1 Dipel® toxin gene cloned by Whiteley and colleagues. In general, the region extending from base pairs 850 to 2,200 exhibits the greatest sequence divergence, although some protoxin genes are nearly identical in this region as well (e.g., the sequenced toxin genes of HD-73 and HD-244). Deletion studies involving several of these P1-type genes have emphasized the importance of sequences in the N-terminal region for insecticidal activity (2,19,100); a minimum toxic fragment mapping from amino acid residues 29–37 to 601–607 has recently been identified for the cloned *berliner* 1715 gene (*bt2*) (52).

Probing of various BT strains with a radioactively labeled restriction fragment derived from the intragenic region of the cloned HD-1 Dipel® toxin gene has been used to investigate the organization and relationship of P1-type protoxin genes in several different subspecies (65,69,109). Kronstad et al. (69) identified multiple crystal-protein genes in strain HD-1 using such an approach. Three *HindIII* restriction fragments (measuring 6.6, 5.3, and 4.5 kb) derived from HD-1

DNA hybridized to the intragenic probe and represented three distinct protoxin genes. Although these *HindIII* fragments do not contain the entire gene, they have served to classify these HD-1 toxin genes as well as related genes in other BT strains. Data obtained in such experiments indicate that multiple P1-type toxin genes exist in BT, that these genes may be present in the same cell on different plasmids and possibly on the bacterial chromosome, and that some strains contain either related genes or multiple copies of the same gene.

Hofte et al. (53) have demonstrated that monoclonal antibodies can be used to distinguish the various classes of P1-type crystal proteins. These monoclonal antibodies could also detect, in certain BT strains, P1-type (bipyramidal) crystal proteins that differed from the classic 4.5, 5.3, and 6.6 types. Likewise, other examples of lepidopteran-active bipyramidal crystal-protein genes that differ dramatically in their N-terminal DNA sequence from the described 4.5, 5.3, or 6.6 P1-type genes have been reported. These include the *Spodoptera*-active *Bta* gene isolated from BT var. *aizawai* 7.29 (89,90) and the *cryA4* gene from BT var. *thuringiensis* HD-2 (7). Interestingly, there appear to be several short stretches of the N-terminal amino acid sequence that are conserved between these newly described crystal proteins and the 4.5, 5.3, and 6.6 P1-type crystal proteins.

Dipteran-active crystal-protein genes

The most well characterized dipteran-active BT is BT var. *israelensis* (BTI), an unusual variety of BT toxic to mosquito and blackfly larvae that was first isolated from a mosquito-breeding site in Israel in 1977 (36). BTI produces an irregularly shaped crystal that is composed of major proteins of about 130 (doublet), 72, and 28 kDa (58,101,113). These BTI crystal proteins are encoded by a 75-MDa transmissible plasmid (38). The genes encoding the major crystal proteins have been isolated, and both the 130-kDa and 72-kDa crystal proteins have been shown to be mosquitocidal (3,6,15,22,58,94,107,108,114). Likewise, the gene encoding the 28-kDa crystal protein has been isolated, and this too has been reported to be mosquitocidal (104). Other reports in the literature indicate that the 28-kDa protein has little or no mosquitocidal activity (13,14,47,56,57,103). Thus, the exact contribution of the individual BTI crystal proteins to the collective insecticidal activity of the strain is not yet clear. Synergistic effects on mosquitocidal activity from mixing of individual BTI crystal proteins have been reported, and all three proteins may contribute significantly (15,111). DNA sequence determinations have shown that the genes encoding these BTI crystal proteins are unique and as different from each other as they are from the lepidopteran-active P1-type toxin genes.

Recent studies of the novel P2 protein of BT var. *kurstaki* have shown that it is toxic to both dipteran *and* lepidopteran larvae (23,113). The 71 kDa P2 protein (59,112) forms a cuboidal crystal that is often found associated with P1 bipyramidal crystals in BT *kurstaki* isolates. Cloning and characterization of the gene (*cryB1*) encoding P2 from BT strain HD-263 has enabled a clear determination of the protein's lepidopteran and dipteran activity as well as its unique amino acid sequence (23). The sequence of the *cryB1* gene is distinct from those reported for P1-type toxin genes or the genes encoding the 28-kDa and 130-kDa BTI crystal proteins. The N-terminal amino acid sequence of the P2 crystal protein does, however, exhibit homology with the 72-kDa mosquitocidal protein from BTI (a stretch of 215 amino acids in CryB1 was 30% homologous to 211 amino acids of the 72-kDa BTI protein). Interestingly, the N-terminal amino acid sequence of the CryB1 protein also exhibits homology with P1-type toxin proteins (a stretch of 100 amino acids is 37% homologous) (23).

A large plasmid (110 MDa) encodes the genes for P1-type crystal proteins as well as P2 in BT var. *kurstaki* strain HD-263 (23). Furthermore, *cryB1*-related sequences have recently been identified in certain BT strains (23,110). Perhaps there exists a family of P2-related crystal proteins in much the same way as there is a family of P1-type crystal proteins (i.e., 4.5, 5.3, and 6.6 classes).

Coleopteran-active crystal-protein genes

The most recent significant finding with regard to BT commercial development has been the discovery of coleopteran-active (beetle-active) BT strains. BT var. *tenebrionis* (BTT), isolated from a *Tenebrio molitor* pupa, was first reported in 1983 to be coleopteran-active (67). BTT produces flat, rectangular crystals comprised of an insecticidal protein with activity against the Colorado potato beetle (*Leptinotarsa decemlineata*) but not lepidopteran larvae. Several other coleopteran-active BT strains have since been reported (e.g., BT var. *san diego* and BT strain EG2158), and the genes encoding these coleopteran-active crystal proteins have been cloned and characterized (24,48,49). Interestingly, all three coleopteran-active strains encode a crystal protein of approximately 73 kDa, and all three proteins have identical amino acid sequences. In contrast to BTT, however, strain EG2158, which was isolated from grain dust, produces both a rhomboid crystal (comprised of the coleopteran-active protein) and a flat diamond crystal. Thus, distinct isolates obtained from varying sources share the iden-

TABLE 2   Sequence Similarities of BT Crystal Proteins

Protein pair	Extent of similarity[b]	Identity (%)
CryB1:P1(6.6)	162–256:146–244	39
CryC:P1(6.6)	99–643:64–608	35
CryC:CryB1	131–219:115–199	36
CryC:CryD	107–230:76–189	33
CryB1:CryD	61–275:45–255	30
CryD:P1(4.5)	No significant identity	

[a]The sequences of the crystal proteins were deduced from their respective gene sequences as detailed for *cryB1*, the 6.6-class P1-type gene from HD-73, the 4.5-class P1-type gene from HD-1, *cryC*, and *cryD* in references 23, 2, 93, 24, and 22, respectively.
[b]The numbers listed in this column for each protein indicate the distance in amino acids from the NH$_2$-terminal methionine.

tical coleopteran-active crystal protein. Certain amino acid sequences within the N-terminal region of the coleopteran-active protein were also shown to be related (approximately 30% homologous) to the N-terminal regions of P1-type toxin genes, to *cryB1*, and to the 130- and 72-kDa BTI proteins (Table 2).

In summary, there is an extensive natural diversity of BT strains and BT insecticidal crystal proteins. There are many lepidopteran-active BT strains as well as some dipteran-active BT varieties, and recently, several BT strains with coleopteran or beetle activity have been isolated.

A given BT strain can contain more than one type of crystal, and a single crystal can be comprised of several related but distinct insecticidal proteins, as in the case of the P1 crystal. These insecticidal proteins are typically encoded by genes present on one or more plasmids.

In view of the fact that BT strains often contain multiple toxin genes, it is likely that the collective insecticidal activity of a strain containing several toxin genes would be influenced by their relative levels of expression and toxicity. Thus, the insecticidal activity of a given strain may be governed not only by the number and nature of toxin genes present, but also by the manner in which these genes (or gene products) interact with one another.

Mechanism of Action

The mechanism of action of P1-type δ-endotoxins, toxin specificity, and the potential development of insect resistance to BT are interconnected. In this review, the term "toxin specificity" is used to describe both the variable toxicity of a particular toxin toward various insect hosts and the differential toxic effects of highly related yet distinct toxins in the same host. The presence of toxin specificity indicates that different insects have varying susceptibilities to different BT toxins. The molecular basis for these differences is closely related to the mechanism of action of P1 toxins and may help define mechanisms for resistance development. An understanding of the mechanism(s) of action, toxin specificities, and potential for development of resistance should permit rational design of new BT products with increased effectiveness and prolong the usefulness of this important class of insecticides. The following discussion will briefly review the mechanism of action of P1-type δ-endotoxins and the potential for resistance development.

BT toxin is an insect stomach poison that must be ingested in order to be effective. Its toxicity, therefore, is related to factors that affect ingestion as well as biochemical changes that ultimately result in the death of the insect. In this review, a series of three events or processes necessary for toxicity are discussed. The first area concerns the effects of insect behavior on ingestion of crystalline protoxin. Second, the toxin is solubilized and proteolytically activated or processed in the insect's midgut. Third, the toxin affects some biochemical process(es) in the midgut epithelial cells. Since all these steps are required for BT toxin to be effective, they all have the potential for involvement in toxin specificity and resistance development.

Little is known about the mechanisms of action of P2 and coleopteran-active toxins. Considerable work has been done on the mechanism of action of the P1-type toxins and mosquitocidal toxins. For information on the mechanism of action of mosquitocidal toxins, the reader is referred to the publications cited in the following references: 3, 6, 15, 22, 58, 94, 107, 114. In this section, the mechanism of action of lepidopteran-active δ-endotoxins (P1-type) will be reviewed, and then information relating to the potential for development of resistance to BT insecticides will be presented. Finally, potential uses of this information in commercialization strategies to enhance the efficacy of BT insecticides will be discussed.

Toxin ingestion

Effect of larval size.   In order to obtain optimal insect control with BT insecticides, the target insect should receive a lethal dose of toxin when any part of the treated plant is ingested—i.e., the first bites of food should contain lethal amounts of toxin regardless of the location of the insect on the plant. Delivery of a lethal dose of toxin can depend not only on the feeding behavior of the larvae before and after toxin ingestion but also on larval size and stage of development.

It is generally assumed that early larval instars of all insects are physiologically more fragile than later instars. This has been documented for some insects, e.g., black cutworm (45) and *Heliothis zea* (83) larvae. Therefore, it is expected that a lethal dose of toxin for the earlier instars would be less on a weight basis than for larvae in later instars. This difference has been demonstrated in laboratory bioassay where LC_{50} measurements on early instar larvae were less than for later instars (25).

The stage of larval development could also directly affect toxin ingestion if larval feeding behavior changes as development progresses. The extent to which feeding behavior is different in different developmental stages is not well-documented. However, differences in distribution patterns of *Plutella xylostella* on *Brassica rapa* suggest different behavioral patterns for different instars (16), and as discussed by Burkett et al. (9), differences in feeding behavior of different larval instars of *H. zea* have been observed by several investigators.

Effects of feeding behavior before toxin ingestion. For the most effective control of insects with BT, the target insect should feed indiscriminately at all times during its life cycle. Such an ideal target insect probably does not exist. Under field or greenhouse conditions, feeding on the plant by insect larvae is behaviorally controlled, and there is a great deal of diversity in the feeding behavior of different lepidopterous larvae. Even within a single species (e.g., neonates of *H. zea*), feeding behavior can vary, probably in response to environmental factors (9). Differences in feeding behavior have also been shown to be related to predator-avoidance strategies (45), and feeding behavior may vary for different stages of larval development. For example, the neonates and earlier instars of many larvae tend to feed in protected areas of the plant possibly because they are more susceptible to physical damage (82) and environmental stresses, such as desiccation and temperature (60). Also, both early and late instars may feed intermittently and continue to search for more satisfying or nutritionally ideal sources of food. This searching behavior (nutrient self-selection) may be related to the tendency of the insect to ingest a nutritionally satisfying diet (17).

Even though it may be difficult to precisely define larval feeding behavior, some insects are considered to be foliage feeders, e.g., gypsy moth and tobacco hornworm, while others prefer certain parts of their host plant throughout their life cycle, e.g., tobacco budworm and bollworm, and still others feed from different parts of the plant at different stages in their life cycle, e.g., European corn borer. When indiscriminate feeding occurs, the chances of effectively using a BT insecticide are increased because in these situations the larvae will usually ingest all the foliage in the immediate area before moving to a different location of the plant. Thus, the opportunity to ingest a lethal dose is maximized. However, this type of feeding usually occurs during the latter stages of development when the larvae are hardier and hence larger doses of toxin are required for lethality.

Because BT toxin must be ingested to be effective, the feeding behavior of the insect must be considered in devising delivery strategies. Those strategies that are able to exploit the feeding behavior of the target insect are expected to be more effective than those strategies that do not address feeding behavior or are unable to do so. Some aspects of formulations and delivery are discussed in more detail below.

Effects of feeding behavior after toxin ingestion.   Larval feeding behavior after toxin ingestion may also have an effect on the efficacy of BT insecticides. A common symptom of BT intoxication is cessation of feeding (29). The reason for this toxin-induced feeding inhibition (TIFI) is not known. As described by Retnakaran et al. (85), after toxin ingestion and before the onset of TIFI, a larva has either ingested a lethal dose or it has obtained a sublethal dose and will eventually recover. If the larva recovers, it will be developmentally delayed but will continue to consume its host plant.

The probability of ingesting a lethal dose of toxin is expected to be less for an insect whose natural tendency is to ingest small amounts and to wander between feedings. Given this type of feeding behavior, ingestion of a small, sublethal amount of toxin during feeding could result in TIFI during the wandering stage. This would allow the larva to recover and resume feeding. Conversely, if the nature of the insect is to feed continuously from an initial site, then it is more likely that a lethal dose will be ingested before the onset of TIFI.

Not enough is known about the feeding behavior of most insects, and less is known about the behavioral alterations that occur after toxin ingestion. The behavioral parameters relating to toxicity that were discussed above are likely to be interactive and complex, and they are expected to vary for different insects and for different developmental stages of the same insect. Knowledge in these areas could be extremely useful for the design of novel delivery systems and formulations to increase the effectiveness of the toxin on the target insects.

Toxin processing

Toxin processing involves the solubilization of crystal protein and subsequent proteolytic digestion of protoxin to an active toxin fragment of about 65 kDa (29). Alkaline pH and the presence of reductant greatly facilitate the solubilization of P1 toxins in vitro (73). Thus, the midguts of most lepidopterous larvae, which are alkaline (pH 8 to 9.5) and contain proteases and reducing equivalents (54), provide a good environment for toxin solubilization and activation in vivo. Lower-molecular-weight peptides can be produced by digestion of protoxin with gut juice or trypsin, but there is little evidence that these fragments are toxic by ingestion (29).

Fast (29) points out that different toxins solubilize at different rates with partially purified midgut proteases or crude gut extracts from different insects and speculates that these differences could affect toxicity and be related to toxin specificity. Proteases might also affect toxin specificity by virtue of their substrate specificities and the amounts present (29,42,71). These differences may result in proteolytic processing at different sites within the toxin. Since proteolytic activation occurs in or near the variable region of the protoxin (52) and this region is presumed to contribute to toxin specificity, it is very likely that the manner in which the toxin is activated in the midgut will affect its activity and specificity.

In fact, the results of Haider et al. (42) demonstrate a relationship between toxin specificity and protease activation. Toxin from BT var. *colmeri*, which is active on both a lepidopteran (*Pieris brassicae*) and a dipteran (*Aëdes aegypti*) insect, was activated with gut extracts from the two insects. Toxin activated with *A. aegypti* gut extract was toxic to *A. aegypti* and not to *P. brassicae*, and toxin activated with *P. brassicae* gut extract was toxic to larvae of both insects. Since polypeptides with slightly different molecular weights were produced with the two gut extracts, it was suggested that the observed activities could be attributed to the manner in which the protoxin was proteolytically processed.

Alteration of biochemical processes in the insect

There are two major populations of cells (columnar and goblet) present in the insect midgut epithelia. BT toxin binds exclusively to the highly convoluted brush border membranes of the columnar cells (74). Recent data of Hofmann et al. (50), using brush border membranes from *P. brassicae*, indicate that activated toxin binds reversibly and then irreversibly to brush border membranes. In a subsequent study using brush border membranes from *Manduca sexta* and *P. brassicae*, it was shown (51) that different P1 toxins attached to different receptors with different affinities and that the number of available receptors for a given P1 toxin varied.

Biochemical studies show that BT toxin inhibits K^+-dependent amino acid transport in membrane vesicles prepared from brush borders (88) and alters the net flux of K^+ ions in isolated midguts (40,43). Evidence (18,50) suggesting that the toxin may be acting as a K^+

channel or may function to create a K^+ channel in the membrane has been reported.

Recently, a phosphatase activity was identified in the brush border membranes of midgut epithelium from _Heliothis virescens_ (28). The activity is present under alkaline conditions and is reversibly inhibited by DET (K_i = 64.7nM). It was suggested that the alkaline phosphatase inhibitory ability of DET is indicative of the ability of the toxin to interact with at least one class of membrane-bound proteins and perhaps anchor to the membrane at this site prior to opening a K^+ channel.

Development of insect resistance to BT toxin

BT products have been used for over 30 years, and there are only isolated reports of reduced effectiveness of BT insecticides (63,75,78). There is a limited amount of data suggesting that resistance to BT may not develop as readily in nature as it has to most conventional chemical pesticides. In several laboratory studies, it was shown that resistance to BT did not develop to a significant extent for Mediterranean flour moth (115), diamondback moth (21), and Egyptian cotton leafworm (96) after prolonged selection at high or moderately high selective doses of BT insecticide. In contrast, development of resistance to chemical pesticides can usually be readily demonstrated in the laboratory, and loss of effectiveness of many chemical pesticides has occurred in the field (35).

Recent studies have shown that BT-resistant colonies of Indianmeal moth (76,77) and _H. virescens_ (99) can be selected in the laboratory. These examples demonstrate that resistance to BT insecticides can be selected under laboratory conditions. If use of BT products should increase dramatically in the future, it is assumed that this would increase the potential for development of resistance (87). Because of these concerns, it is important to understand how resistance has developed in selected laboratory populations and to determine the genetic potential of different insect species to develop resistance.

Knowledge of insect feeding behavior and an understanding of the molecular basis of the mechanism of action of BT toxins helps to define steps where changes could occur that would decrease toxicity or result in a less susceptible insect. For example, a biochemical or behavioral change that caused TIFI to occur more rapidly after toxin ingestion could prevent a lethal dose from being ingested at an early developmental stage. Likewise, difficulties in solubilizing the toxin in the midgut or alterations in the protease composition of the midgut could alter the stability of the toxin in the midgut and affect toxicity. In addition, changes in the receptors on the brush border membranes and other subtle membrane changes that might affect the interaction of the toxin with its target sites on the membrane could affect toxicity. Changes such as these might have little effect on the overall fitness of the insect population.

Some insects may differ in their susceptibility to BT because they differ in their genetic potential to develop resistance. This is expected because a number of insect species are unaffected by BT, and thus, species with marginal sensitivity presumably exist. It is also expected that resistance could be more readily selected from marginally sensitive species than it could from extremely susceptible insect species.

Another factor that is expected to be related to resistance development is the degree of inbreeding within populations. Within highly inbred populations, a significant subpopulation can carry resistance determinants, particularly if the population has ever been exposed to selection by the pesticide. In populations that are not highly inbred, introduction of susceptible genotypes tends to decrease homozygosis and the frequency of resistance determinants.

Reports on susceptibility of natural populations of Indianmeal moth (*Plodia interpunctella*) to BT insecticides are consistent with its inbred nature and the likelihood that it is marginally sensitive to BT insecticides. Different populations of Indianmeal moth were reported to have substantially different susceptibilities to BT (62), some differing by as much as 42-fold (34). Further, the low slopes of the dose-response curves reported suggest the presence of significant subpopulations with marginal susceptibilities to BT (62). Limited genetic diversity of the selected population and heavy use of BT insecticides has been suggested to be a contributing factor to a reported low efficacy of BT against diamondback moth in the Philippines (63).

Many strains of natural isolates of BT produce more than one toxin protein. Many of the highly active lepidopteran strains contain at least two P1-type genes and one P2-type gene (69). Both P1 and P2 gene products are active on lepidopteran insects and, because they have limited amino acid sequence homology, it is hypothesized that they may have some differences in their mechanisms of action. If so, then strains producing more than one toxin may be less likely to generate resistance in target insect populations than those strains expressing a single toxin. Furthermore, resistance that developed to one toxin might not reflect resistance to the other (cross-resistance). The existence of toxins with different modes of action also supports the hypothesis that resistance to BT may occur less frequently than it has with conventional chemical pesticides because the latter generally disrupt a single very specific biochemical reaction or process.

The data of McGaughey and Johnson (79) may provide an example of lack of cross-resistance among different toxins. They showed that

Indianmeal moth selected for resistance to Dipel® containing HD1, serotype *kurstaki*, had increased resistance to 18 other *kurstaki* and 12 other *thuringiensis* isolates but not to strains from 6 other serotypes. It cannot be determined if there is a lack of cross-resistance in these experiments because the number of toxin proteins present in the strains used and their activities were not reported.

One of the few cases (99) of selection in the laboratory in which a significant, high level of resistance to BT toxin has been obtained involved prolonged selection of *H. virescens* (tobacco budworm) with a single toxin protein. It is possible that resistance was attainable because a single toxin was used in the selection and therefore only a single biochemical change was required, e.g., an alteration in the site of toxin reception on brush border. Data on the susceptibility of this insect to P2 and other P1 toxins would be valuable in assessing the potential for resistance development in this insect.

The rapidity with which field resistance may develop in the future will likely depend on (1) the genetic potential of target insects to develop resistance, e.g., their inherent susceptibility to BT insecticides, and (2) how well we use information, such as the understanding of the mechanism of action of BT toxicity, to guide our future use of BT-based insecticides.

The potential for resistance development to BT insecticides has recently been addressed by a group composed of representatives of companies interested in commercial applications of BT. The objectives of this BT Management Working Group are to encourage and fund academic research aimed at understanding the mode of action of BT and at determining whether resistance to BT developed in the laboratory insect populations is predictive of resistance development in the field. The group plans to raise funds from the participating companies and solicit research proposals for funding. Once the program is established, the primary function of the group will be to serve as a focus for the dissemination of information relating to BT use and management. Researchers funded by the group will be encouraged to publish their results in major international journals, and the group will work toward developing guidelines within the industry for implementing measures to preserve the continued effectiveness of BT (P. Marrone, personal communication, committee chairperson, Monsanto).

Commercialization Strategies for Increasing the Effectiveness of BT Insecticides

Commercialization strategies

There are two primary objectives, of relatively equal importance, for any successful commercialization strategy for BT-based insecticides.

The first is to identify or genetically engineer a toxin or a combination of toxins that are highly active on the target insect(s). (In this review, the phrase "genetically engineer" will refer to activities involving recombinant DNA techniques.) The second objective is to translate this activity to the field using a delivery system that will maximize product stability and efficacy. This delivery system should take advantage of any aspects of either insect behavior or biochemistry to deliver maximal doses of toxin to the target insect as effectively as possible.

Two commercialization strategies are currently being actively pursued for BT-based bioinsecticide products. One strategy is to improve the conventional spray-on formulation of BT crystal protein. This strategy can be pursued by altering the active ingredient (i.e., BT strain or crystal protein) and/or changing aspects of the formulation itself.

The ability to improve BT strains has been greatly aided by our increased knowledge of the molecular genetic basis of BT's insecticidal activity. The diversity of BT-toxin-encoding genes, especially as related to differences in insecticidal activity, coupled with the ability of certain toxin-encoding plasmids to conjugally transfer, provides the basis for the directed development of BT strains through the application of microbial genetics. Novel combinations of δ-endotoxin genes can be derived without the use of recombinant DNA techniques via selective plasmid curing and conjugal transfer. "Selective plasmid curing" refers to the isolation of BT strain derivatives that have lost one or more plasmids carrying toxin genes of low activity. Selective plasmid curing can yield partially cured derivatives (strains harboring a limited number of toxin plasmids) that can subsequently be used as recipients in conjugal transfer. Conjugal transfer of toxin plasmids into partially cured derivatives can lead to BT strains with elevated activities against specific insect pests as well as strains with a broader spectrum of insecticidal activity.

An example of a BT strain with a broader spectrum of activity that was derived via conjugal transfer is a strain developed by Ecogen Inc. for use as a potato insecticide product. In addition to the Colorado potato beetle, an important coleopteran pest of potatoes, there are also lepidopteran pests such as the European corn borer that are significant pests in certain geographical areas. Thus, a BT strain with acceptable activities against both insects was constructed as follows: A BT strain with insecticidal activity against the Colorado potato beetle was isolated, and the coleopteran-active crystal protein was found to be encoded by a transmissible plasmid. This transmissible plasmid was then introduced via conjugation into a cured derivative of another BT strain having potent lepidopteran activity, specifically against the European corn borer. In this manner, a new strain was created with insecticidal activity against both pests. In field tests, this strain has shown performance as a broader-spectrum potato insecticide that is equivalent to chemical pesticide standards.

This conjugal approach to BT strain development is limited, however, to those toxin plasmids that are transmissible. Certain highly active toxin genes are located on plasmids that are not easily transferred, and some toxin-plasmid combinations are unstable due to incompatibility (J. M. Gonzalez, Jr., unpublished). It is also possible that potent insecticidal crystal proteins are poorly produced in native strains due to inefficient promoters and the like. Under these circumstances, the development of genetically manipulated strains of BT having novel activities against specific insects will be greatly facilitated by the application of recombinant DNA technology. Recombinant DNA technology enables the isolation and alteration of specific toxin genes of interest in vitro. These alterations could include changes in the control regions of the toxin genes as well as specific alterations in the coding regions of these genes, using site-specific mutagenesis techniques. Altered genes could be introduced into either BT or related *Bacillus* hosts, such as *Bacillus megaterium*, in order to develop spray-on BT products.

A second current commercialization strategy is to introduce BT toxin genes into the genome of plants or plant-associated microorganisms. When specific promoters are used that allow toxin gene expression in the plant or plant-associated microorganism, insect control can be achieved in laboratory tests. Researchers at Agrigenetics and Plant Genetic Systems have engineered various BT toxin genes into tobacco and developed insect-resistant plants (1,102). Likewise, Monsanto researchers have introduced BT toxin genes into tomato plants and into a root-colonizing pseudomonad (32,80). Recently, Monsanto has announced the successful introduction of BT toxin genes into cotton plants (33). Crop Genetics International has engineered the plant endophyte *Clavibacter xyli* subsp. *cynodontis* to express a BT var. *kurstaki* crystal protein, and introduced the plant endophyte into corn (66). Thus, recombinant DNA technology enables the use of BT toxin genes in a variety of host-vector systems and for various applications.

Both commercialization strategies for development of improved BT-based products share three general steps:

1. Selection of the active ingredient and maximization of its potency
2. Production of the insecticide and its delivery to the field in an effective form
3. Registration of the product

In addition, the potential for resistance development must be considered for each delivery strategy. Although the two strategies have certain technical aspects in common, they differ in other important aspects. In the following discussion, each commercialization strategy will be discussed in relation to the three steps presented above.

Development of a conventional BT product for foliar application

Selection of the active ingredient.   The first step in a commercialization strategy for strain construction of a conventional BT spray-on product is to identify the toxin or combination of toxins that have maximal activity on the target insect(s). Selection of the highly active toxins is performed in laboratory bioassays by placing larvae of the target insect on artificial diet or plant material to which known amounts of crystalline toxin have been added by either surface contamination or diet incorporation. Although this type of laboratory bioassay measures inherent toxicity of a particular δ-endotoxin, it does not necessarily correlate with field performance because the assay does not take into account either effects of insect behavior on toxicity or environmental effects on residual activity.

Many highly active BT strains that are active against lepidoptera have been isolated from nature and contain multiple toxin genes (69). In order to obtain strains that are maximally active on a single target insect, it is necessary to determine the activities of the individual toxins and then combine the most active toxins in the final strain. The combinations of toxins chosen should take advantage of synergistic interactions where possible. Testing the individual toxins for activity is usually achieved either by curing strains of toxin plasmids or by transferring toxin plasmids to recipient strains that do not produce toxin (10). In some cases, it may be necessary to clone and characterize individual toxin genes from a complex background in order to determine the most potent δ-endotoxin against a specific target insect.

It is often desirable to control more than one insect species on a particular crop. In many cases, a single toxin will not be highly active on all target insects—e.g., tobacco budworm and bollworm on cotton, and Colorado potato beetle and European corn borer on potatoes. Even though many strains isolated from nature contain multiple toxin genes, they may not necessarily contain the optimum combination of toxin genes to effectively control a specific insect complex. In addition to producing novel combinations of toxin genes via conjugation as previously described, it may be possible in the future, as more information is obtained on the mechanism of action of endotoxin, to genetically engineer toxins to increase their effectiveness on desired insect targets. If this goal is achieved, production of a single toxin by a strain could be sufficient to achieve control of more than one target insect.

Since spores can influence the toxicity of BT against some insects (44), it must be determined whether their presence or absence is desirable in the final product. If spores contribute significantly to the activity of the insecticide, then it is likely that they would be included. The choice to include or eliminate spores in the insecticide will depend on the situation.

Production and field delivery.   BT insecticides are typically produced by large-scale liquid fermentation of the selected strain. The fermentation broth is concentrated and then either formulated into an aqueous flowable product or dried and formulated into a wettable powder or an oil-based flowable formulation. Even though 20 to 30% of the cellular dry weight accumulates as protoxin protein (73), improvements are still desirable. This is particularly true if more than one protoxin is being produced to control multiple insects because the activity of each toxin on its specific target is diluted by the presence of the second toxin directed toward the other target insect. Thus, the genetic construction of a single gene for a protoxin directed against multiple insects could benefit the economics of production if the strain produced as much protoxin on a per cell basis as a strain expressing two protoxin genes.

In the short term, improvements in the performance and economics of production for spray-on BT products will be achieved by identifying the most active naturally occurring toxin(s) for the target insect, combining these toxins in a single strain, and producing stable formulations that can be efficiently applied. Although some improvement can clearly be made by choosing the most potent toxin for the target insect, significant increases in efficacy can also be made through formulation improvements.

An ideal formulation of a BT insecticide should have the following properties: (1) It should prevent inactivation of the toxin under diverse environmental conditions; (2) it should be able to be easily and efficiently delivered to the plant surfaces; and (3) its application should either be directed to the plant part(s) preferred by the target insect or it should alter or somehow exploit the normal feeding preference of the insect to promote toxin ingestion.

The first two factors (environmental stability and efficient delivery) involve physicochemical properties of the formulation. These problems must be overcome for all insecticides, and have typically been addressed by including additives in the formulation that prevent, for example, inactivation at pH extremes or by UV light, and by the inclusion of additives that promote adherence of the formulation to the plant (spreaders, stickers, emulsifiers, etc.). Recently a BT gene has been expressed in a strain of *Pseudomonas fluorescens*, which is subsequently killed before delivery to the field. This product (MCAP®, Mycogen Inc.), which is currently under development, also appears to be effective in reducing environmental inactivation (31). Other chemical encapsulation methodologies—e.g., starch encapsulation (27)—are being developed for delivery of BT insecticides.

Feeding preference is important for BT insecticides: since the insecticide must be ingested, feeding preference can contribute to or detract from product efficacy. In almost all cases, traditional spray-on formulations strive to achieve uniform coverage of the plant. This may result in poor performance because much of the treated surface of the plant may be ignored by insects that have a feeding preference—i.e., the preferred feeding site may be small relative to the treated area. The problem is exacerbated if the target area of the plant is fast-growing and continually producing new, untreated tissue. In addition, if the insect does not ingest a lethal dose shortly after spraying, it is less likely to do so at a later time because the active ingredients of the insecticide will degrade principally because of photo-inactivation of the toxin (81) and the spores (68). Thus, repeated applications of the formulated product may be required to achieve control.

Some work has been done on the effects of feeding stimulants (mixtures of sugars, vegetable oils, and flours) on the efficacy of currently available BT products. Although benefits can be observed in laboratory studies (86), improvement in product performance under field conditions is variable (5,61,86,97). In one case, crop yield was not enhanced by application of BT insecticide with a feeding stimulant, although the number of larvae was significantly reduced (5). In another study (61), significant decreases in the number of larvae and significant yield increases were observed in some treatments in which feeding stimulants were included. Even though chemical insecticides were superior to BT products in these studies, the results indicate that feeding stimulants can improve field performance of BT insecticides. More recently, it has been recognized that droplet size and particle distribution of the insecticide on the plant surface can affect performance of the pesticide for certain insects (8,30). Clearly, further work will be required to develop more potent feeding stimulants. Novel methods of formulating BT insecticides with feeding stimulants could increase the effectiveness of spray-on BT insecticides.

Registration.   A concern relative to the commercialization of BT products in the United States is the ability to receive approval for field testing. Regulation of spray-on products is administered by the EPA. Currently, the ability to field-test and register a new spray-on type BT product that has not been genetically engineered has been demonstrated (Table 1). Undoubtedly, the relatively short time required to register these spray-on products was due, at least in part, to the fact that there were several BT products already on the market and the active ingredients consist of naturally occurring strains. After a natural isolate of BT has been proven by extensive field testing to be effective in controlling the target insect, the time required for registration depends primarily on the time required to perform and pass Tier I toxicology testing. If the strain, the technical-grade active ingredient, and the formulated end product pass the various required toxicological tests, the insecticide can be registered by the U.S. Environmental Protection Agency (EPA) without further testing.

Approvals for field testing genetically altered strains (transconjugants or plasmid-cured derivatives) and five Environmental Use Permits (EUPs) have been obtained by Ecogen Inc. for cotton, vegetable crops, forest trees, and potatoes. Because BT products containing natural isolates are already registered, it is anticipated that registration of products containing genetically altered BT strains that are not derived via recombinant DNA techniques will be possible by following procedures similar to those used for the registration of natural isolates.

Development of BT-based products with novel delivery systems

The approaches that utilize either a live, genetically engineered, plant-associated bacterium or a genetically engineered plant as a carrier of the BT toxin have essentially the same end result: i.e., the toxin is expressed in the plant itself and probably in all plant tissues. Certain aspects concerned with selection of the active ingredient, the production and delivery of the toxin to the field, and approval to commercialize the product are different for this strategy than in the strategies for development of a conventional spray-on product.

Selection of the active ingredient.   It is desirable to select the toxin gene that codes for a toxin with maximal activity on the target insect regardless of whether the toxin is expressed within the plant itself or applied as a spray-on product. The toxicity of a plant that produces BT toxin would be solely due to the activity of the toxin itself and could not be potentiated by the presence of spores. In addition, it may be necessary in certain cases to introduce multiple toxin genes into the recombinant organism. For example, if either (1) synergy between two different toxins was important for activity or (2) it was desirable to control more than one insect and a single toxin could not be identified that had good activity on both, then more than one toxin gene would have to be introduced.

Production and field delivery.   The most obvious difference between the two commercialization strategies lies in the production and delivery of the toxin to the field. When a genetically engineered plant or plant-associated bacterium is used, toxin is produced in the plant tissue. Thus, there is no need to formulate the toxin, as there is in conventional spray-on BT products, or to continually apply it throughout the growing season.

The regulation of toxin synthesis in a genetically engineered plant is not well understood. Ideally, the toxin gene could be fused to specific promoters that would either limit expression to specific target organs of the plant or would be expressed only during insect feeding, e.g., by wound response. However, controlled expression of this nature has not been reported. Toxin gene expression in a plant-associated endophyte would presumably be constitutive, and toxin would be produced in all parts of the plant. However, for *C. xyli*, expression of toxin is lowest in rapidly growing parts of the plant (84). It has not been reported whether this level of expression is sufficient to achieve effective control.

Registration. Regulation of genetically engineered plants producing BT toxin is administered by the U.S. Department of Agriculture (USDA), and regulation of plant-associated bacteria producing BT toxin may be administered by both the USDA and the EPA. (The USDA is involved only in those cases where the recombinant organism is a plant pathogen.) There are currently no genetically engineered organisms for production of BT toxin that have been commercialized. Only recently have permits been issued for testing genetically engineered plants, and permission for small-scale testing of the xylem-inhabiting bacterium *C. xyli* was recently obtained. In the latter trials, elaborate precautions were taken to prevent environmental spread of the organism (66). Because permits on genetically engineered organisms expressing BT toxin have been obtained, there is optimism that such similarly derived products may eventually be commercialized.

Comparison of development strategies

The major differences between the two commercialization strategies, a conventional spray-on BT product and a genetically engineered plant as an example of a novel host-vector system, are highlighted in Table 3. These differences are discussed in more detail below.

Selection of the active ingredient. Selection of the most active toxin for both commercialization strategies is performed in laboratory insect bioassays as previously described, but the molecular form of the toxin used in the bioassays may be different. For conventional spray-on BT

TABLE 5 Differences between Strategies for Development of BT Insecticides for Delivery as a Conventional Spray-On Product or Delivery in a Genetically Engineered Plant

Parameter compared	Comments	
	Spray-on	Plant genetic engineering
Toxin selection:		
Selection of most active toxin(s) against target insect	Standard bioassay with crystalline protoxin(s).	Truncated toxin introduced into plant via recombinant DNA technology. Insect bioassays may not accurately measure toxicity of expressed toxin.
Ease of combining desired toxin genes	Novel combinations of toxin genes may be obtained via conventional microbial genetic approaches.	Novel combinations of toxin genes technically more difficult to obtain.
Effect of spores on activity	Easily incorporated.	Absence may greatly reduce toxicity of final product against certain insects.
Production, formulation, and delivery:		
Delivery of multiple toxins	Construct bacterial strains by conjugation and plasmid curing.	Introduction of multiple toxin genes using recombinant DNA techniques has not been demonstrated but should be possible.
Product application	Production and application costs are comparable to conventional pesticides.	Reduced application costs because toxin is continually synthesized.
Effects of feeding behavior on product effectiveness	Feeding behavior of target insect is critical for good performance.	Feeding behavior of target insect may have less effect on product efficacy.
Environmental stability	Dependent on the formulation and possibly the strain.	Stability may be of little concern because toxin is continually synthesized.
Resistance development:		
Potential for resistance development	Resistance development less likely to occur under normal conditions.	May promote more rapid resistance development unless special precautions are taken.
Other considerations:		
Registration with the EPA	May be possible within 1 year of the registration application date.	Uncertain status but possibly longer than for a spray-on product.
Time required for product development	Should be possible in 2–4 years after desirable toxins are identified.	Uncertain status but probably longer than for a spray-on product.

133

products, the toxin is ingested by the target insect in a crystalline form both in laboratory bioassays and in the field. In this delivery strategy, spores are generally present, unless they have been specifically omitted. Thus, toxicity measurements made on individual toxins, mixtures of toxins, and mixtures of spores and toxins are likely to be an indicator of the potential toxicity (which must be achieved by proper formulation) of a spray-on product applied in the field.

In contrast, the toxicity of a genetically engineered plant producing one or more toxins may be less accurately reflected by the toxicity of the respective toxins in laboratory bioassays. Truncated toxin genes have been introduced into plants, and it is difficult or impossible to use exactly the same truncated form of the toxin protein(s) in laboratory bioassay. In the worst case, a gene producing a highly active toxin in laboratory bioassay might code for a relatively nontoxic protein in the plant. Although genetically engineered plants are qualitatively toxic to insects, the quantitative activity of the toxin (i.e., LC_{50}) has not been reported.

The two commercialization strategies clearly differ in their abilities to deliver spores, which could increase toxicity of the product against certain insects. The inclusion of spores in the product can be easily accomplished for a spray-on product but is not possible when the toxin is expressed in the plant tissue itself. In contrast, if it is desirable to produce a spore-free product, the goal is easily accomplished if the toxin is produced in the plant. A spray-on product with reduced or no live spores could also be achieved by several different approaches, e.g., production in non-spore-forming strains other than BT (31), isolation of a sporulation mutant of BT (105,106), or treatment to inactivate spores (Toaro CT®, Toa Gosei, Japan), etc.

Production and field delivery.   The active ingredient for use in a spray-on BT insecticide is produced by large-scale fermentation of the selected BT strain, whereas BT toxin delivered to the field by a genetically engineered plant is produced in the plant itself. Because many BT toxins can be introduced into or removed from the BT strain of interest, it is usually possible to genetically construct a bacterial strain that will express multiple toxin genes with the desired specificities. The introduction and expression of multiple genes is potentially much more difficult for genetically engineered plants. This is in part due to the complexities of plant transformation, as compared to standard bacterial genetic manipulation techniques. Furthermore, introduction of multiple toxin genes in the same plant has not yet been reported, and to do so and maintain stability may at present be technically more difficult than introducing a single gene. Plant genetic engineering is further complicated by the need to regenerate and commercially propagate the engineered plant. To propagate large numbers of genetically engineered plants for delivery to the field, it would be desirable if the BT gene(s) were stably maintained during meiosis. Currently, transmission to the F1 progeny has been achieved (102). Information on the stability of the toxin gene(s) beyond this stage has not been published.

The introduction of BT toxin genes into a variety of plant species is clearly limited by the availability of efficient plant tissue culture techniques for the target plant. Although many species of plants can be grown in tissue culture, few species can be regenerated into plants from callus tissue and still fewer can be regenerated from either callus tissue or single cells.

The delivery of BT toxin by a genetically engineered plant may have some advantages over conventional spray-on approaches because the toxin is produced continuously by the plant cells. Obvious advantages include the elimination of application and formulation costs. There would be less concern about the environmental stability of the toxin because the toxin would either be continually synthesized during all stages of vegetative growth or synthesized in response to wounding. Furthermore, it is thought that expression of the toxin within the plant would be advantageous for control of those insects that prefer a very specific part of the plant, particularly those that burrow into the plant and are thus very difficult to control with spray-on BT products.

A requirement of any BT product is to deliver enough toxin to the target insect to achieve good control. In spray-on products, this is accomplished and can be improved by developing more stable formulations and by using multiple applications. Products involving genetically engineered plants achieve this goal as a result of continual production of the toxin within the plant.

Effective control of certain target insects has clearly been achieved using spray-on BT products, and improvements in production and formulation technologies should make these products even more versatile. Currently, it appears that delivery of BT toxin in a genetically engineered organism may increase efficacy and have some other advantages over a spray-on product for commercialization of BT-based insecticides. However, there is little information on the ability of genetically engineered plants to deliver enough toxin to the entire plant or to specific target organs of the plant (particularly to rapidly growing areas) during the entire period of time when the plants are likely to be attacked by insects. Also, it is unknown whether environmental conditions adversely affect toxin production in the plant. If gene expression within the plant is difficult to control, problems may arise that could severely limit the effectiveness of larval control.

Resistance development.   There is little information available on the potential for resistance development in nature when insect populations are exposed to lethal doses of spray-on BT insecticides over long periods of time. There is even less information available on the potential for resistance development when the toxin is expressed in the target plant itself. However, our understanding of how resistance develops to chemical pesticides suggests that resistance development could occur more rapidly when BT toxin is expressed in a genetically engineered plant, particularly if the plant produces a single toxin. Additional work will be required to assess the potential for resistance development when BT is delivered by these two strategies.

One potential problem with expression of the toxin in the *plant* is that delivery may be too efficient and provide too strong a selection for resistance development. If all crop plants grown over large acreages were producing BT toxin, then the probability that the surviving insects would carry some level of resistance would be increased, particularly if there were no alternative hosts. Mating of this selected population would tend to rapidly increase the resistance rate. This problem might be overcome by growing some proportion of plants that do not produce BT toxin. This would provide a reservoir on which susceptible individuals could survive. Mating of susceptible and resistant individuals would tend to delay the development of widespread resistance.

Use of spray-on BT products may not cause a comparable problem because target plant coverage is never complete. Thus, there are always untreated areas to support susceptible individuals. In addition, the toxin is continually inactivated on the surface of the plant and can fall to sublethal amounts if not applied repeatedly. This provides an additional opportunity for the survival of susceptible individuals.

Summary

Bacillus thuringiensis, or BT as it is commonly known, is an important biological control agent used since the late 1960s for the control of lepidopteran target pests. Recent novel BT isolates have extended its utility to include the control of certain dipteran and coleopteran pest species. Significant research in the area of the molecular genetic basis of insecticidal activity has shown that the genes encoding BT insecticidal crystal proteins are most often carried by plasmids, certain of which have the ability to transfer via conjugation. Numerous toxin genes have been isolated and characterized with respect to their DNA sequence. These analyses have yielded important information regarding the diversity of BT toxin genes and set the stage for future gene structure and function studies.

Recent research has begun to focus on the mechanism of action of BT toxin. It is clear that the action of BT toxin entails a complex process that must begin with toxin ingestion by the target insect and proteolytic processing within the insect gut. Alteration of certain, not yet well understood, biochemical processes leads to a disruption of osmoregulation in the insect midgut and death of the insect.

Two commercialization strategies for BT-based insecticide products are currently being pursued. One strategy is to improve the conventional spray-on formulation of BT crystal protein either by altering the active ingredient (e.g., novel BT strain derived via conjugation) or by improving aspects of the formulation. A second commercialization strategy involves the incorporation of BT toxin genes into novel delivery systems, such as a crop plant, via plant molecular biology technology. Both of these strategies have unique considerations with respect to product development, field delivery, and registration as well as the potential for the development of insect resistance to BT.

Acknowledgments

We thank our colleagues at Ecogen Inc., especially B. C. Carlton, W. P. Donovan, L. English, D. Avé, and J. L. McIntyre, for helpful discussion and critical reading of the manuscript. A special thanks to G. Gheorghiou for unpublished material.*

References

1. Adang, M. J., Firoozabady, E., Klein, J., DeBoer, D., Sekar, V., Kemp, J. D., Murray, E., Rocheleau, T. A., Rashka, K., Staffeld, G., Stock, C., Sutton, D., and Merlo, D. J. 1987. Expression of a *Bacillus thuringiensis* insecticidal crystal protein gene in tobacco plants. In *Molecular Strategies for Crop Protection*, C. J. Arntzen and C. Ryan (eds.). Alan R. Liss, New York, pp. 345–353.
2. Adang, M. J., Staver, M. J., Rocheleau, T. A., Leighton, J., Barker, R. F., and Thompson, R. V. 1985. Characterized full-length and truncated plasmid clones of the crystal protein of *Bacillus thuringiensis* subsp. *kurstaki* HD-73 and their toxicity to *Manduca sexta*. *Gene* 36:289–300.
3. Angsuthanasombat, C., Chungjatupornchai, W., Kertbundit, S., Luxananil, P., Settasatian, C. Wilairat, P. and Panyim, S. 1987. Cloning and expression of the 130-kd mosquito-larvicidal delta-endotoxin gene of *Bacillus thuringiensis* var. *israelensis* in *Escherichia coli*. *Mol. Gen. Genet.* 208:384–389.
4. Aronson, A. I., Beckman, W., and Dunn, P. 1986. *Bacillus thuringiensis* and related insect pathogens. *Microbiol. Rev.* 50:1–24.
5. Bell, M. R., and Romine, C. L. 1980. Tobacco budworm field evaluation of microbial control in cotton using *Bacillus thuringiensis* and a nuclear polyhedrosis virus with a feeding adjuvant. *J. Econ. Entomol.* 73:427–430.
6. Bourgouin, C., Klier, A., and Rapoport, G. 1986. Characterization of the genes encoding the haemolytic toxin and the mosquitocidal delta-endotoxin of *Bacillus thuringiensis israelensis*. *Mol. Gen. Genet.* 205:390–397.
7. Brizzard, B. L., and Whiteley, H. R. 1988. Nucleotide sequence of an additional crystal protein gene cloned from *Bacillus thuringiensis* subsp. *thuringiensis*. *Nucleic Acids Res.* 16:2723–2734.

---

* While this chapter was in press, a new nomenclature and classification scheme for BT crystal proteins were proposed, as discussed in Ref. 53a.

8. Bryant, J. E., and Yendol, W. G. 1988. Evaluation of the influence of droplet size and density of *Bacillus thuringiensis* against gypsy moth larvae (Lepidoptera: *Lymantriidae*). *J. Econ. Entomol.* 81:130–134.

9. Burkett, G. R., Schneider, J. C., and Davis, F. M. 1983. Behavior of the tomato fruitworm, *Heliothis zea* (Boddie) (Lepidoptera: Noctuidae), on tomato. *Environ. Entomol.* 12:905–910.

10. Carlton, B. C. 1988. Development of genetically improved strains of *Bacillus thuringiensis*. In *Biotechnology of Crop Protection*, P. A. Hedrin, J. J. Menn, and R. M. Hollingworth (eds.). American Chemical Society, Washington, D.C., pp. 261–279.

11. Carlton, B. C., and Gonzalez, J. M., Jr. 1985. The genetics and molecular biology of *Bacillus thuringiensis*. In *The Molecular Biology of the Bacilli*, D. A. Dubnau (ed.). Academic, New York, pp. 211–249.

12. Chapman, J. S., and Carlton, B. S. 1985. Conjugal plasmid transfer in *Bacillus thuringiensis*. In *Plasmids in Bacteria*, D. R. Helenski, S. N. Cohen, D. B. Clewell, D. A. Jackson, and A. Hollaender (eds.). Plenum, New York, pp. 453–467.

13. Cheung, P. Y. K., Buster, D., and Hammock, B. D. 1987. Lack of mosquitocidal activity by the cytolytic protein of the *Bacillus thuringiensis* subsp. *israelensis* parasporal crystal. *Curr. Microbiol.* 15:21–23.

14. Cheung, P. Y. K., and Hammock, B. D. 1985. Separation of three biologically distinct activities from the parasporal crystal of *Bacillus thuringiensis* var. *israelensis*. *Curr. Microbiol.* 12:121–126.

15. Chilcott, C. N., and Ellar, D. J. 1988. Comparative toxicity of *Bacillus thuringiensis* var. *israelensis* crystal proteins *in vivo* and *in vitro*. *J. Gen. Microbiol.* 134:2551–2558.

16. Chua, T. H., and Lim, B. H. 1979. Distribution pattern of diamondback moth, *Plutella xylostella* (L.) (*Lepidoptera: Plutellidae*) on choy-sum plants. *A. Ang. Entomol.* 88:170–175.

17. Cohen, R. W., Waldbauer, G. P., Friedman, S., and Schiff, N. M. 1987. Nutrient self-selection by *Heliothis zea* larvae: a time-lapse film study. *Entomol Exp. Appl.* 44:65–73.

18. Crawford, D. N., and Harvey, W. R. 1988. Barium and calcium block *Bacillus thuringiensis* subspecies *kurstaki* delta-endotoxin inhibition of potassium current across isolated midgut of larval *Manduca sexta*. *J. Exp. Biol.* 137:277–286.

19. Dean, D. A., Sabourin, J. R. and McLinden, J. H. 1986. Comparative DNA sequencing and restriction site analysis of *Bacillus thuringiensis* insecticidal crystal protein genes In *Fundamental and Applied Aspects of Invertebrate Pathology*, R. A. Samson, J. M. Vlak, and D. Peters (eds.). Foundation of the Fourth International Colloquium of Invertebrate Pathology, Wageningen, The Netherlands, p. 394.

20. de Barjac, H. 1981. Identification of H-serotypes of *Bacillus thuringiensis*. In *Microbial Control of Pests and Plant Diseases, 1970–1980*, H. D. Burgess (ed.). Academic, New York, pp. 35–43.

21. Devriendt, M., and Martouret, D. 1967. Absence de resistance a *Bacillus thuringiensis* chez la teigne des cruciferes, *Plutella maculipennis* (Lep.: Hyponomeutidae). *Entomophaga.* 21:189–199.

22. Donovan, W. P., Dankocsik, C., and Gilbert, M. P. 1988. Molecular characterization of a gene encoding a 72-kilodalton mosquito-toxic crystal protein from *Bacillus thuringiensis* subsp. *israelensis*. *J. Bacteriol.* 170:4732–4738.

23. Donovan, W. P., Dankocsik, C. C., Gilbert, M. P., Gawron-Burke, M. C., Groat, R. G., and Carlton, B. C. 1988. Amino acid sequence and entomocidal activity of the P2 crystal protein, an insect toxin from *Bacillus thuringiensis* var. *kurstaki*. *J. Biol. Chem.* 263:561–567.

24. Donovan, W. P., Gonzales, J. M., Jr., Gilbert, M. P., and Dankocsik, C. 1988. Isolation and characterization of EG2158, a new *Bacillus thuringiensis* toxic to coleopteran larvae, and nucleotide sequence of the toxin gene. *Mol. Gen. Genet.* 214:365–372.

25. Dulmage, H. T. 1973. Assay and standardization of microbial insecticides. *Ann. N.Y. Acad. Sci.* 217:187–199.

26. Dulmage, H. T., and cooperators. 1981. Insecticidal activity of isolates of *Bacillus thuringiensis* and their potential for pest control. In *Microbial Control of Pests and Plant Diseases, 1970–1980*, H. D. Burgess (ed.). Academic, New York, pp. 193–222.

27. Dunkle, R. L., and Shasha, B. S. 1988. Starch-encapsulated *Bacillus thuringiensis*: a potential new method for increasing environmental stability of entomopathogens. *Environ. Entomol.* 17:120–126.

28. English, L. H., and Readdy, T. L. 1989. Delta endotoxin inhibits a phosphatase in midgut epithelial membranes of *Heliothis virescens*. *Insect Biochem.* (in press).

29. Fast, P. G. 1981. The crystal toxin of *Bacillus thuringiensis*. In *Microbial Control of Pests and Plant Diseases, 1970–1980*, H. D. Burgess (ed.). Academic, London, pp. 223–248.

30. Fast, P. G., and Sundaram, A. 1988. The relationship of foliar deposit of *Bacillus thuringiensis* to the mortality of Eastern spruce budworm, *Choristoneura fumiferana* Clem. *Proceedings XVIIIth International Congress of Entomology, Vancouver, British Columbia, Canada*, (abstr.), p. 428.

31. Finlayson, M., and Gaertner, F. 1989. Alternative hosts for *Bacillus thuringiensis* delta-endotoxin. *J. Cell. Biochem.* 13A(suppl.):152 (abstr.).

32. Fischoff, D. A., Bowdish, K. S., Perlak, F. J., Marrone, P. G., McCormick, S. M., Niedermeyer, J. G., Dean, D. A., Kusano-Kretzmer, K., Mayer, E. J., Rochester, D. E., Rogers, S. G., and Fraley, R. T. 1987. Insect tolerant tomato plants. *Bio/Technology* 5:815–917.

33. Fishbein, G. W. 1989. *Gen. Eng. Lett.* 9:4.

34. Georghiou, G. P. 1988. Implications of potential resistance to biopesticides. In *Biotechnology, Biological Pesticides, and Novel Plant-Pest Resistance for Insect Pest Management*, Insect Pathology Resource Center (ed.). Boyce Thompson Institute for Plant Research at Cornell University, Ithaca, New York (in press).

35. Georghiou, G. P., and Saito, T. (eds.). 1983. *Pest Resistance to Pesticides*. Plenum, New York.

36. Goldberg, L. J., and Margalit, J. 1977. A bacterial spore demonstrating rapid larvicidal activity against *Anopheles sergentii, Uranotaenia unguiculata, Culex univitattus, Aedes aegypti* and *Culex pipiens. Mosquito News* 37:355–358.

37. Gonzalez, J. M., Jr., Brown, B. J., and Carlton, B. C. 1982. Transfer of *Bacillus thuringiensis* plasmids coding for delta-endotoxin among strains of *B. thuringiensis* and *B. cereus. Proc. Natl. Acad. Sci. USA* 79:6951–6955.

38. Gonzalez, J. M., Jr., and Carlton, B. C. 1984. A large transmissible plasmid is required for crystal toxin production in *Bacillus thuringiensis* var. *israelensis. Plasmid* 11:28–38.

39. Gonzalez, J. M., Jr., Dulmage, H. T., and Carlton, B. C. 1981. Correlation between specific plasmids and delta-endotoxin production in *Bacillus thuringiensis. Plasmid* 5:351–365.

40. Gupta, B. L., Dow, J. A. T., Hall, T. A., and Harvey, W. R. 1985. Electron probe X-ray microanalysis of the effects of *Bacillus thuringiensis* var *kurstaki* crystal protein insecticide on ions in an electrogenic K^+-transporting epithelium of the larval midgut in the lepidopteran, *Manduca sexta, in vitro. J. Cell Sci.* 74:137–152.

41. Haider, M. Z., and Ellar, D. J. 1988. Nucleotide sequence of a *Bacillus thuringiensis aizawai* ICI entomocidal crystal protein gene. *Nucleic Acids Research* 16:10927.

42. Haider, M. Z., Knowles, B. H., and Ellar, D. J. 1986. Specificity of *Bacillus thuringiensis* var *colmeri* insecticidal delta-endotoxin is determined by differential proteolytic processing of the protoxin by larval gut proteases. *Eur. J. Biochem.* 156:531–540.

43. Harvey, W. R., and Wolfersberger, M. G. 1979. Mechanisms of inhibition of active potassium transport in isolated midgut of *Manduca sexta* by *Bacillus thuringiensis* endotoxin. *J. Exp. Biol.* 83:293–304.

44. Heimpel, A. M., and Angus, T. A. 1959. The site of action of crystalliferous bacteria in lepidoptera larvae. *J. Insect Pathol.* 1:152–170.

45. Heinrich, B. 1979. Foraging strategies of caterpillars, leaf damage and possible predator avoidance strategies. *Oecologia* 42:325–337.

46. Held, G. A., Bulla, L. A., Jr., Ferrari, E., Hoch, J. A., Aronson, A. I., and Minnich, S. A. 1982. Cloning and localization of the lepidopteran protoxin gene of *Bacillus thuringiensis* subsp. *kurstaki*. *Proc. Natl. Acad. Sci. USA* 79:6065–6069.

47. Held, G. A., Huang, A. Y., and Kawanishi, C. Y. 1986. Effect of removal of the cytolytic factor of *Bacillus thuringiensis* subsp. *israelensis* on mosquito toxicity. *Biochem. Biophys. Res. Commun.* 141:937–941.

48. Herrnstadt, C., Gilroy, T. E., Sobieski, D. S., Bennett, B. D., and Gaertner, F. H. 1987. Nucleotide sequence and deduced amino acid sequence of a coleopteran-active delta-endotoxin gene from *Bacillus thuringiensis* subsp. *san diego*. *Gene* 57:37–46.

49. Herrnstadt, C., Soares, G. G., Wilcox, E. R., and Edwards, D. L. 1986. A new strain of *Bacillus thuringiensis* with activity against coleopteran insects. *Bio/Technology* 4:305–308.

50. Hofmann, C., Luthy, P., Hutter, R., and Pliska, V. 1988. Binding of the delta endotoxin from *Bacillus thuringiensis* to brush-border membrane vesicles of the cabbage butterfly (*Pieris brassicae*). *Eur. J. Biochem.* 173:85–91.

51. Hofmann, C., Vanderbruggen, H., Hofte, H., Van Rie, J., Jansens, S., and Van Mellaert, H. 1988. Specificity of *Bacillus thuringiensis* delta-endotoxin is correlated with the presence of high-affinity binding sites in the brush border membrane of target insect midguts. *Proc. Natl. Acad. Sci. USA* 85:844–848.

52. Hofte, H., de Greve, H., Seurinck, J., Jansens, S., Mahillon, J., Ampe, C., Vandekerckhove, J., Vanderbruggen, H., van Montagu, M., Zabeau, M., and Vaeck, M. 1986. Structural and functional analysis of a cloned delta endotoxin of *Bacillus thuringiensis berliner* 1715. *Eur. J. Biochem.* 161:273–280.

53. Hofte, H., Van Rie, J., Jansens, S., Van Houtven, A., Vanderbruggen, H., and Vaeck, M. 1988. Monoclonal antibody analysis and insecticidal spectrum of three types of lepidopteran-specific insecticidal crystal proteins of *Bacillus thuringiensis*. *Appl. Environ. Microbiol.* 54:2010–2017.

53a. Hofte, H. and Whiteley, H. 1989. Insecticidal crystal proteins of *Bacillus thuringiensis*. *Microbiol. Rev.* 53:242–255.

54. House, H. L. 1965. Digestion. In *The Physiology of Insects*, M. Rockstein (ed.). Academic, New York, pp. 815–858.

55. Huber, E., and Luthy, P. 1981. *Bacillus thuringiensis* delta endotoxin: composition and activation. In *Pathogenesis of Invertebrate Microbial Diseases*, E. W. Davidson (ed.). Allenheld, Osmun, Totowa, New Jersey, pp. 209–234.

56. Hurley, J. M., Bulla, L. A., Jr., and Andrews, R. E., Jr. 1987. Purification of the mosquitocidal and cytolytic proteins of *Bacillus thuringiensis* subsp. *israelensis*. *Appl. Environ. Microbiol.* 53:1316–1321.

57. Hurley, J. M., Lee, S. G., Andrews, R. E., Jr., Kloweden, M. J., and Bulla, L. A., Jr. 1985. Separation of the cytolytic and mosquitocidal proteins of *Bacillus thuringiensis* subsp. *israelensis*. *Biochem. Biophys. Res. Commun.* 126:961–965.

58. Ibara, J. E., and Federici, B. 1986. Isolation of a relatively nontoxic 65 kDa protein inclusion from the parasporal body of *Bacillus thuringiensis* subsp. *israelensis*. *J. Bacteriol.* 165:527–533.

59. Iizuka, T., and Yamamoto, T. 1983. Possible location of mosquitocidal protein in the crystal preparation of *Bacillus thuringiensis* subsp. *kurstaki*. *FEMS Microbiol. Lett.* 19:187–192.

60. Jackson, D. M. 1982. Searching behavior and survival of 1st instar codling moths. *Ann. Entomol Soc. Am.* 75:284–289.

61. Johnson, D. R. 1982. Suppression of *Heliothis* spp. on cotton by using *Bacillus thuringiensis, Baculovirus heliothis*, and two feeding adjuvants. *J. Econ. Entomol.* 75:207–210.

62. Kinsinger, R. A., and McGaughey, W. H. 1979. Susceptibility of populations of Indianmeal moth and almond moth to *Bacillus thuringiensis*. *J. Econ. Entomol.* 72:346–349.

63. Kirsch, K., and Scmutterer, H. 1988. Low efficacy of a *Bacillus thuringiensis* (Berl.) formulation in controlling the diamondback moth *Plutella xylostella* (L.), in the Philippines. *J. Appl. Entomol.* 105:249–255.

64. Klier, A., Fargette, F., Ribier, J., and Rapaport, G. 1982. Cloning and expression of the crystal protein genes from *Bacillus thuringiensis* strain *berliner* 1715. *EMBO J.* 1:791–799.
65. Klier, A., Lereclus, D., Ribier, J., Bourgouin, C., Menou, G., Lecadet, M., and Rapoport, G. 1985. Cloning and expression in *Escherichia coli* of the crystal protein gene from *Bacillus thuringiensis* strain *aizawai* 7-29 and comparison of the structural organization of genes from different serotypes. In *Molecular Biology of Microbial Differentiation*, J. A. Hoch and P. Setlow (eds.). American Society for Microbiology, Washington, D.C., pp. 217–224.
66. Kostka, S. J., Tomasino, S. F., Turner, J. T., and Reeser, P. W. 1988. Field release of a transformed strain of *Clavibacter xyli* subsp. *cynodontis* (cxc) containing a delta-endotoxin gene *Bacillus thuringiensis* subsp. *kurstaki* (BT). Abstracts of 1988 American Phytopathological Society Meeting. p. 221. The American Phytopathological Society, St. Paul, MN; Nov. 13–17, 1988
67. Krieg, A., Huger, A. M., Langenbruch, G. A., and Schnetter, W. 1983. *Bacillus thuringiensis* var. *tenebrionis*, a new pathotype effective against larvae of Coleoptera. *Z. Angew Entomol.* 96:500–508.
68. Krieg, V. A., Groner, A., and Matter, M. 1980. Uber die Wirkung von mittel- und langwelligen ultravioletten Strahlen (UV-B und UV-A) auf insektenpathogene Bakterien und Viren und deren Beeinfussung durch UV-Schutzstoffe. *Nachrichtenbl. Deut. Pflanzenschutzd.* (Braunsxheweig). 32:100–105.
69. Kronstad, J. W., Schnepf, H. E., and Whiteley, H. R. 1983. Diversity of locations for the *Bacillus thuringiensis* crystal protein genes. *J. Bacteriol.* 154:419–428.
70. Krywienczyk, J., and Angus, T. M. 1967. A serological comparison of several crystalliferous insect pathogens. *J. Invertebr. Pathol.* 9:126–128.
71. Lecadet, M-M., and Martouret, D. 1987. Host specificity of the *Bacillus thuringiensis* delta-endotoxin toward lepidopteran species: *Spodoptera littoralis* Bdv. and *Pieris brassicae* L. *J. Invertebr. Pathol.* 49:37–48.
72. Li, R. S., Jarrett, P., and Burgess, H. D. 1987. Importance of spores, crystals, and delta-endotoxins in the pathogenicity of different varieties of *Bacillus thuringiensis* in *Galleria mellonella* and *Pieris brassicae*. *Invertebr. Pathol.* 50:277–284.
73. Lilley, M., Ruffell, R. N., and Somerville, H. J. 1980. Purification of the insecticidal toxin in crystals of *Bacillus thuringiensis*. *J. Gen. Microbiol.* 118:1–11.
74. Luthy, P., Jaquet, F., Hofmann, C., Huber-Lukac, M., and Wolfersberger, M. G. 1986. Pathogenic actions of *Bacillus thuringiensis* toxin. *Zentralbl. Bakteriol. Hyg.* 15(suppl.):161–166.
75. McGaughey, W. H. 1985. Evaluation of *Bacillus thuringiensis* for controlling Indianmeal moths in farm grain bins and elevator silos. *J. Econ. Entomol.* 78:1089–1094.
76. McGaughey, W. H. 1985. Insect resistance to the biological insecticide *Bacillus thuringiensis*. *Science* 229:193–195.
77. McGaughey, W. H., and Beeman, R. W. 1988. Resistance to *Bacillus thuringiensis* in colonies of Indianmeal moth and almond moth (Lepidoptera: *Pyralidae*). *J. Econ. Entomol.* 81:28–33.
78. McGaughey, W. H., and Dicke, E. B. 1980. Methods of applying *Bacillus thuringiensis* to stored grain for moth control. *J. Econ. Entomol.* 73:228–229.
79. McGaughey, W. H., and Johnson, D. E. 1987. Toxicity of different serotypes and toxins of *Bacillus thuringiensis* to resistant and susceptible Indianmeal moths (Lepidoptera: *Pyralidae*). *J. Econ. Entomol.* 80:1122–1126.
80. Obukowicz, M. G., Perlak, F. J., Kusano-Kretzmer, K., Mayer, E. J., and Watrud, L. 1986. Integration of the delta-endotoxin gene of *Bacillus thuringiensis* into the chromosome of root colonizing strains of pseudomonads using Tn5. *Gene* 45:327–331.
81. Pozsgay, M., Fast, P., Kaplan, H., and Carey, P. R. 1987. The effect of sunlight on the protein crystals from *Bacillus thuringiensis* var *kurstaki* HD1 and NRD12: a Raman spectroscopic study. *J. Invertebr. Pathol.* 50:246–253.

82. Reese, J. C. 1981. Insect dietetics: complexities of plant-insect interactions. In *Current Topics in Insect Endocrinology and Nutrition*, G. Bhaskaran, S. Friedman, and J. G. Rodriguez (eds.). Plenum, New York.

83. Reese, J. C. 1983. Nutrient-allelochemical interactions in host plant resistance. In *Plant Resistance to Insects*, P. A. Hedin (ed.). American Chemical Society, Washington, D.C., pp. 231–243.

84. Reeser, P. W., and Dostka, S. J. 1988. Population dynamics of *Clavibacter xyli* subsp. *cynodonitis* (CXC) and a CXC/*Bacillus thuringiensis* subsp. *kurstaki* (BT) recombinant in corn (*Zea mays*). Abstracts of 1988 American Phytopathological Society Meeting, p. 221. The American Phytopathological Society, St. Paul, MN; Nov. 13–17, 1988.

85. Retnakaran, A., Lauzon, H., and Fast, P. 1983. *Bacillus thuringiensis* induced anorexia in the spruce budworm, *Choristoneura fumiferana*. *Entomol. Exp. Appl.* 34:233–239.

86. Richter, A. R., and Fuxa, J. R. 1984. Preference of five species of *Noctuidae* for feeding-stimulant adjuvants. *J. Georgia Entomol. Soc.* 19:383–387.

87. Rowe, G. E., and Margaritis, A. 1987. Bioprocess developments in the production of bioinsecticides by *Bacillus thuringiensis*. *CRC Crit. Rev. Biotechnol.* 6:87–127.

88. Sacchi, V. F., Parenti, P., Hanozet, G. M., Giordana, B., Luthy, P., and Wolfersberger, M. G. 1986. *Bacillus thuringiensis* toxin inhibits K^+-gradient dependent amino acid transport across the brush border membrane of *Pieris brassicae* midgut cells. *FEBS Lett.* 204:213–218.

89. Sanchis, V., Lereclus, D., Menou, G., Chaufaux, J., and Lecadet, M-M. 1988. Multiplicity of delta-endotoxin genes with different insecticidal specificities in *Bacillus thuringiensis aizawai* 7.29. *Mol. Microbiol.* 2(3):393–404.

90. Sanchis, V., Lereclus, D., Menou, G., Chaufaux, J., Guo, S., and Lecadet, M.-M. 1989. Nucleotide sequence and analysis of the N-terminal coding region of the Spodoptera active delta endotoxin gene in *Bacillus thuringiensis aizawai* 7-29. *Mol. Microbiol.* 3:229–238.

91. Schnepf, H. E., and Whiteley, H. R. 1981. Cloning and expression of the *Bacillus thuringiensis* crystal protein gene in *Escherichia coli. Proc. Natl. Acad. Sci. USA* 78:2893–2897.

92. Schnepf, H. E., and Whiteley, H. R. 1985. Delineation of a toxin-encoding segment of a *Bacillus thuringiensis* crystal protein gene. *J. Biol. Chem.* 260:6273–6280.

93. Schnepf, H. E., Wong, H. C., and Whiteley, H. R. 1985. The amino acid sequence of a crystal protein from *Bacillus thuringiensis* deduced from the DNA base sequence. *J. Biol. Chem.* 260:6264–6272.

94. Sekar, V., and Carlton, B. C. 1985. Molecular cloning of the delta-endotoxin gene of *Bacillus thuringiensis* var. *israelensis. Gene* 33:151–158.

95. Shibano, Y., Yamagata, N., Nakamura, N., Iizuku, T., Sugisaki, H., and Takanami, M. 1985. Nucleotide sequence coding for the insecticidal fragment of the *Bacillus thuringiensis* crystal protein. *Gene* 34:243–251.

96. Sneh, B., and Schuster, S. 1983. Effect of exposure to sublethal concentrations of *Bacillus thuringiensis* Berl. ssp. *entomocidus* on the susceptibility to the endotoxin of subsequent generations of the Egyptian leafworm *Spodoptera littoralis* Boisd. (Lep.: *Noctuedae*). *A. Ang. Entomol.* 96:425–428.

97. Southern, P. S., and Jackson, D. M. 1984. Control of *Heliothis virescens* on flue-cured tobacco using *Bacillus thuringiensis* and a cottonseed flour feeding stimulant. *Tob. Sci.* 186:40–43.

98. Stahly, D. P., Dingman, D. W., Bulla, L. A., Jr., and Aronson, A. I. 1978. Possible origin and function of the parasporal crystals in *Bacillus thuringiensis* subsp. *kurstaki. Biochem. Biophys. Res. Com.* 84:581–588.

99. Stone, T. B., Sims, S. R., and Marrone, P. G. 1989. Selection of tobacco budworm for resistance to a genetically engineered *Pseudomonas fluorescens* containing the delta-endotoxin of *Bacillus thuringiensis* subsp. *kurstaki. J. Invertebr. Pathol.* 53:228–234.

100. Thorne, L., Gaduno, F., Thompson, T., Decker, D., Zounes, M., Wild, M., Walfield, A. M., and Pollock, T. J. 1986. Structural similarity between the Lepidoptera and

Diptera-specific insecticidal endotoxin genes of *Bacillus thuringiensis* subspp. *kurstaki* and *israelensis*. *J. Bacteriol.* 166:801–811.

101. Tyrell, D. J., Bulla, L. A., Andrews, R. E., Kramer, K. J., Davidson, L. I., and Nardin, P. 1981. Comparative biochemistry of entomocidal parasporal crystals of selected *Bacillus thuringiensis* strain. *J. Bacteriol.* 145:1052–1062.

102. Vaeck, M., Reynaerts, A., Hofte, H., Jansens, S., De Beuckeleer, M., Dean, C., Zabeau, M., Van Montagu, M., and Leemans, J. 1987. Transgenic plants protected from insect attack. *Nature* 327:33–37.

103. Visser, B., van Workum, M. E. S., Dullemans, A., and Waalwijk, C. 1986. The mosquitocidal activity of *Bacillus thuringiensis* var. *israelensis* is associated with M_r 230, 000 and 130,000 crystal proteins. *FEMS Microbiol. Lett.* 30:211–214.

104. Waalwijk, C., Dullemans, A. M., van Workum, M. E. S., and Visser, B. 1985. Molecular cloning and the nucleotide sequence of the M_r 28,000 crystal protein gene of *Bacillus thuringiensis* subsp. *israelensis*. *Nucleic Acids Res.* 13:8207–8217.

105. Wakisaka, Y., Masaki, E., Koizumi, K., Nishimoto, Y., Endo, Y., Nishimura, M. S., and Nishiitsutsuji-Uwo, J. 1982. Asporogenous *Bacillus thuringiensis* mutant producing high yields of delta-endotoxin. *Appl. Environ. Microbiol.* 43:1498–1500.

106. Wakisaka, Y., Masaki, E., and Nishimoto, Y. 1982. Formation of crystalline delta-endotoxin or poly-B-hydroxy-butyric acid granules by asporogenous mutants of *B. thuringiensis*. *Appl. Environ. Microbiol.* 43:1473–1480.

107. Ward, E. S., and Ellar, D. J. 1987. Nucleotide sequence of a *Bacillus thuringiensis* var. *israelensis* gene encoding a 130 kDa delta-endotoxin. *Nucleic Acids Res.* 15:7195.

108. Ward, E. S., and Ellar, D. J. 1988. Cloning and expression of two homologous genes of *Bacillus thuringiensis* subsp. *israelensis* which encode 130-kilodalton mosquitocidal proteins. *J. Bacteriol.* 170:727–735.

109. Whiteley, H. R., Schnepf, H. E., Kronstad, J. W., and Wong, H. C. 1984. Structural and regulatory analysis of a cloned *Bacillus thuringiensis* crystal protein gene. In *Genetics and Biotechnology of the Bacilli*, A. T. Ganesan and J. A. Hoch (eds.). Academic, New York, pp. 375–386.

110. Widner, W. R., and Whiteley, H. R. 1988. Two highly related insecticidal crystal proteins of *Bacillus thuringiensis* subsp. *kurstaki* possess different host range specificities. *J. Bacteriol.* 171:965–974.

111. Wu, D., and Chang, F. N. 1985. Synergism in mosquitocidal activity of 26 and 65 kDa proteins from *Bacillus thuringiensis* subsp. *israelensis* crystal. *FEBS Lett.* 190:232–236.

112. Yamamoto, T., and McLaughlin, R. E. 1981. Isolation of a protein from the parasporal crystal of *Bacillus thuringiensis* var. *kurstaki* toxic to the mosquito larva, *Aedes taeniorhynchus*. *Biochem. Biophys. Res. Commun.* 103:414–421.

113. Yamamoto, T., Iizuka, T., and Aronson, J. N. 1983. Mosquitocidal protein of *Bacillus thuringiensis* subsp. *israelensis*: identification and partial isolation of protein. *Curr. Microbiol.* 9:279–284.

114. Yamamoto, T., Watkinson, I. A., Kim, L., Sage, M. V., Stratton, R., Akande, N., Li, Y., Ma, D-P., and Roe, B. A. 1988. Nucleotide sequence of the gene coding for a 130 kDa mosquitocidal protein of *Bacillus thuringiensis israelensis*. *Gene* 66:107–120.

115. Yamvrias, C. 1962. Contribution a l'etude du mode d'action de *Bacillus* Berliner vis-a-vis de la teigne de la farine *Anagasta* (*Ephestia*) *kuekniella* Zeller (Lep.). *Entomophaga* 7:101–159.

Molecular Genetics of Nitrogen Fixation in Plant-Bacteria Symbioses

James M. Ligon

Ciba-Geigy Corporation
Agricultural Biotechnology Research Unit
P.O. Box 12257
Research Triangle Park, North Carolina 27709

Introduction

Nitrogen, an element that is essential to all life, is abundant on the earth and composes nearly 80% of the atmosphere. Despite this fact,

nitrogen is one of the major nutrients that limits the growth and productivity of plants in many ecosystems. This is a result of the inability
of plants and other eukaryotes to directly utilize atmospheric nitrogen, or dinitrogen, to meet their biological requirements for this element. Plants can utilize only combined forms of nitrogen such as ammonia and nitrate as sources of nitrogen. Ultimately, combined forms
of nitrogen are derived from atmospheric nitrogen by the process of
nitrogen fixation. Nitrogen fixation includes the reduction of
dinitrogen to oxides of nitrogen by physical processes, both naturally
occurring and synthetic, and the biologically catalyzed reduction to
ammonia. Biological nitrogen fixation is a property found only in bacteria and on a worldwide scale accounts for approximately 2×10^8
tons of fixed nitrogen per year (Burris 1980), or about two-thirds of the
total.

In the past 25 years there has been intense interest in nitrogen-
fixing bacteria that are capable of forming symbiotic relationships
with plants. The plants provide a reduced environment and carbohydrates for the bacterial endosymbionts to support their metabolism,
and the bacteria in turn fix atmospheric nitrogen that is used by the
plant to meet its biological needs for this element. Due to this relationship, plants capable of forming these symbioses are generally able
to grow in nitrogen-poor soils that would not otherwise support vigor-
ous plant growth. The result of decades of research in this area has
been a greatly enhanced understanding of the biochemistry, genetics,
and regulation of biological nitrogen fixation. The goal of this chapter
is to review the current state of our understanding of the genetics and
regulation of biological nitrogen fixation in three important bacterial-
plant symbiotic systems: *Rhizobium-Bradyrhizobium*–leguminous
plants, *Frankia*–actinorhizal plants, and *Anabaena*–azolla, and to
suggest avenues for future approaches that may lead to enhancement
of nitrogen fixation in these symbioses.

Rhizobia and frankiae are soil bacteria that are capable of inducing
the formation of nitrogen-fixing root nodules on their respective host
plants. Rhizobia are Gram-negative rods that form a symbiotic association with leguminous plants and are classified in two genera:
Rhizobium, which includes the fast-growing species, and *Bradyrhizobium*, which includes slow-growing species. *Frankia* is an
unspeciated bacterial genus in the family *Actinomycetales* that forms
a similar association with a diverse group of actinorhizal plants that
includes species in eight families and 21 genera (Lechevalier 1983,
Akkermans and Houwers 1979) that are mostly small trees and woody
shrubs. *Frankia* have a septated hyphal morphology, and most strains
produce sporangia with numerous nonmotile spores. *Frankia* have
only recently been studied in depth, due in part to their slow growth

under laboratory conditions and to the fact that the ability to grow them in pure culture has only recently been acquired. (Callaham et al. 1978). The physiology and genetics of *Frankia* have been reviewed recently (Normand and Lalonde 1986, Tjepkema et al. 1986).

Like rhizobia and frankiae, the heterocystous cyanobacterium *Anabaena* forms a nitrogen-fixing symbiosis with azolla, which is an aquatic fern. However, the *Anabaena*–azolla symbiosis does not include the formation of root nodules. Instead, *Anabaena* occupies a cavity in azolla located at the base of each dorsal leaf lobe. A colony of vegetative *Anabaena* filaments is closely associated with the apical meristem of each azolla stem, and as the developing bilobed leaves differentiate, these filaments are partitioned into, and enclosed by, the newly formed cavities. The endophyte is associated with the megasporocarp and acts as an inoculum for the developing azolla sporophyte, thus ensuring the survival of the symbiotic association from one generation to the next (Calvert et al. 1979). This association has been used for centuries in the Far East as a green manure in the cultivation of rice, and with the introduction of new, high-yield varieties of rice that require higher levels of nitrogen for growth, it has gained increased importance.

In all three of these symbioses, the bacterial endosymbiont undergoes some degree of differentiation to attain its nitrogen-fixing form (see Figure 1). In the case of rhizobia, each cell transforms in the interior of the nodule into an enlarged spherical bacteroid that has different membrane and physiological properties from free-living cells (reviewed by Appleby 1984). *Frankia* and *Anabaena* differ from rhizobia since only a portion of the cells differentiate into specialized nitrogen-fixing cells, known as "vesicles" and "heterocysts," respectively. Each of these have thickened cell walls whose apparent function is to act as a diffusion barrier to oxygen, which is toxic to nitrogenase, the enzyme that catalyzes the nitrogen fixation reaction. In *Anabaena*, the O_2-generating photosynthetic apparatus, "photosystem II," is inactive within heterocysts. Rhizobia and legumes manage the oxygen paradox by jointly synthesizing the oxygen-binding compound leghemoglobin (reviewed by Appleby 1984). Unlike their rhizobial counterparts, *Frankia* and *Anabaena* are capable of free-living growth using N_2 as a source of nitrogen through nitrogen fixation. However, several strains of *Bradyrhizobium* and *Azorhizobium sesbaniae* (a rhizobial species that nodulates the nonlegume *Parasponia*) are capable of fixing nitrogen in the free-living state under carefully defined conditions, but they are incapable of using this nitrogen for their own growth. Nevertheless, these strains have been useful as models for the study of the regulation of nitrogen fixation in symbiotic bacteria.

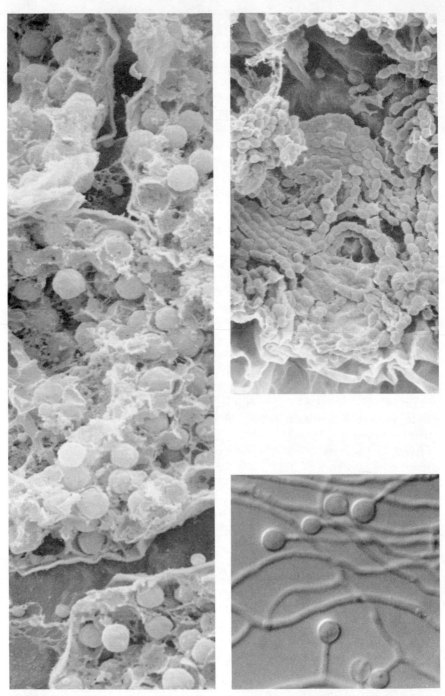

Figure 1 (*A*) Electron micrograph of the interior of a peanut root nodule showing *Bradyrhizobium* sp. bacteroids within cavities in the nodules (*courtesy of R. Steece*). (*B*) A similar photograph of the *Anabaena* endophyte from leaf cavities of azolla (*courtesy of R. Rowland*). In each of the photographs, the endophyte cells are approximately 1 μ in diameter. (*C*) Free-living cells of *Frankia* grown under nitrogen-fixing conditions (*courtesy of M. Stowers*). The vesicles are the spherical cells seen at the end of stalks that branch from the hypha and are approximately 5 μ in diameter.

Much of the current understanding of the biochemistry and genetics of nitrogen fixation has been a product of research using *Klebsiella pneumoniae* as a model. This organism fixes nitrogen in the free-living state under anaerobic, nitrogen-limiting conditions and has been a good model organism because of its great genetic manipulability. Knowledge of the processes involved in nitrogen fixation in *K. pneumoniae* has been invaluable in gaining a better understanding of this process in symbiotic organisms. Therefore, many comparisons will be made throughout this review between nitrogen fixation in this organism and in the plant-symbiotic nitrogen-fixing organisms.

Nitrogenase

Nitrogenase is a complex, oxygen-labile enzyme that is composed of two protein components: dinitrogenase and dinitrogenase reductase. Dinitrogenase consists of two dissimilar polypeptides, with a conformation of $\alpha_2\beta_2$, and a molecular mass of 200 to 250 kDa (Tso 1974, Nakos and Mortenson 1971). The α and β subunits of dinitrogenase are encoded by the *nifD* and *nifK* genes, respectively. Active dinitrogenase contains two iron-molybdenum cofactors (FeMoco) and four [4Fe—4S] clusters. In addition, dinitrogenase contains 2 Mo and 24 to 32 nonheme Fe atoms (Eady and Postgate 1974, Mortenson and Thorneley 1979, Lowe et al. 1980). Mossbauer analysis of the dinitrogenase proteins from *Azotobacter vinelandii, K. pneumoniae*, and *Clostridium pasteurianum* suggests 33 ±3 Fe atoms per molecule (Smith 1983, Orme-Johnson and Munck 1980), and sulfide analysis indicates approximately the same number of S atoms (Mortenson and Thorneley 1979). There are no known species differences with respect to metal content.

Dinitrogenase reductase consists of two identical polypeptides that are encoded by the *nifH* gene. It has a total molecular mass of approximately 60 kDa and contains a single [4Fe—4S] cluster that functions as a one-electron donor. The role of dinitrogenase reductase in nitrogenase catalysis is the transfer of electrons to dinitrogenase, which contains the active site where the reduction of nitrogen occurs (Hageman and Burris 1980, Mortenson and Thorneley 1979). The pathway of electron transport of the oxidized form of dinitrogenase reductase has been well established for *K. pneumoniae* (Shah et al. 1983). An oxidoreductase, encoded by the *nifJ* gene, catalyzes the two-electron oxidation of pyruvate to acetyl-S-CoA and CO_2 with the two electrons used to reduce two molecules of a flavodoxin, the product of the *nifF* gene (Nieva-Gomez et al. 1980). This flavodoxin operates between the semiquinone and hydroquinone states in the transfer of one electron to oxidized dinitrogenase reductase (Klugkist et al. 1985,

Shah et al. 1983). Prior to the reduction of N_2, reduced dinitrogenase reductase, dinitrogenase, and two molecules of Mg-ATP combine to form a ternary complex. Within this complex, the hydrolysis of two molecules of ATP is coupled to the transfer of one electron from dinitrogenase reductase to dinitrogenase (Eady et al. 1978, Hageman et al. 1980), with a concomitant one-electron reduction of the substrate.

The rate of electron transfer in the active dinitrogenase complex is relatively rapid compared to the flux of electrons through dinitrogenase during turnover (Thorneley et al. 1975). The slowest, and rate limiting, step in the cycle of nitrogenase catalysis occurs after electron transfer between the two dinitrogenase proteins. It is the dissociation of the oxidized dinitrogenase reductase from dinitrogenase (Thorneley and Lowe 1983). It has been suggested that the dissociation of the two protein components is necessary because dinitrogenase reductase cannot be reduced while it is bound to dinitrogenase (Hageman and Burris 1980). Since 6 mol of electrons are required to reduce 1 mol of N_2 to 2 mol of NH_3, dinitrogenase must complete the cycle of binding, transfer of one electron, release from dinitrogenase, and reduction by the flavodoxin a total of six times to complete the reduction of 1 mol of N_2. Accordingly, nitrogenase reaction rates for N_2 reduction are usually low.

There has been a long-standing feeling among researchers that FeMoco is bound near the substrate binding site and is involved in catalysis (Smith 1977). Details of the structure of FeMoco are few. However, it is generally agreed that it contains no amino acids, it has an organic component, and it has a molecular weight of about 5,000. Recently, Hoover et al. (1987) demonstrated that the organic portion of FeMoco is homocitrate. A system for the in vitro synthesis of FeMoco was developed by Shah et al. (1986); it requires the nifB, nifN, and nifE gene products, as well as a low-molecular-weight factor produced in the presence of the nifV gene. Hoover et al. (1987) isolated the nifV factor and identified it as homocitrate. They demonstrated that the addition of homocitrate to in vitro FeMoco biosynthetic reactions could replace the requirement for the nifV factor and complements the Nif$^-$ phenotype of nifV mutants of K. pneumoniae (Hoover et al. 1988). Therefore, they suggested that nifV codes for homocitrate synthase.

In K. pneumoniae 17 genes have been identified that play a role in nitrogen fixation (Table 1). They are organized in eight transcription units within a 24-kb region of the K. pneumoniae genome (Orme-Johnson 1985, Ausubel and Cannon 1981, Roberts and Brill 1980, Cannon et al. 1985). In addition to the structural nitrogenase genes (nifH, nifD, and nifK) and genes that have a role in electron transfer

TABLE 1 **The** *nif* Genes of *Klebsiella pneumoniae* and the Function of Their Protein Products

Gene	Protein function	Reference
nifQ	Mo uptake	Imperial et al. 1984
nifB	Required for the synthesis of FeMoco	St. John et al. 1975, Shah and Brill 1977, Roberts et al. 1978, Orme-Johnson 1985
nifA	Activates expression of *nif* operons	Roberts et al. 1978, Dixon et al. 1977, Dixon et al. 1980, Houmard et al. 1980, Buchanan-Wallaston et al. 1981*a*
nifL	Inhibits activation of *nif* operons	Hill et al. 1981, Kennedy et al. 1981, Merrick et al. 1982, Buchanan-Wallaston et al. 1981*b*
nifF	Ferredoxin, electron transport to dinitrogenase reductase	St. John et al. 1975, Roberts et al. 1978, Hill and Kavanagh 1980, Nieva-Gomez et al. 1980
nifM	Processing of active dinitrogenase reductase	Roberts et al. 1978
nifV	Synthesis of homocitrate, a component of FeMoco	McLean and Dixon 1981, Hoover et al. 1987, Hoover et al. 1988
nifS	Maturation of nitrogenase complex	Roberts et al. 1978
nifU	Unknown	
nifX	Unknown	
nifN	Required for the synthesis of FeMoco	Roberts et al. 1978, Orme-Johnson 1985
nifE	Required for the synthesis of FeMoco	Roberts et al. 1978, Orme-Johnson 1985
nifY	Unknown	
nifK	Structural gene for the β subunit of dinitrogenase	Roberts et al. 1978
nifD	Structural gene for the α subunit of dinitrogenase	Roberts et al. 1978
nifH	Structural gene for dinitrogenase reductase	Roberts et al. 1978, Dixon et al. 1977
nifJ	Electron transport to ferredoxin coded by *nifF*	Hill and Kavanagh 1980, Nieva-Gomez et al. 1980, Bogusz et al. 1981

(*nifJ* and *nifF*) or the biosynthesis of an active FeMoco (*nifE, nifN, nifB*, and *nifV*), other *nif* genes have been identified, and the functions of the products of some of these are known. The *nifQ* gene product is required for the uptake of molybdenum (Imperial et al. 1984), and the *nifM* gene is involved in processing of the *nifH* polypeptides to yield an active protein (Roberts et al. 1978). The role of the *nifS* protein is not well-characterized, but there is some evidence that suggests it is required for the maturation of a fully functional nitrogenase enzyme complex. Studies with *K. pneumoniae* have been unable to resolve whether it functions in the processing of dinitrogenase, or of dinitrogenase reductase, or in the assembly of these two proteins to form the nitrogenase enzyme complex (Orme-Johnson 1985, Roberts et al. 1978). *Bradyrhizobium japonicum nifS* mutants are not completely Fix⁻, and exhibit reduced levels of nitrogen fixation in the nodule (Ebeling et al. 1987). The *nifA* and *nifL* genes encode the positive and negative regulatory proteins, respectively, that regulate transcription of all *nif* gene operons (Buchanan-Wollaston et al. 1981*a*, Buchanan-Wallaston et al. 1981*b*, Dixon et al. 1977, Dixon et al. 1980, Hill et al. 1981, Houmard et al. 1980, Kennedy et al. 1981, Merrick et al. 1982, Roberts et al. 1978). The functions of the *nifU*, *nifX*, and *nifY* genes are unknown.

Regulation of Nitrogenase Activity

ntr regulatory genes

The mechanisms of *nif* gene regulation have been extensively studied in *K. pneumoniae*, and studies with *Rhizobium meliloti, B. japonicum*, and other symbiotic diazotrophs indicate that the mechanisms are very similar in these bacteria. The major difference in the regulation of nitrogenase between free-living and symbiotic nitrogen-fixing bacteria lies in the control of the *nifA* gene and its protein product (NifA). In most, if not all, nitrogen-fixing bacteria *nif* gene expression is dependent on NifA and the *ntrA* gene product, NtrA (Buchanan-Wallaston et al. 1981*a*, Fischer et al. 1986, Powlowski et al. 1987, Schetgens et al. 1985, Szeto et al. 1984). However, a higher level of control is exerted by the nitrogen regulatory system (*ntr*), which is sensitive to cellular levels of nitrogen. A proposed regulatory model adapted from Gussin et al. (1986) is presented in Figure 2, and it will serve as a framework for the following discussion of the regulation of nitrogen metabolism and fixation. The nomenclature used in describing genes involved in nitrogen fixation in symbiotic organisms makes a distinction between "*nif*" genes, those genes that are homologous to corresponding genes in *K. pneumoniae*, and "*fix*" genes, those genes that are not homologous to known *K. pneumoniae nif* genes but are required for normal symbiotic nitrogen fixation.

ntrA.   The *ntr* genes are organized in two operons, *glnA-ntrBC* and *ntrA*. NtrA and the *ntrC* gene product (NtrC) are required for the activation of all *ntr*-regulated gene systems, including glutamine synthetase (*glnA*), arginine transport, proline and histidine transport and utilization, and *nifLA*, the *nif* regulatory gene operon (Ausubel

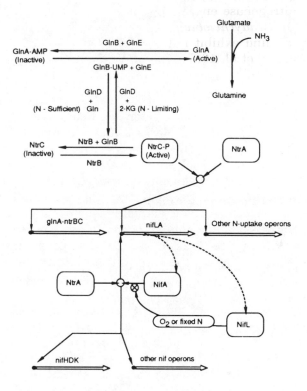

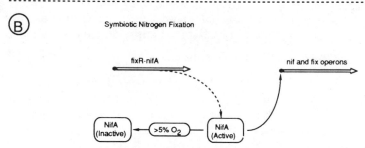

Figure 2 A model depicting (A) the regulation of transcription of nitrogenase genes and the activity of their protein products in the free-living *Klebsiella pneumoniae* system and (B) how the regulation of the symbiotic nitrogenase systems differs from this. See the text for a detailed description.

1984, Alvarez-Morales et al. 1984, Magasanik 1982, Merrick 1983). There is much evidence that indicates that NtrA is a *nif*-specific σ factor. Although an NtrA–RNA polymerase core enzyme complex has not been isolated, partially purified NtrA binds stoichiometrically with *E. coli* RNA polymerase in vitro (Hunt and Magasanik 1985), and in the presence of RNA polymerase core enzyme it exhibits DNA-binding activity (Hunt and Magasanik 1985, Hirschman et al. 1985). Also, the migration of NtrA on SDS-polyacrylamide gels and its high content of acidic amino acids, as deduced from the *ntrA* gene sequence, are properties common to other known σ factors (de Bruijn and Ausubel 1983, Hunt and Magasanik 1985, Merrick and Gibbins 1985). The *A. vinelandii ntrA* gene was sequenced by Merrick et al. (1987), and this sequence was compared to the sequence of the *K. pneumoniae* and *R. meliloti ntrA* genes. All three genes shared regions of homology, but were not homologous to the genes of other known sigma factors. A region in the center of the predicted amino acid sequence of the NtrA proteins is homologous to a similar region in other σ factors, and the C-terminal regions are homologous to the β subunit of RNA polymerase (RpoC). Among the three NtrA proteins, two other conserved regions were found, one of which is homologous to known DNA-binding motifs (Merrick et al. 1987).

Transcriptional fusions of the *ntrA* promoter and *lacZ* or tetracycline resistance genes have shown that *ntrA* gene expression is not controlled by the level of ammonia (Castano and Bastaracchea 1984, de Bruijn and Ausubel 1983). This suggests that cellular *ntr* gene regulation is not mediated by the transcriptional control of *ntrA*. However, when *ntrA* was introduced into an *E. coli ntrA-lacZ* background on a multicopy plasmid, the *ntrA::lacZ* gene fusion was overexpressed (Merrick and Stewart 1985), indicating that *ntrA* may be negatively regulated by a repressor that can be titrated out with excess copies of an *ntrA* operator.

ntrC. The NtrC protein is encoded by the *glnA-ntrBC* operon, whose expression is regulated from three promoters. Two of these are tandem promoters, *glnA*p1 and *glnA*p2, that are located upstream from the *glnA* gene (Reitzer and Magasanik 1985). A third promoter is located in the intergenic region between *glnA* and *ntrB* (Reitzer and Magasanik 1983, Ueno-Nishio et al. 1984). The *glnA*p1 and *ntrB* promoters are relatively weak and are expressed under conditions of excess ammonia in order to provide low levels of GlnA, NtrB, and NtrC. Under these conditions the *ntrB* promoter is necessary due to the presence of a transcriptional terminator between the *glnA* and *ntrB* genes that would otherwise reduce the levels of NtrB and NtrC derived from the *glnA*p1 promoter alone to less than one molecule per cell (Reitzer and Magasanik 1985). When cellular ammonia levels become limiting, NtrC and NtrA act in concert to activate transcription of the *glnA-ntrBC* operon from the strong *glnA*p2 promoter (Reitzer and Magasanik 1985).

In addition to activating the *glnA*p2 promoter, NtrC has been shown to repress transcription from the *glnA*p1 and *ntrB* promoters (reviewed by Kustu et al. 1986). Since NtrC appears to act as both a negative and positive regulator of *glnA-ntrBC* transcription, it has been suggested that NtrC is a DNA-binding protein. This has been demonstrated for the *E. coli ntrB* promoter with DNase footprinting and methylation protection experiments (Ueno-Nishio et al. 1984). Subsequently, five ntrC binding sites have been identified in a region upstream of the *glnA*p2 promoter (Hirschmann et al. 1985, Reitzer and Magasanik 1986). Studies by Reitzer and Magasanik (1986) have shown that the NtrC binding sites farthest upstream of the *glnA*p2 transcription start site are the most important of the five such sites identified. They demonstrated that deletion of the NtrC binding sites farthest upstream resulted in the inability to activate the *glnA-ntrBC* operon under low NtrC concentrations and in an 80% reduction of activation with high levels of NtrC. Surprisingly, activation of the *glnA-ntrBC* operon under conditions of low NtrC concentration could occur even when the two upstream binding sites were moved more than 1,000 bp upstream, or 2,000 bp downstream, from their normal location. This result strengthens the proposal that NtrC acts as a DNA-binding protein in conjunction with NtrA–RNA polymerase to activate transcription of *ntr*-regulated promoters.

ntrB. According to the model, the ability of NtrC to act as a transcriptional activator is modulated by NtrB in response to the nitrogen status of the cell (Alvarez-Morales et al. 1984). Under conditions of low glutamine concentration NtrB is required for the NtrC-mediated activation of the *glnA*p2 promoter. The NtrC-activation activity of NtrB is in turn activated, or inactivated, by the *glnB* gene product, GlnB. Uridylylated GlnB does not interfere with the NtrB-mediated activation of NtrC, while in the presence of deuridylylated GlnB, NtrB inactivates NtrC. The uridylylation-deuridylylation of GlnB is controlled by the *glnD* gene product (GlnD) in response to the intracellular ratios of glutamine and 2-ketoglutarate. Through this series of interactions, high glutamine concentrations, reflecting a nitrogen-sufficient environment, result in the inactivation of NtrC by NtrB, and *ntr*-regulated operons, including the *nifLA* operon, are not expressed. Conversely, low glutamine levels, which reflect a nitrogen-deficient environment, do not trigger the NtrB-mediated inactivation of NtrC, and the *nifLA* operon and other similarly regulated operons are activated.

This model is supported by several lines of evidence. First, *glnA* mutants do not respond to high levels of ammonia, but to high concentrations of glutamine, indicating that glutamine, and not ammonia, influences the regulation of the *glnA-ntrBC* operon (Krajewska-Grynkiewicz and Kustu 1983). Second, the effects of mutations in the *glnB*, *glnD*, and *ntrB* genes on levels of glutamine synthetase are consistent with those predicted by this model (Bloom et al. 1977, Bueno et al. 1985, Foor et al. 1980, Reuveny et al. 1981). Third, it has been demonstrated that in vitro transcription of *glnA* is activated by NtrC only in the presence of NtrB (Hunt and Magasanik 1985). Recent evidence has demonstrated that NtrB has kinase-phosphatase activity that phosphorylates NtrC under nitrogen-limiting conditions and dephosphorylates it when nitrogen is sufficient (Ninfa and Magasanik 1986).

Role of NifA and NifL

The nitrogenase gene system in *K. pneumoniae* is but one of the gene systems regulated by NtrC that play a role in nitrogen acquisition and metabolism. We have seen how *ntr* gene regulation is sensitive to cellular nitrogen levels as reflected by the glutamine/2-ketoglutarate ratio. If cellular nitrogen levels are low, such that NtrC concentration is elevated and the *nifLA* operon is expressed, a second level of control that is sensitive to oxygen and to intermediate levels of nitrogen ultimately controls the expression of *nif* genes. Expression of the *K. pneumoniae* regulatory *nif* operon, *nifLA*, requires NtrC (Drummond et al. 1983, Ow and Ausubel 1983) and NtrA (Hirschman et al. 1985, Hunt and Magasanik 1985). As discussed previously, the level of NtrC derived from the *ntrB* promoter under conditions of high nitrogen availability is not sufficient to activate the *nifLA* operon. When cellular nitrogen levels decline and the *glnA*p2 promoter is activated, the level of active NtrC increases until the *nifLA* operon is transcribed. The proteins encoded by the *nifL* (NifL) and *nifA* (NifA) genes have opposing functions. NifA is required in addition to NtrA for the activation of the structural *nif* gene operon, *nifHDK*, as well as all other operons containing *nif* genes. NifL interacts with NifA to prevent NifA-mediated gene activation in the presence of oxygen or intermediate levels of ammonia (Sibold and Elmerich 1982, Dixon et al. 1980, Hill et al. 1981, Merrick et al. 1982). NifL affects NifA-mediated, but not NtrC-mediated, gene activation, probably by direct interaction with NifA that leads to its inactivation (Cannon et al. 1985). The mechanisms by which NifL senses oxygen and ammonia levels are unknown. The mechanism of NifA-mediated activation of *nif* promoters is less well understood than that of the NtrC-mediated activation of the *glnA*p2 promoter. This is a result of the lack of a system for the transcription of *nif* promoters in vitro or for detecting binding of NifA to *nif* promoters. Multicopy plasmids containing *nif*-specific promoter sequences have been shown to inhibit expression of chromosomal *nif* genes, presumably by titration of NifA protein (Buck and Cannon 1987). Regulation of *nif* genes by NifA-NifL results in their inactivation under conditions that are detrimental to the nitrogenase enzyme complex (high oxygen concentrations) or that do not favor nitrogen fixation as a means of acquiring nitrogen (sufficient levels of fixed nitrogen).

There is still some confusion about the mechanism of inhibition of *nif* gene expression due to the effects of oxygen. In addition to the effect of oxygen as a modulator of NifL-mediated inactivation of NifA, some reports indicate that oxygen inhibits the expression of the *nifLA* operon (Dixon et al. 1980, Kong et al. 1986), while others have found no such effect (Cannon et al. 1985, Collins and Brill 1985). It is clear that oxygen has no direct effect on NifA activity since transcription of *nif* genes is not affected by oxygen when *nifA* is expressed constitutively from strong antibiotic resistance promoters (Buchanan-Wallaston 1981a).

Several lines of evidence suggest that NtrC and NifA are structurally and functionally related. First, NifA and NtrC both require NtrA for activity (Merrick 1983, Ow and Ausubel 1983). Second, NifA and NtrC are both capable of activating the *K. pneumoniae nifLA* promoter (Merrick 1983, Ow and Ausubel 1983, Drummond et al. 1983, Sundaresan et al. 1983). Third, NifA can activate several promoters that are normally activated by NtrC, including the *glnA* promoter (Merrick 1983, Ow and Ausubel 1983). Finally, the *nifA* and *ntrC* genes of *K. pneumoniae* and the *R. meliloti nifA* gene are structurally homologous and share strongly conserved regions in the central and C-terminal domains (Buikema et al. 1985, Drummond et al. 1986).

Evidence supporting the regulatory model in symbiotic systems

The *K. pneumoniae* model for the regulation of nitrogen metabolism and nitrogen fixation has served as a useful foundation for the investigation of similar regulatory systems in the symbiotic nitrogen-fixing bacteria. A regulatory gene that is structurally homologous to the *ntrC* gene in enteric bacteria was recently identified by Szeto et al. (1987) in *R. meliloti*. This gene shares sequence homology with the *ntrC* gene from *K. pneumoniae* (Drummond et al. 1986) and a similar gene identified in *Azorhizobium sesbaniae*. In addition, Szeto and Ausubel (1987) have demonstrated the existence of an *ntrC* homologue in *B. japonicum*. As in the *K. pneumoniae* system, an *ntrB*-like gene was found by Drummond et al. (1986) to be located immediately upstream from the *ntrC* gene in *R. meliloti*. As would be expected of an *ntrC* gene, the *R. meliloti* gene is expressed only under nitrogen-limiting conditions and it is required for growth on nitrate and for the *ex planta* transcription of several *R. meliloti nif* genes (Szeto et al. 1987). However, root nodules formed by *ntrC* mutants fixed nitrogen as well as wild-type *R. meliloti*. Similarly, in *B. japonicum* the *nifH* and *nifD* promoters are activated by the *nifA* protein alone (Alvarez-Morales and Hennecke 1985), so NtrC appears not to be required for *nif* gene activation. This suggests that in these bacteria the *ntrC* gene is not involved in the regulation of *nif* genes *in planta*. Since nitrogen fixation usually occurs only *in planta* in these bacteria, NtrC may not be involved in the regulation of nitrogen fixation genes at all, and its function may be restricted to the regulation of other nitrogen uptake systems.

An *ntrB*-like protein from *A. sesbaniae* has been shown to share homology with regulatory proteins from other species of bacteria (Nixon et al. 1986). The gene that encodes this protein is located adjacent to a region that is homologous to the *K. pneumoniae ntrC* gene.

A *nifA* gene has been identified in several rhizobial species, including *R. meliloti* (Szeto et al. 1984, Weber et al. 1985), *B. japonicum* (Fischer et al. 1986), *Rhizobium leguminosarum* (Schetgens et al. 1985), and *Azorhizobium sesbaniae* strain ORS571 (Pawlowski et al. 1987). The *R. meliloti nifA* gene is structurally and functionally similar to the *nifA* gene of *K. pneumoniae* (Buikema et al. 1985). The DNA sequence of the *K. pneumoniae nifA* gene was recently published, and it shares sequence homology with the *R. meliloti nifA* gene (Drummond et al. 1986). The *nifA* gene in this organism is located directly downstream from the *fixABC* operon, and it has been demonstrated that at least 50% of the transcripts derived from it in the nodule are derived from read-through transcription from the *fixABC* promoter, which is itself activated by NifA (Buikema et al. 1985, Kim et al. 1985, 1987). This suggested the possibility that the *nifH* and *fixA* promoters are activated by the *R. meliloti* NtrC-like protein and that *nifA* transcription could be activated by its own gene product that is produced initially from the *fixA* promoter. In order to test this hypothesis, Szeto et al. (1987) cloned the *R. meliloti ntrC* gene using interspecies homology with the *E. coli ntrC* gene. They constructed *ntrC::Tn5* mutants that were unable to activate the *nifH* or *fixA* promoters in response to nitrogen starvation during free-living growth, while the wild-type was competent in this regard (Sundaresan et al. 1983). The *ntrC::Tn5 R. meliloti* mutants formed effective nodules and were capable of activating *nifH-lacZ* and *fixA-lacZ* fusions to levels comparable to the wild type in the nodule (Szeto et al. 1987). Szeto et al. (1984) had previously demonstrated that in *R. meliloti*, NifA is required for *nif* gene activation during symbiotic growth but not for activation of *nif* promoters in the free-living state. Therefore, it appears that both NtrC and NifA are capable of activating the *R. meliloti nifH* promoter, but NtrC appears to operate in the free-living state while NifA is active in the nodule. Since the level of NtrC-mediated activation of *nifH* during free-living conditions is only 10% of the level obtained by NifA activation in symbiosis, and since acetylene reduction activity has never been demonstrated in free-living *R. meliloti*, the significance of the NtrC-mediated *nif* gene activation is unclear.

Through the use of in vitro transcription-translation systems, Beynon et al. (1988) recently determined the translational initiation point for the *R. meliloti nifA* gene. It was demonstrated that four distinct polypeptides were produced from the *nifA* gene. One is probably derived from a fortuitous translation start site outside the *nifA* region and is probably not functional. Of the other three, only the largest polypeptide derived from the first translation start site demonstrated the ability to activate *nif* gene–*lacZ* fusions. The *R. meliloti nifA* gene product was capable of activating *nif* promoters from *K. pneumoniae* and rhizobia strains (Beynon et al. 1988). Furthermore, overproduction of a modified NifA in which the N-terminal domain was deleted resulted in elevated transcriptional activity from *nif* promoters. Therefore, the N-terminal domain of NifA appears to hinder the NifA gene-activation function. It has been suggested that the N-terminal domain is involved in the regulation of NifA activity.

A *nifA* gene was identified in *B. japonicum*, and its location mapped to the region immediately upstream from the *fixA* gene (Fischer et al. 1986). Partial DNA sequencing of this gene demonstrated that it shares homology with the *K. pneumoniae* and *R. meliloti nifA* genes. It was further demonstrated by Fischer et al. (1986) that *lacZ* gene fusions with *nif* gene promoters from *K. pneumoniae* and *R. meliloti* were activated by the protein product of the *B. japonicum nifA* gene. In an *E. coli* background, the *B. japonicum* NifA-mediated transcriptional activation of *nif* genes was dependent on NtrA in accordance with the expectations of a typical NifA protein (Fischer et al. 1986). Site-directed insertion and deletion-replacement mutagenesis revealed that the *nifA* gene in *B. japonicum* is the gene distal to the promoter within a multigenic operon (Fischer et al. 1986). NifA⁻ mutants produced Fix⁻ nodules and were pleiotropic in that they failed to produce several *nif* proteins, including NifH, NifD, and NifK, and they produced an altered nodulation profile on plants, producing many small, widely distributed nodules in which the bacteroids deteriorated at an early stage in nodule development (Fischer et al. 1986). This suggests that in addition to its role in *nif* gene regulation, the *nifA* gene is required for the establishment of a determinate nitrogen-fixing nodule.

The effect of truncations of the *B. japonicum nifA* gene on the activation of a *nifD-lacZ* fusion was recently studied by Fischer and Hennecke (1987a). The results demonstrated that up to 241 amino acid residues out of a total of 605 could be deleted from the N terminus without loss of the transcriptional activation function, whereas deletion of only 2 amino acids from the C terminus resulted in a 90% reduction of this activity. This result is consistent with the findings discussed earlier, which indicate that the N-terminal domain of the *nifA* gene in *R. meliloti* is not required for the *nif* gene activation function.

At the present time, no *nif* genes that have regulatory functions have been identified in *Anabaena* or *Frankia*. However, initial evidence suggests that there is a functional *nifA* gene in *Frankia*. A genomic library of DNA from a *Frankia* alder isolate was transferred to a Nif⁻ *K. pneumoniae nifA* deletion mutant, and Nif⁺ *K. pneumoniae* transconjugants were generated that, unlike the mutant, were capable of growth on a nitrogen-free medium (Ligon 1986).

Structure of *nif* gene promoters

Examination of the DNA sequences in the promoter regions and transcription initiation sites of several different NtrC- and NifA-activated genes has revealed that these promoters share a similar structure (reviewed by Ausubel 1984, Ausubel et al. 1985, Cannon et al. 1985) that is characterized by the sequence CTGGYAYR-N4-TTGCA in the position approximately -26 to -10, relative to the transcription initiation site (Y = pyrimidine, R = purine) (Beynon et al. 1983, Buchanan-Wollaston et al. 1981b, Ow et al. 1983, Sundaresan et al. 1983). This sequence constitutes what has come to be known as a "consensus *nif* promoter" and differs from the canonical *E. coli* promoter at -35 to -10. The consensus *nif* promoter is found in the promoters of the *K. pneumoniae nif* operons examined to date, as well as in several rhizobia *nif* gene promoters (Alvarez-Morales and Hennecke 1985) and preceding the *nif* genes of other diazotrophic organisms, such as *A. vinelandii* (Brigle et al. 1985). Studies by Ow et al. (1985) and Buck et al. (1985) using site-directed or oligonucleotide-directed mutagenesis suggest that NifA and NtrC recognition of *nif* promoters may involve interactions at or near the -12 base.

The conservation of NifA- and NtrC-activated promoters is not inclusive of all *nif* genes. The *R. meliloti nifA* promoter does not resemble the *nif* consensus promoter (Buikema et al. 1985), nor do the *nifH* and *glnA* promoters of *Anabaena* 7120, which have the sequence TCTAC at −14 to −10, replacing TTGCA (Haselkorn et al. 1983, Tumer et al. 1983). This is not surprising in the case of *R. meliloti* since *nifA* appears to be regulated in a different manner than the other *nif* genes. In *Anabaena*, this difference may be due to the fact that the activation of nitrogen fixation is coordinated with the development of heterocysts and a unique DNA rearrangement near the *nif* genes (discussed later). Recent scrutiny of the *nifH* gene of *Clostridium pasteurianum* indicates that this gene also does not contain a typical *nif* consensus promoter but has instead the canonical *E. coli* promoter (Chen et al. 1986).

Recently, the *nif* consensus promoter sequence has been found in the promoter region of genes with functions not normally associated with nitrogen fixation. The promoter for the *xylABC* operon in *Pseudomonas putida*, involved in the utilization of xylose, contains sequences that are homologous to the conserved *nif* consensus promoter (Dixon 1986). Dixon (1986) constructed *xylA-lacZ* fusions and demonstrated that in *E. coli* the activation of this fusion is dependent on the presence of NtrA and either NtrC, NifA, or XylR, the product of the regulatory gene for the *xylABC* operon in *Pseudomonas*. Conversely, the *K. pneumoniae nifL* and *nifH* promoters were not activated by NtrA and XylR. The significance of the structural and functional similarity between the *xylA* promoter of *P. putida* and the *nif* consensus promoter is unknown.

A recent development in the study of *nif* promoters is the identification of upstream activator sequences that have a role in the activation of *nif* genes. Conserved sequences located in the −103 to −153 region have been identified in 19 different *nif* promoters from a variety of species, including the *nifH, nifU,* and *nifB* promoters of *K. pneumoniae*, the *nifH* and *nifD* promoters of *B. japonicum* (Alvarez-Morales et al. 1986), and 10 rhizobia and 2 *A. vinelandii nif* promoters (Buck et al. 1986). The consensus upstream activator sequence contains two conserved sequences, 5-TGT-N_{10}-ACA-3 and 5-TGTCG-N_6-CRACA-3. Like the NtrC binding sequence, both of these sequences are symmetric, suggesting that they may be recognition sites for a binding protein, for which NifA is the logical candidate. These sequences appear to be necessary for NifA-mediated, but not NtrC-mediated, gene activation, although there is some evidence that they may be required for the NtrC-mediated activation of one of the two *glnA* promoters in *E. coli* (Reitzer and Magasanik 1986).

Buck et al. (1987) examined the DNA sequence in the region between the *nifF* and *nifL* genes of *K. pneumoniae*, which are in separate operons and are transcribed in opposite directions. These investigators did not find a typical upstream activator sequence for the *nifF* gene in this region. However, a consensus upstream activator sequence was identified in the *nifL* coding region at +59. In order to determine if this activator sequence was affecting the expression of the *nifL* promoter in the unusual position 3' to the *nifL* promoter, they cloned a typical upstream activator sequence in the same position relative to the *nifH* promoter that lacked its indigenous upstream activator, and found that this configuration was not transcriptionally active in the presence of NifA. Subsequently, they demonstrated that the upstream activator sequence in the *nifL* coding region modulates expression of the *nifF* promoter at a distance of 263 bp from the transcription start site of this gene, a much greater distance than is found for other such sequences.

In most cases studied thus far, deletion of the upstream activator sequence of a *nif* promoter results in an approximately 30-fold reduction of the NifA-dependent promoter activity (Buck et al. 1986). Deletions of the upstream activator regions of the *B. japonicum nifH* and *nifD* promoters resulted in a drastic reduction in the activation of these promoters in the presence of the *K. pneumoniae* NifA protein in *E. coli* (Alvarez-Morales et al. 1986). Characteristics of deletion derivatives of the upstream activators of the *nifH* promoter of *R. meliloti* were studied by Better et al. (1985). These also demonstrated reduced *K. pneumoniae* NifA-dependent promoter activity in *E. coli*. However, the deletions had no effect on the expression of *nifH* in bacteroids in the nodule. Later, Earl (1986) demonstrated that these deletion derivatives were capable of wild-type levels of expression in *E. coli* in the presence of the *R. meliloti* NifA protein. These results are contrary to those concerning the role of upstream activators in other *nif* promoters.

Changing the position or orientation of the upstream activators of *nif* genes in *K. pneumoniae* had varying effects on gene expression. Moving the upstream activator sequence more than 1,200 bp upstream of its normal position reduced promoter activity to 10% of the wild-type level, while inverting its orientation had no effect on activity (Buck et al. 1986).

In addition to the confusion concerning the role of the upstream activator sequences of the *R. meliloti nifA* gene, there are other cases in which the roles of these sequences are in doubt. For instance, the *K. pneumoniae nifL* promoter lacks both an NtrC binding site and an upstream activator sequence, yet it is activated by NtrC and NifA. However, it has been demonstrated that deletion of regions upstream and very distant from the *nifL* promoter (farther than −160) result in a fourfold decrease in NtrC- and NifA-dependent activation (Drummond et al. 1983), suggesting that, like the upstream activator of the *nifF* promoter, the *nifL* upstream activator may be located further upstream than usual. There is a sequence in this region that differs from the strong NtrC binding site by only three bases and may serve as a weak NtrC binding site.

The preponderance of evidence to date suggests that both the *nif* consensus promoter (−26 to −12) and the upstream activator sequence are required for maximal expression of *nif* promoters. Mutational analysis of the *nifH* promoter by Brown and Ausubel (1984) demonstrates this point. They used the phenomenon called "*nif* inhibition" that is exhibited by high-copy-number plasmids containing strong *nif* promoters (Buchanan-Wallaston et al. 1981*b*, Riedel et al. 1983) to study the effect of mutations in the *nifH* promoter and upstream activator regions. Plasmids containing the *nifH* promoter in high copy number strongly inhibit the expression of chromosomal *nif* genes, presumably by titration of a limited number of positive regulator molecules (NifA). Point mutations at two sites in the consensus sequence of the *nifH* promoter relieved *nif* inhibition, as did a similar mutation at a site (−136) in the upstream activator sequence. A deletion of 112 bp from −184 to −72 prevented *nif* inhibition altogether.

The mechanism by which the upstream activator sequences modulate transcriptional activity is unknown. Nor is it known how binding of NtrC to these sequences facilitates the activation of Ntr-specific promoters. Several possible mechanisms are discussed by Gussin et al. (1986).

Role of oxygen in the regulation of symbiotic nitrogen fixation

For many years it has been suggested that there must exist a plant signal that elicits the activation of nitrogenase genes in the bacterial endosymbionts of plant-bacterial nitrogen-fixing symbioses. This seemed reasonable, especially in the case of rhizobia and bradyrhizobia, which are not generally capable of free-living nitrogen fixation. However, evidence is accumulating that indicates that the signal that activates symbiotic nitrogen fixation is an environmental condition rather than a specific signal derived from the plant host. The environmental stimulus appears to be the microaerobic environment that develops in the interior of root nodules. Ditta et al. (1987) constructed *nifA-lacZ* fusions in *R. meliloti* and demonstrated that they were transcriptionally activated in vegetative cells under reduced levels of oxygen. In addition, *nifA* expression and NifA function were not affected by the availability of fixed nitrogen under these conditions, nor did they require NtrC. It was also demonstrated that the activity of the *R. meliloti* NifA protein is greatly reduced in the presence of high concentrations of oxygen. Furthermore, they showed that the NifA N-terminal domain, which has been proposed to possess regulatory functions, can be deleted without affecting the oxygen-induced reduction in NifA activity. Evidence presented by Szeto et al. (1987) indicates that in *R. meliloti* the *ntrC* gene is not required for symbiotic nitrogen fixation, whereas in *K. pneumoniae* it is required for the activation of *nifLA*, thus supporting the proposition that regulation of nitrogen fixation in *R. meliloti* responds to different stimuli.

Evidence for the influence of oxygen on the regulation of nitrogen fixation in *B. japonicum* has come from the work of Fischer et al. (1986) and Regensburger et al. (1986). They demonstrated that expression of the *nifH* and *nifDK* operons in *B. japonicum* is sensitive to the level of oxygen and is activated under microaerobic conditions. In the presence of a constitutively expressed *B. japonicum* or *K. pneumoniae nifA* gene, Fischer and Hennecke (1987a) found that a *nifD-lacZ* fusion was expressed at a high level in both *E. coli* and *B. japonicum* under microaerobic conditions. However, under aerobic conditions this fusion was poorly expressed in the presence of *B. japonicum* NifA, whereas it was expressed at a level equal to that of microaerobic conditions in the presence of *K. pneumoniae* NifA. These results indicate that oxygen blocks the activation of *nif* promoters in *B. japonicum* by directly affecting the NifA protein. However, in *K. pneumoniae* oxygen is known to indirectly influence the activity of NifA by interacting with NifL (Merrick et al. 1982).

Recently, Thöny et al. (1987) determined that the *B. japonicum nifA* gene is transcribed from the promoter of the *fixR* gene that is located immediately 5' to the *nifA* gene. The *fixRnifA* operon was shown to be expressed under the following conditions: (1) in soybean nodules, (2) under anaerobic culture with nitrate as a terminal electron acceptor, (3) under microaerobic conditions, and (4) under aerobic culture. They further demonstrated that NifA is not itself an activator of the *fixR* promoter and neither is NtrC since no activation was seen in *E. coli* in the presence of *K. pneumoniae* NtrC and because *ntrC* mutants of *Azorhizobium sesbaniae* and *R. meliloti* are Fix$^+$ (Pawlowski et al. 1987, Szeto et al. 1987). Because the *fixRnifA* operon in *B. japonicum* is expressed under aerobic conditions, it is possible that the cell expresses this operon at a low level during aerobic growth so that NifA protein will be available should the environment become microaerobic, favoring nitrogen fixation activity. This type of mechanism might seem wasteful; however, a similar mechanism operates in *K. pneumoniae* during anaerobic nitrogen fixation, when NifL is synthesized in case a rapid repression of *nif* genes is required. It is also possible that a new transcriptional factor may control expression of the *fixRnifA* operon. A better understanding of the regulation of nitrogen fixation in rhizobial species depends upon understanding how expression of the *fixRnifA* operon is regulated.

An examination of the half-life of *nifA* mRNA from *B. japonicum* in aerobic and anaerobic conditions by Fischer et al. (1988) demonstrated that *nifA* mRNA has the same half-life in the presence of oxygen as in its absence. They further demonstrated that the NifA protein is irreversibly inactivated by oxygen. When free-living cells were induced to fix nitrogen under microaerobic conditions and subsequently switched to an aerobic environment, nitrogenase activity was rapidly eliminated and was not restored upon the reinstitution of a microaerobic environment in the presence of chloramphenicol. Fischer et al. (1988) also compared the sequences of the *nifA* genes from *B. japonicum, R. meliloti, R. leguminosarum*, and *K. pneumoniae* and reported that the N-terminal domain is poorly conserved among all organisms, but that there is high conservation in the central domain and moderate sequence conservation in the C-terminal domain. They found a short region encoding 30 amino acids between the central and C-terminal domains in the rhizobial species that was absent in *K. pneumoniae*, which they have called the "interdomain linker" (IDL). The independent conversion of two conserved Cys residues in the IDL to Ser resulted in an inactive NifA, but similar replacements of amino acid residues between these Cys residues had no effect on activity. Similar results were obtained for two conserved Cys residues in the C-terminal domain. In addition, reducing or increasing the number of amino acids between the Cys residues in the IDL resulted in complete or severe reduction in NifA activity, indicating that the spacing between the Cys residues is important.

Strong iron chelators were found to inhibit *B. japonicum* NifA-dependent *nif* gene activation but did not inhibit *K. pneumoniae* NifA activity (Fischer et al. 1988). Furthermore, the inhibitory effect of chelators on *B. japonicum* NifA was reversed by the addition of Fe^{2+} or Mn^{2+}. This has led these workers to hypothesize that the Cys residues in the IDL and C-terminal regions of NifA from rhizobial species chelate Fe^{2+} to create the proper configuration for NifA activity. In the presence of oxygen, the Fe^{2+} is oxidized to Fe^{3+}, inducing a conformational change in NifA that results in its inactivation.

A cowpea *Bradyrhizobium* species, strain 32H1, synthesizes nitrogenase, ammonium transport proteins, and an electrogenic K^+–H^+ antiporter in vitro under conditions of low oxygen concentration ($\leq 0.2\%$) (Gober and Kashket 1987). This phenomenon was observed only under conditions of adequate K^+ (8 to 12mM) and not when K^+ was limiting (50μM). When cells were grown with low K^+

and oxygen, the addition of K^+ led to the development of phenotypic properties associated with bacteroids, including nitrogenase activity (acetylene reduction), induction of an ammonium transport system and the K^+–H^+ antiporter, as well as increased synthesis of two heme biosynthetic enzymes, δ-aminolevulinate synthase and δ-amino-levulinate dehydrogenase. Further, addition of K^+ caused the repression of glutamine synthetase and of capsular polysaccharide synthesis. Similar results of K^+-mediated regulation of symbiotic functions were also observed in *B. japonicum*, suggesting that in addition to oxygen, potassium may also have a role in the regulation of these functions.

There are two plausible mechanisms for oxygen mediation of the activation of *nif* promoters by NifA. NifA may itself be modified by oxygen at specific amino acid residues or at a prosthetic group as proposed by Fischer et al. (1988). It is also possible that the NifA protein is modified by another protein or proteins that are sensitive to oxygen levels, similar to NifL in *K. pneumoniae*. However, there is no evidence of a *nifL* homologue in *B. japonicum* (Fischer and Hennecke 1987*b*, Fischer et al. 1986) so this hypothesis appears unlikely to be correct.

There is a growing body of evidence that indicates a correlation between the activities of enzymes that control the tertiary structure of DNA and gene expression in bacteria (reviewed by Drlica 1984 and McClure 1985). A relationship between DNA supercoiling and the activation of genes regulated by oxygen has recently been demonstrated. Novak and Maier (1987) have demonstrated that DNA gyrase is required for hydrogenase expression in *B. japonicum* under microaerobic conditions, and Kranz and Haselkorn (1986) have shown that the induction of nitrogenase genes in *Rhodopseudomonas capsulata* and *K. pneumoniae* is blocked in the presence of inhibitors of DNA gyrase. Evidence indicating that expression from the *K. pneumoniae* *nifL* promoter is dependent on DNA gyrase activity, while expression from other *nif* promoters is not, was recently reported by Dimri and Das (1988). Clearly, further research is required in order to better understand the mechanisms of how oxygen affects the expression of nitrogen fixation and other symbiotic functions in symbiotic nitrogen-fixing bacteria.

Regulation of dicarboxylic acid transport

It has long been felt that succinate or other dicarboxylic acids serve as the rhizobial symbiont's source of carbon and energy in the nodule (reviewed by Guerinot and Chelm 1987). Recent support for this hypothesis has been derived from mutations in *Rhizobium* that eliminate the ability to transport dicarboxylic acids and result in ineffective symbiotic nitrogen fixation (Bolton et al. 1986, Engelke et al. 1987, Finan et al. 1983, Ronson et al. 1981). Bolton et al. (1986) isolated *R. meliloti* mutants that were incapable of growth on succinate and produced Fix⁻ nodules. Similarly, Engelke et al. (1987) demonstrated by inhibition studies that in free-living *R. meliloti*, succinate, fumarate, and malate are transported by a common active transport system that is induced by succinate. They isolated seven *Tn5* insertion mutants that were incapable of dicarboxylic acid transport (*dct*), and of these, five induced nodule formation but demonstrated only 50% of the wild-type nitrogen fixation levels. Bacteroids isolated from nodules induced by these mutants transported succinate and malate at about 30 to 50% of the level of wild-type bacteroids. The other two *dct* mutants formed small, white, ineffective, or Fix,⁻ nodules, and transport of succinate and malate could not be detected in bacteroids from these nodules.

The structural gene for C_4-dicarboxylic acid transport, *dctA*, was isolated from *R. leguminosarum* by Ronson and Astwood (1985), and it was found to contain a *nif* consensus promoter sequence in the region 5' to the coding sequence. This led to the suspicion that the *dct* genes may be under the control of the *ntr* regulatory system. Subsequently, Ronson and Astwood (1985) isolated the *dct* regulatory gene, *dctD*, and demonstrated that it codes for a protein that is closely related to NtrC and NifA. In subsequent work, Ronson et al. (1987) cloned an *ntrA*-like gene from *R. meliloti* and showed that its protein product was required for C_4-dicarboxylate transport, the assimilation of nitrate, and symbiotic nitrogen fixation. Comparison of the DNA sequence of this gene with that of the *K. pneumoniae ntrA* gene showed that the *R. meliloti* gene shared 38% homology with the *K. pneumoniae* gene. Watson et al. (1988) have demonstrated that genes involved in C_4-dicarboxylate transport in *R. meliloti* are located on a megaplasmid that also carries genes that have a role in the biosynthesis of exopolysaccharides.

Further experiments that demonstrate the involvement of *dct* genes in nitrogen fixation have been reported by Birkenhead et al. (1988), who transferred *R. meliloti dct* genes to a *B. japonicum* strain. Expression of the *R. meliloti dct* genes in *B. japonicum* resulted in increased growth rates in media containing dicarboxylic acids as the sole carbon source and increased succinate uptake under aerobic conditions. Under free-living nitrogen-fixing conditions, the *B. japonicum* strain containing the *R. meliloti dct* genes demonstrated increased levels of nitrogenase activity over the parent strain without these genes. From this work it is unclear whether the increase in nitrogenase activity is due to the introduction of a more effective *dct* system than may be present in *B. japonicum*, or to an increase in *dct* expression through increased gene copy number.

Organization of *nif* Genes in Symbiotic Bacteria

Rhizobium meliloti

In the *R. meliloti* strains studied thus far, the *nif* genes that have been mapped are located on a symbiotic megaplasmid (Ausubel et al. 1985). Mutations that map in the *R. meliloti* chromosome or to a second megaplasmid have been shown to result in Fix⁻ phenotypes (Finan et al. 1986, Forrai et al. 1983, Hynes et al. 1986), indicating that some genes required for symbiotic nitrogen fixation may be located there. The organization of the *R. meliloti*, *B. japonicum*, and *Anabaena nif* genes identified to date is shown in Figure 3, which should be used as a reference in this discussion. A cluster of *nif* structural genes has been identified in *R. meliloti* by homology with the *K. pneumoniae* *nifHDK* genes (Corbin et al. 1982). The *R. meliloti* *nifA* gene was mapped approximately 5.0 kb upstream of the *nifHDK* gene cluster (Szeto et al. 1984). In a central region of the protein encompassing 200 amino acids, it shares 50% homology with both the *K. pneumoniae* *nifA* and *ntrC* genes, which have been shown to share this same region of homology (Buikema et al. 1985). Since the growth of *R. meliloti* strains with mutations in the *nifA* gene region is not impaired on media containing proline, histidine, arginine, or glutamine as sole nitrogen source, the *nifA* gene is likely to be the *R. meliloti* regulatory gene that is specific for the *nif* gene system similar to *K. pneumoniae* *nifA*. By transposon mutagenesis, a *R. meliloti* gene essential for nitrogen fixation was identified directly downstream of the *nifA* gene (Buikema et al. 1987). The DNA sequence of this gene region was determined and compared to the sequence of *K. pneumoniae* *nifBQ* genes since they are located in a similar location relative to *nifA*. The *R. meliloti* gene downstream of the *nifA* gene shares about 50% DNA base homology with the *K. pneumoniae* *nifB* gene and is therefore likely to be the *nifB* gene of *R. meliloti*. No homology was found with the *K. pneumoniae* *nifQ* gene in the region downstream from the *R. meliloti* *nifB* gene, indicating that the analogous gene in *R. meliloti* is not located in this region.

Another cluster of genes in *R. meliloti* that is required for symbiotic nitrogen fixation was identified and mapped in the region between the *nifHDK* and *nifAB* gene clusters and is called *fixABC* (Pühler et al. 1984). Genetic analyses of the *fixABC* genes indicate that the genes

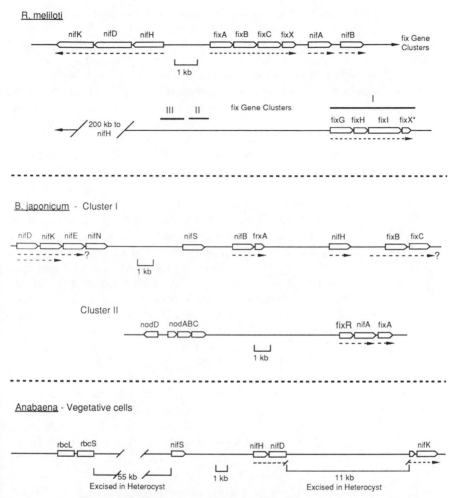

Figure 3 Organization of genes involved in symbiotic nitrogen fixation in *Rhizobium meliloti, Bradyrhizobium japonicum*, and *Anabaena*. The broken arrows below genes indicate the transcriptional relationships of the genes where known.

comprise a single operon (Better et al. 1983, Earl et al. 1987, Ruvkun et al. 1982) and that the *fixA* promoter contains a characteristic *nif* consensus promoter sequence (Better et al. 1983) and is activated by *R. meliloti* NifA (Pühler et al. 1984). The role of these genes is unknown. However, the DNA sequence of the *fixC* gene predicts that the FixC protein contains a hydrophobic region for membrane insertion (Earl et al. 1987). The entire *fixABC* operon has been sequenced, and no homology has been detected between these genes and any *K.*

pneumoniae nif gene (Earl et al. 1987). However, *fixABC* homology was detected in all rhizobial species examined, including *R. leguminosarum, R. trifolii, B. japonicum,* and *Bradyrhizobium* from *Parasponia* (Earl et al. 1987). Therefore, it has been suggested that these genes have functions that are important during symbiotic nitrogen fixation but are not required for free-living fixation as in *K. pneumoniae.* However, homology to the *R. meliloti fixABC* genes has been detected in the free-living nitrogen-fixing bacteria *Azotobacter vinelandii* (Gubler and Hennecke 1986, Earl et al. 1987) and *Azotobacter chroococcum* (Evans et al. 1988), indicating that the *fixABC* genes may have functions that are more general in nature than originally believed. Fix⁻ *R. meliloti* transposition mutants were generated with the indigenous transposable element *ISRm2* (Dusha et al. 1987). One of these insertions mapped immediately downstream of the *fixC* gene, and DNA sequence analysis of this region identified a short open reading frame (ORF) in this area. It has been designated *fixX*, and its protein product contains a cluster of cysteine residues that are characteristic of ferredoxins. It has been suggested by Dusha et al. (1987) that the function of the *fixX* gene product is the transfer of electrons to nitrogenase, similar to the *nifF* gene in *K. pneumoniae.* A *fixX* gene has now been found in a similar location in other fast-growing rhizobia including *R. leguminosarum* (Gronger et al. 1987) and *R. trifolii* (Iismaa and Watson 1987). The *fixX* gene in *R. meliloti* appears to be a part of the *fixABC* operon (Earl et al. 1987).

Several groups have examined the megaplasmid pSym of *R. meliloti* by mutagenesis in order to find other *fix* genes that may be required for symbiotic nitrogen fixation. One such *fix* gene region containing three *fix* complementation groups was discovered and mapped to an area of the Sym plasmid approximately 200 kb from the *nifH* gene (Batut et al. (1985a,b). Kahn et al. (1987) have demonstrated that these three *fix* gene clusters are in a 12-kb region and that a 5-kb region of one cluster is duplicated elsewhere in the *R. meliloti* genome and contains functional *fix* genes. In addition, the *fix* genes within this duplicated region are transcribed only during symbiosis and independently of NifA. These genes were also shown to be highly conserved among fast-growing rhizobia (Kahn et al. 1987). Through a more detailed examination of the new *fix* gene cluster, Kahn et al. (1987) have identified four *fix* genes in this cluster. They have named these *fixG, H, I,* and *X*, and all are on a single transcriptional unit, which has been sequenced, and these data suggest that they encode membrane-integrated proteins that may be involved in a redox process that is specific to symbiotic nitrogen fixation. Since this gene cluster is distinct from the *fixABCX* gene cluster near *nifHDK*, consideration should be given to renaming the *fixX* gene in this new cluster to avoid confusion.

Bradyrhizobium japonicum

Unlike the fast-growing rhizobia, the *nif* and *fix* genes of *B. japonicum* are located on the chromosome, not on plasmids. Many of the genes required for symbiotic nitrogen fixation in *B. japonicum* have been mapped to two unlinked gene clusters (Figure 3) that are located in a highly specialized region of the chromosome that does not contain genes essential for growth (Adams et al. 1984, Kaluza et al. 1985). At least five other regions of the *B. japonicum* chromosome contain genes with functions that affect symbiotic nitrogen fixation and have been identified by random *Tn5* mutagenesis (Regensberger et al. 1986).

Symbiotic gene cluster I contains the structural *nif* genes in addition to other *nif* genes. The organization of the *nif* structural genes in *B. japonicum* differs from that found in *R. meliloti* as the *nifH* and *nifDK* genes are separated by 17 kb of intervening DNA (Fischer and Hennecke 1984, Kaluza et al. 1983). The DNA between the *nifDK* and *nifH* genes has been extensively mutagenized in order to identify other genes in this area that are required for nitrogen fixation (Ebeling et al. 1987, Noti et al. 1986). This region has also been analyzed for interspecies homology to *nif* genes from *K. pneumoniae* (Ebeling et al. 1987). Through these means, the *nifE*, *nifN*, *nifS*, and *nifB* genes were identified in cluster I, and mutations in each of these resulted in Nif⁻, Fix⁻ phenotypes, except in the case of *nifS*, which proved to be leaky.

The *nifE* gene in *B. japonicum* is located immediately downstream from the *nifK* gene, and Ebeling et al. (1987) have determined that it is expressed from the *nifD* promoter. However, mRNA corresponding to *nifDK*, and not including *nifE*, was detected, so there may be transcription termination signals between *nifK* and *nifE* that result in less *nifDKE* message than *nifDK* message (Hennecke et al. 1987). It is not known whether the *nifN* gene that is located 3 to *nifE* is also expressed from the *nifD* promoter. The *nifS* gene is located approximately 6 kb 3' to the *nifN* gene, and the *nifB* gene is found about 1 kb downstream from it (Ebeling et al. 1987). The promoter and transcription start site for the *nifS* gene have been identified and mapped. The *nifS* promoter contains a perfect *nif* consensus promoter and is preceded by a single putative upstream activator sequence (Ebeling et al. 1987). Examination of the DNA sequence immediately downstream of the *nifB* gene revealed the presence of a small ORF that has the capability to encode a polypeptide of 74 amino acids (Ebeling et al.

1988). A high degree of homology was found between this ORF and ferredoxin genes from several diverse bacteria, and, therefore, it has been designated *"frxA."* A *nifB-frxA-lacZ* translational fusion containing the *nifB* promoter was shown to be activated in the presence of *K. pneumoniae NifA* in *E. coli*, indicating that *frxA* is cotranscribed with *nifB* from the *nifB* promoter. Mutagenesis of *frxA* demonstrated that it is not absolutely required for nitrogen fixation in bacteroids, but is required for optimal levels of fixation in free-living cells. Therefore, it appears to have a role in nitrogen fixation, but the nature of its role is unknown.

Also located on cluster I are the *fixB* and *fixC* genes. These genes are located adjacent to each other approximately 2.5 kb downstream from the *nifH* gene and are organized on a single transcriptional unit (Hennecke et al. 1987). The *fixA* gene is located apart from the *fixBC* genes in gene cluster II (Hennecke et al. 1987), and a consensus *nif* promoter sequence has been found in the *fixA* promoter region (Fuhrmann et al. 1985). Gubler and Hennecke (1988) recently identified the *fixB* promoter and demonstrated the presence of a consensus *nif* promoter sequence within it. They also conducted transcriptional mapping of the *fixBC* and *fixA* operons and demonstrated that the transcription of both is dependent on the presence of the *nifA* gene and on low concentrations of oxygen. The *fixA*, *fixB*, and *fixC* genes in *B. japonicum* and in *Azorhizobium sesbaniae* are indispensable not only for symbiotic fixation, but also for free-living fixation under microaerobic conditions (Gubler and Hennecke 1986, Donald et al. 1986). Homology to *fixA*, *fixB*, and *fixC* has thus far been detected only in rhizobial strains and in nonrhizobial diazotrophs that fix nitrogen aerobically or microaerobically such as *Azospirillum brasilense* (Fogher et al. 1985) and *Azotobacter vinelandii* (Gubler and Hennecke 1986, Earl et al. 1987). No such homology has been detected with DNA from *K. pneumoniae*, an organism that fixes nitrogen only in anaerobic environments. Therefore, it is likely that these genes have functions that involve the transport of electrons to nitrogenase that are unique to organisms that fix nitrogen in the presence of oxygen.

In addition to the *fixA* gene, the cluster II gene region contains the *fixRnifA* operon (Fischer et al. 1986, Hennecke et al. 1987). The *fixR* gene encodes a polypeptide consisting of 278 amino acids, whose function is unknown (Thöny et al. 1987). However, a deletion of the *fixR* gene that permitted transcription of *nifA*, reduced symbiotic fixation by 50%. The *fixR* promoter region contains a consensus *nif* promoter sequence and a region (at -50 to -148) that is required for maximal expression. Also located on the cluster II symbiotic gene region are the common nodulation genes that are required for the formation of nodules by *B. japonicum* (Fischer et al. 1986, Hennecke et al. 1987).

Anabaena

Employing the interspecies homology among *nif* genes, Rice et al. (1982) isolated clones that contained the *nifH, nifD, nifK,* and *nifS* genes from vegetative cells of *Anabaena* strain 7210. Subsequent analysis of these genes indicated that the *nifH* and *nifD* genes were adjacent to each other and the *nifK* gene was located about 11 kb to the 3' end of the *nifD* gene. At a later time, the same genes were isolated from purified heterocysts, the specialized cells where nitrogenase activity is found. In this case it was demonstrated that the *nifH, nifD,* and *nifK* genes were organized together in a single transcriptional unit that was expressed from a promoter located 5' to the translational start site of the *nifH* gene (Golden et al. 1985). Further investigation of this discrepancy has led to the discovery that the 11 kb of DNA separating the *nifD* and *nifK* genes is excised during the differentiation of vegetative cells to nitrogen-fixing heterocysts. DNA sequence analysis of the border regions on each end of the excised 11 kb DNA fragment revealed that in vegetative DNA this segment of DNA interrupts the coding sequence of the *nifD* gene such that a small portion of the 3' end of the *nifD* gene is located adjacent to the 5' end of the *nifK* gene (Golden et al. 1985). Therefore, the excision of this 11 kb fragment restores the integrity of the *nifHDK* operon. It was noticed that plasmids containing a 17 kb vegetative DNA fragment from *Anabaena* that includes the *nifK* gene, the 11 kb excision fragment, and part of the *nifD* gene underwent excision of the 11 kb element in a *recA E. coli* (Lammers et al. 1986). This was used to construct a genetic system to study the excision process in *E. coli*. A *lacZ* gene was cloned into the 11 kb excision element so that excision would lead to a β-galactosidase-negative phenotype. Mutagenesis of the 11 kb excision element led to the creation of mutants that were incapable of excising the 11 kb DNA element, thus demonstrating that the genetic information required for the excision event was encoded within the 11 kb element. This led to the identification of the *xisA* gene that is responsible for the excision event, and the location of this gene was determined to be 240 bp from the recombination site near the *nifK* gene (Golden 1987, Lammers et al. 1986). An 11 bp direct repeat has been found at the recombination site on each end of the 11 kb excision element.

The *nifS* gene of *Anabaena* has been mapped to a region upstream from the *nifHD* genes, and surprisingly, a second DNA rearrangement has recently been discovered that occurs near the 5' end of the *nifS* gene and results in the placement of the *rbcL* and *rbcS* genes about 10 kb from the reconstituted *nifHDK* operon (Golden 1987). The size of the excision in this case is 55 kb, and the break points have been cloned and sequenced and bear no resemblance to those involved in the *nifD-nifK* excision event (Golden 1987). This suggests that this rearrangement is catalyzed by a site-specific recombination enzyme other than that encoded by *xisA*. Golden et al. (1987) have also presented evidence that indicates transcription of the *nifS* gene is dependent upon excision of the 55-kb element, which suggests that this rearrangement results in a promoter sequence being juxtaposed with the *nifS* gene.

A second copy of the *nifH* gene, called *nifH**, has been identified in *Anabaena*, but it is not linked to the *nifHD-nifK* gene region, nor is it linked to a second copy of *nifD* (Haselkorn 1986). There is an additional ORF between *nifH** and its promoter, but the function of this potential gene is unknown. The regulation of expression of the *nifH** gene appears to be identical to that of the *nifHDK* operon.

Frankia

Due to the lack of genetics systems in *Frankia* and the difficulty of working with these bacteria, very little is known about the genes involved in nitrogen fixation in this group of organisms. In a survey of interspecies homology of DNA from a wide range of diazotrophic bacteria to the structural *nif* genes from *K. pneumoniae*, Ruvkun and Ausubel (1980) demonstrated that there was homology to these genes in DNA from a *Frankia* strain. Simonet et al. (1986) utilized this homology to demonstrate the presence of DNA sequences homologous to the *nifHDK* gene fragment from *K. pneumoniae* in chromosomal DNA and on a 190-kb plasmid in a *Frankia* strain isolated from *Alnus rugosa*. This indicates that the *nifHDK* genes in this strain may be reiterated, or that they are split, with one or more of the genes on the plasmid and the others on the chromosome. Ligon and Nakas (1987) recently reported the cloning and mapping of four *nif* genes from a *Frankia* isolate. These included the *nifD* and *nifK* genes, which are contiguous and appear to be cotranscribed with *nifD* being promoter-proximal to *nifK* (Ligon and Nakas 1988). The *nifH* gene was shown not to be closely linked to the *nifDK* gene cluster. They also probed the region surrounding the *nifDK* gene cluster, as well as total genomic DNA, with gene probes from *K. pneumoniae* that collectively included all other known *nif* genes from that organism in order to locate other *Frankia nif* genes. Only a DNA fragment containing the *nifE* and *nifN* genes demonstrated homology to *Frankia* DNA. The *nifE* and *nifN* genes were mapped and have a similar location in the region immediately downstream from *nifK* as in *K. pneumoniae, A. vinelandii, B. japonicum,* and other nitrogen-fixing bacteria. Contrary to these findings, Simonet and coworkers have identified two *nif* gene clusters by homology with *K. pneumoniae* genes in a different *Frankia* strain isolated from *Alnus* (personal communication). One of these clusters contains the *nifH, nifD,* and *nifK* genes that are contiguously located, and the second cluster appears to contain the *nifB* and *nifA* genes that also map next to each other. Recently, Prakash and Cummings (1988) demonstrated that the *nifHDK* gene fragment of *K. pneumoniae* hybridizes to a single *BamHI* restriction fragment from the DNA of a *Frankia* isolate, thus indicating that in this strain the *nifH, nifD,* and *nifK* genes are clustered. Similar evidence for a *nifH, nifD, nifK* gene cluster in several different *Frankia* isolates was recently reported by Normand et al. (1988). Therefore, it appears likely that, as in the organization of the *nif* structural genes in different rhizobia, *Frankia nif* genes may be clustered in most strains but may not be clustered in other strains.

Progress in resolving the lack of genetic transfer mechanisms and the concomitant difficulty in studying the genetics of nitrogen fixation in *Frankia* may come from a unique approach reported recently by Prakash and Cummings (1988). Through protoplast fusion between *Frankia* and *Streptomyces,* a closely related actinomycete for which there are well-developed systems of genetic transfer, they selected fusants that were capable of rapid growth, a characteristic of *Streptomyces,* on a nitrogen-free medium, a characteristic of *Frankia.* Fusants with these characteristics were isolated and all demonstrated the ability to reduce acetylene, with a few also capable of inducing effective nodules on the proper actinorhizal host. They presented evidence suggesting that these clones had *Streptomyces* genetic backgrounds and may, therefore, be derived from a process in which all of the genes required for nitrogen fixation in *Frankia* have stably recombined into the *Streptomyces* chromosome. If these results are verified, this approcah might be used to study the genetics of nitrogen fixation in *Frankia* in a *Streptomyces* genetic background.

Future Prospects for Increasing Symbiotic Nitrogen Fixation

Symbiotic nitrogen fixation is widely recognized as a trait that is beneficial to the plant host, and many of those plants capable of forming this type of symbiosis have agronomic and/or ecological importance. It is therefore only natural that mechanisms for increasing nitrogen fixation within the symbiotic relationship, or extending the ability to fix nitrogen to plants outside of symbiotic relationships, are of tremendous interest. With the development of new genetic techniques over the past decade and the increased understanding of the processes involved in symbiotic nitrogen fixation that have resulted from their judicious application, these goals, though still untested, do not appear as elusive as they once seemed. Possible approaches for increasing plant-associated nitrogen fixation will now be discussed in more detail.

The most promising avenues for increasing symbiotic nitrogen fixation in the short term are those that focus on the process of nitrogen fixation in the bacterial partner, since this is the aspect of the symbiosis about which there is a greater understanding, and which is most easily manipulated. One such approach that is close to being tested is the attainment of increased symbiotic nitrogen fixation by increasing the number of nitrogen fixation genes in the endosymbiont with a concomitant increase in the number of active nitrogenase molecules. A company with interests in agricultural biotechnology has recently received approval to field-test a *R. meliloti* strain that has been genetically manipulated to contain increased copies of *nif* genes. They have reported that in greenhouse studies the modified *R. meliloti* strain demonstrates elevated levels of symbiotic nitrogen fixation activity. Increasing the copy number or the expression of the *nifH* gene in particular may be the most successful of these types of approaches. Since the rate-limiting step in nitrogen fixation appears to be the cycle of binding reduced dinitrogenase reductase, the *nifH* gene product, to dinitrogenase followed by a one-electron transfer, it has been proposed that increased copies of the *nifH* gene and its protein product may result in increasing the turnover rate of nitrogenase. An excess of the *nifH* gene product has been shown to be required for maximal nitrogen fixation activity in *K. pneumoniae* (Thorneley and Lowe, 1984). This may be the basis for the presence of more than one copy of the *nifH* gene in some diazotrophs such as *Azotobacter vinelandii* (Jacobson et al. 1986), *Rhizobium phaseoli* (Quinto et al. 1982, 1985), and *Azorhizobium sesbaniae* (Norel and Elmerich 1987). However, *B. japonicum*, *Anabaena*, *Frankia*, and other symbiotic diazotrophs appear to contain a single copy of *nifH*, and these may benefit most from this strategy.

Nitrogen fixation in the *Rhizobium*-legume symbiosis is thought to be limited by the amount of plant-derived photosynthate available to bacteroids (Hardy and Havelka 1975, Ryle et al. 1979). The work of Birkenhead et al. (1988) suggests that increasing the ability of the endosymbiont to utilize photosynthate in the nodule may lead to increased fixation rates. These investigators transferred *R. meliloti dct* genes to *B. japonicum* and demonstrated a 50% increase in the uptake of C_4-dicarboxylic acids, increased growth rates on succinate as the sole carbon source, and a corresponding increase in the levels of symbiotic nitrogen fixation. These results suggest that increases in nitrogen fixation may result from an increase in the number of *dct* genes, or from the introduction of *dct* genes from other species that are inherently more effective than the indigenous *dct* system.

Introducing a more efficient nitrogenase enzyme into symbiotic diazotrophs is another approach with some merit. The specific activity of nitrogenase in vivo, and of purified components in vitro, differs among diazotrophs. Nitrogenase activities in free living *A. vinelandii* have been measured by acetylene reduction and are about 80 nanomoles of C_2H_4 per milligram of protein per minute (Upchurch and Mortenson 1980), whereas similar measurements for *B. japonicum* in soybean nodules show much lower rates of about 6 nanomoles per milligram of protein per minute (Wittenberg et al. 1974). This difference cannot be accounted for solely by differences in the genetic or biochemical environments of these nitrogenase systems, since it has been demonstrated that purified nitrogenase from *A. vinelandii* has a specific activity that is approximately threefold higher than that of *B. japonicum* (Burgess et al. 1980, Israel et al. 1974). Therefore, it may be possible to attain increased symbiotic nitrogen fixation by introducing a more active nitrogenase from another diazotroph, or similarly, by replacing the indigenous nitrogenase system with a more active nitrogenase. There is a significant history of the transfer of nitrogenase genes to other bacteria that results in the transfer of nitrogen fixation ability to bacteria that were previously unable to fix nitrogen (reviewed by Robson et al. 1983).

One often mentioned and attractive target for future efforts in this field is the creation of plants capable of fixing nitrogen independent of a symbiotic relationship. This goal will require much more than just the transfer of the genes necessary for nitrogen fixation from a bacterium to a plant. One serious obstacle to this approach is the necessity to provide protection of the oxygen-labile nitrogenase enzyme from oxygen in an aerobic organism. One possible solution to this problem that has been proposed is to introduce the nitrogen fixation apparatus into chloroplasts in which the oxygen-generating activity of photosystem II has been inactivated. This would require that only a portion of the chloroplasts be modified so as not to have a severely detrimental effect on the photosynthetic capabilities of the plant.

Perhaps a more feasible approach is to extend the ability of forming a symbiotic nitrogen-fixing relationship to plant species that presently are unable to do so. Several aspects unique to *Frankia* could make this group of bacteria a logical starting point for this strategy. First, *Frankia* as a group has a much broader host range than do rhizobial species. *Frankia* form a symbiotic relationship with over 19 genera of dicotyledonous plants from seven families (Akkermans and Houwers 1979). The application of breeding techniques and tissue culture methodologies, to introduce the ability to enter into a symbiotic relationship with *Frankia* into plant species is one potential avenue to this end. *Betula* is a plant genus of some economic potential that is closely related to actinorhizal species and so could be a good candidate on which to test this approach. A second feature of *Frankia* that makes it better suited to this approach is that since it deals with the problem of oxygen protection of nitrogenase by the sequestering of nitrogenase in vesicles, potential plant hosts would not have to acquire this function. However, given the limited state of our understanding of the plant-microbe symbiosis in the plant host, it would be naive to think that this goal is readily attainable. This point underscores the great disparity that exists between the level of understanding of various aspects of nitrogen fixation on the bacterial side, compared to the plant side, of the symbiosis. Since the ability to fix nitrogen is a bacterial trait, and bacteria can be manipulated genetically much more easily than plants, our understanding of bacterial processes germane to the symbiosis is much greater than our understanding of contributions of the plant partner. Ultimately, increasing nitrogen fixation in plants by any means will surely require a better understanding and appreciation of the contributions and needs of the plant partner in the symbiosis.

References

Adams, T. H., McClung, C. R., and Chelm, B. K. 1984. Physical organization of the *Bradyrhizobium japonicum* nitrogenase gene region, *J. Bacteriol.* 159:857–862.

Akkermans, A. D. L., and Houwers, A. 1979. Symbiotic nitrogen fixers available for use in forestry. In *Symbiotic Nitrogen Fixation in the Management of Temperate Forests*, J. C. Gordon, C. T. Wheeler, and D. A. Perry (eds.). Oregon State University Press, Corvalis, pp. 23–35.

Alvarez-Morales, A., and Hennecke, H. 1985. Expression of *Rhizobium japonicum nifH* and *nifDK* operons can be activated by the *Klebsiella pneumoniae nifA* protein but not by the product of *ntrC, Mol. Gen. Genet.* 199:306–314.

Alvarez-Morales, A., Betancourt-Alvarez, M., Kaluza, K., and Hennecke, H. 1986. Activation of the *Bradyrhizobium japonicum nifH* and *nifDK* operons is dependent on promoter-upstream DNA sequences, *Nucleic Acids Res.* 14:4208–4227.

Alvarez-Morales, A., Dixon, R., and Merrick, M. 1984. Positive and negative control of the *glnA-ntrBC* regulon in *Klebsiella pneunomiae, EMBO J.* 3:501–507.

Appleby, C. A. 1984. Leghemoglobin and *Rhizobium* respiration, *Annu. Rev. Plant Physiol.* 35:443–478.

Ausubel, F. M. 1984. Regulation of nitrogen fixation genes, *Cell* 37:5–6.

Ausubel, F. M., and Cannon, F. C. 1981. Molecular genetic analysis of Klebsiella pneumoniae nitrogen fixation (nif) genes, *Cold Spring Harbor Symp. Quant. Biol.* 45:487–499.

Ausubel, F. M., Buikema, W. J., Earl, C. D., Klingensmith, J. A., Nixon, B. T., and Szeto, W. W. 1985. Organization and regulation of *Rhizobium meliloti* and *Parasponia Bradyrhizobium* nitrogen fixation genes. In *Nitrogen Fixation Research Progress*, H. J. Evans, P. J. Bottomley, and W. E. Newton (eds.). Martinus Nijhoff, Dordrecht, The Netherlands, pp. 165–171.

Batut, J., Terzaghi, B., Gherardi, M., Huguet, M., Tezaghi, E., Garnerone, A. M., Boistard, P., and Huguet, T. 1985a. Localization of a symbiotic *fix* region more than 200 kilobases from the *nod-nif* region, *Mol. Gen. Genet.* 199:232–239.

Batut, J., Boistard, P., Debellé, F., Dénarié, J., Ghai, J., Huguet, T., Infante, D., Martinez, E., Rosenberg, C., Vasse, J., and Truchet, G. 1985b. Developmental biology of the *Rhizobium meliloti*-alfalfa symbiosis: a joint genetic and cytological approach. In *Nitrogen Fixation Research Progress*, H. J. Evans, P. J. Bottomley, and W. E. Newton (eds.). Martinus Nijhoff, Dordrecht, The Netherlands, pp. 109–115.

Better, M., Ditta, G., and Helinski, D. R. 1985. Deletion analysis of *Rhizobium meliloti* symbiotic promoters, *EMBO J.* 4:2419–2424.

Better, M., Lewis, B., Corbin, D., Ditta, G., and Helinski, D. R. 1983. Structural relationships among *Rhizobium meliloti* symbiotic promoters, *Cell* 35:479–485.

Beynon, J., Cannon, M., Buchanon-Wallaston, V., and Cannon, F. 1983. The *nif* promoters of *Klebsiella pneumoniae* have a characteristic primary structure, *Cell* 34:665–671.

Beynon, J. L., Williams, M. K., and Cannon, F. C. 1988. Expression and functional analysis of the *Rhizobium meliloti nifA* gene, *EMBO J.* 7:7–14.

Birkenhead, K., Manian, S. S., and O'Gara, F. 1988. Dicarboxylic acid transport in *Bradyrhizobium japonicum*: use of *Rhizobium meliloti dct* gene(s) to enhance nitrogen fixation, *J. Bacteriol.* 170:184–189.

Bloom, F. R., Levin, M. S., Foor, F., and Tyler, B. 1977. Regulation of glutamine synthetase formation in *Escherichia coli*: characterization of mutants lacking the uridylyltransferase, *J. Bacteriol.* 134:569–577.

Bogusz, D., Houmard, J., and Aubert, J. 1981. Electron transport to nitrogenase in *Klebsiella pneumoniae*, *Eur. J. Biochem.* 120:421–426.

Bolton, E., Higgisson, B., Harrington, A., and O'Gara, F. 1986. Dicarboxylic acid transport in *Rhizobium meliloti*: isolation of mutants and cloning of dicarboxylic acid transport genes, *Arch. Microbiol.* 144:142–146.

Brigle, K. E., Newton, W. E., and Dean, D. R. 1985. Complete nucleotide sequence of the *Azotobacter vinelandii* nitrogenase structural gene cluster, *Gene* 37:37–44.

Brown, S. E., and Ausubel, F. M. 1984. Mutations affecting regulation of the *Klebsiella pneumoniae nifH* (nitrogenase reductase) promoter, *J. Bacteriol.* 157:143–147.

Buchanan-Wallaston, V., Cannon, M. C., Beynon, J. L., and Cannon, F. C. 1981a. Role of the *nifA* gene product in the regulation of *nif* expression in *Klebsiella pneumoniae*, *Nature* 294:776–778.

Buchanan-Wallaston, V., Cannon, M. C., and Cannon, F. C. 1981b. The use of cloned *nif* (nitrogen fixation) DNA to investigate transcriptional regulation of *nif* expression in *Klebsiella pneumoniae*, *Mol. Gen. Genet.* 184:102–106.

Buck, M., and Cannon, W. 1987. Frameshifts close to the *Klebsiella pneumoniae nifH* promoter prevent multicopy inhibition by hybrid *nifH* plasmids, *Mol. Gen. Genet.* 207:492–498.

Buck, M., Khan, H., and Dixon, R. 1985. Site-directed mutagenesis of the *Klebsiella pneumoniae nifL* and *nifH* promoters and in vivo analysis of promoter activity, *Nucleic Acids Res.* 13:7621–7638.

Buck, M., Miller, S., Drummond, M., and Dixon, R. 1986. Upstream activator sequences are present in the promoters of nitrogen fixation genes, *Nature* 320:374–378.

Buck, M., Woodcock, J., Cannon, W., Mitchenall, L., and Drummond, M. 1987. Positional requirements for the function of *nif*-specific upstream activator sequences, *Mol. Gen. Genet.* 210:140–144.

Bueno, R., Pahel, G., and Magasanik, B. 1985. Role of *glnB* and *glnD* gene products in regulation of the *glnALG* operon of *Escherichia coli*, *J. Bacteriol.* 164:816–822.

Buikema, W. J., Klingensmith, J. A., Gibbons, S. L., and Ausubel, F. M. 1987. Conservation of structure and location of *Rhizobium meliloti* and *Klebsiella pneumoniae nifB* genes, *J. Bacteriol.* 169:1120–1126.

Buikema, W. J., Szeto, W. W., Lemley, P. V., Orme-Johnson, W. H., and Ausubel, F. M. 1985. Nitrogen fixation specific regulatory genes of *Klebsiella pneumoniae* and *Rhizobium meliloti* share homology with the general nitrogen regulatory gene *ntrC* of *K. pneumoniae*, *Nucleic Acids Res.* 13:4539–4555.

Burgess, B. K., Jacobs, D. B., and Stiefel, E. I. 1980. Large-scale purification of high activity *Azotobacter vinelandii* nitrogenase, *Biochem. Biophys. Acta.* 614:196–209.

Burris, R. H. 1980. The global nitrogen budget: science or seance? In *Nitrogen Fixation*, W. E. Newton and W. H. Orme-Johnson (eds.), vol. I. University Park Press, Baltimore, pp. 7–16.

Callaham, D., Del Tredici, P., and Torrey, J. 1978. Isolation and cultivation in vitro of the actinomycete causing root nodulation in *Comptonia, Science* 199:899–902.

Calvert, H., Chaudary, A., and Lalonde, M. 1979. Structure of an unusual root nodule symbiosis in a nonleguminous herbaceous dicotyledon. In *Symbiotic Nitrogen Fixation in the Management of Temperate Forests*, J. C. Gordon, C. T. Wheeler, and D. A. Perry (eds.). Oregon State University Press, Corvallis, Oregon, pp. 474–475.

Cannon, F., Beynon, J., Buchanan-Wollaston, V., Burghoff, R., Cannon, M., Kwiatkowski, R., Laurer, G., and Rubin, R. 1985. Progress in understanding organization and expression of *nif* genes in *Klebsiella.* In *Nitrogen Fixation Research Progress*, H. J. Evans, P. J. Bottomley, and W. E. Newton (eds.). Martinus Nijhoff, Dordrecht, The Netherlands, pp. 453–460.

Castano, I., and Bastaracchea, F. 1984. *glnF-lacZ* fusions in *Escherichia coli*: studies on *glnF* expression and its chromosomal orientation, *Mol. Gen. Genet.* 195:228–233.

Chen, K., Chen, C. -K., and Johnson, J. L. 1986. Structural features of multiple *nifH*-like sequences and very biased codon usage in nitrogenase genes of *Clostridium pasteurianum, J. Bacteriol.* 166:162–172.

Collins, J. J., and Brill, W. J. 1985. Control of *Klebsiella pneumoniae nif* mRNA synthesis, *J. Bacteriol.* 162:1186–1190.

Corbin, D., Ditta, G., and Helinski, D. R. 1982. Clustering of nitrogen fixation (*nif*) genes in *Rhizobium meliloti, J. Bacteriol.* 149:221–228.

de Bruijn, F. J., and Ausubel, F. M. 1983. The cloning and characterization of the *glnF* (*ntrA*) gene of *Klebsiella pneumoniae*: role of *glnF* (*ntrA*) in the regulation of nitrogen fixation (*nif*) and other nitrogen assimilation genes, *Mol. Gen. Genet.* 192:342–353.

Dimri, G. P., and Das, H. K. 1988. Transcriptional regulation of nitrogen fixation genes by DNA supercoiling, *Mol. Gen. Genet.* 212:360–363.

Ditta, G., Virts, E., Palomares, A., and Kim, C. -H. 1987. The *nifA* gene of *Rhizobium meliloti* is oxygen regulated, *J. Bacteriol.* 169:3217–3223.

Dixon, R. A. 1986. The *xylABC* promoter from the *Pseudomonas putida* TOL plasmid is activated by nitrogen regulatory genes in *Escherichia coli, Mol. Gen. Genet.* 203:129–136.

Dixon, R. A., Kennedy, C., Kondorosi, A., Krishnapillai, V., and Merrick, M. 1977. Complementation analysis of *Klebsiella pneumoniae* mutants defective in nitrogen fixation, *Mol. Gen. Genet.* 157:189–198.

Dixon, R., Eady, R. R., Espin, G., Hill, S., Iaccarino, M., Kahn, D., and Merrick, M. 1980. Analysis of regulation of *Klebsiella pneumoniae* nitrogen fixation (*nif*) gene cluster with gene fusions, *Nature* 286:128–132.

Donald, R. G. K., Nees, D. W., Raymond, C. K., Loroch, A. I., and Ludwig, R. A. 1986. Characterization of three genomic loci encoding *Rhizobium* sp. strain ORS571 nitrogen fixation genes, *J. Bacteriol.* 165:72–81.

Drlica, K. 1984. Biology of bacterial deoxyribonucleic acid topoisomerases, *Microbiol. Rev.* 48:273–289.

Drummond, M., Clements, K. J., Merrick, M., and Dixon, R. 1983. Positive control and autogenous regulation of the *nifLA* promoter in *Klebsiella pneumoniae, Nature* 301:302–307.

Drummond, M., Whitty, P., and Wootton, J. 1986. Sequence and domain relationships of *ntrC* and *nifA* from *Klebsiella pneumoniae*: homologies to other regulatory proteins, *EMBO J.* 5:441–447.

Dusha, I., Kovalenko, S., Banfalvi, Z., and Kondorosi, A. 1987. *Rhizobium meliloti* insertion element ISRm2 and its use for identification of the *fixX* gene, *J. Bacteriol.* 169:1403–1409.

Eady, R. R., and Postgate, J. R. 1974. Nitrogenase, *Nature* 249:805–810.

Eady, R. R., Lowe, D. J., and Thorneley, R. N. F. 1978. Nitrogenase of *Klebsiella pneumoniae*: a pre-steady state burst of ATP hydrolysis is coupled to electron transfer between the component proteins, *FEBS Lett.* 95:211–213.

Earl, C. D. 1986. Studies of the structure, regulation and function of the *fixABC* genes of *Rhizobium*. Ph.D. thesis, Harvard University, Cambridge, Massachusetts.

Earl, C. D., Ronson, C. W., and Ausubel, F. M. 1987. Genetic and structural analysis of the *Rhizobium meliloti fixA, fixB, fixC* and *fixX* genes, *J. Bacteriol.* 169:1127–1136.

Ebeling, S., Hahn, M., Fischer, H. -M., and Hennecke, H. 1987. Identification of *nifE*-, *nifN*- and *nifS*-like genes in *Bradyrhizobium japonicum*, *Mol. Gen. Genet.* 207:503–508.

Ebeling, S., Noti, J. D., and Hennecke, H. 1988. Identification of a new *Bradyrhizobium japonicum* gene (*frxA*) encoding a ferredoxin-like protein, *J. Bacteriol.* 170:1999–2001.

Engelke, T., Jagadish, M. N., and Pühler, A. 1987. Biochemical and genetic analysis of *Rhizobium meliloti* mutants defective in C_4-dicarboxylate transport, *J. Gen. Microbiol.* 133:3019–3029.

Evans, D., Jones, R., Woodley, P., and Robson, R. 1988. Further analysis of nitrogen fixation (*nif*) genes in *Azotobacter chroococcum*: identification and expression in *Klebsiella pneumoniae* of *nifS, nifV, nifM* and *nifB* genes and localization of *nifE/N-*, *nifU-, nifA-* and *fixABC* genes, *J. Gen. Microbiol.* 134:931–942.

Finan, T. M., Wood, J. M., and Jordan, D. C. 1983. Symbiotic properties of C_4-dicarboxylic acid transport mutants of *Rhizobium leguminosarum*, *J. Bacteriol.* 154:1403–1413.

Finan, T. M., Kunkel, B., De Vos, F., and Signer, E. R. 1986. Second symbiotic megaplasmid in *Rhizobium meliloti* carrying exopolysaccharide and thiamine synthesis genes, *J. Bacteriol.* 167:66–72.

Fischer, H. -M., and Hennecke, H. 1984. Linkage map of the *Rhizobium japonicum nifH* and *nifDK* operons encoding the polypeptides of the nitrogenase enzyme complex, *Mol. Gen. Genet.* 196:537–540.

Fischer, H. -M., and Hennecke, H. 1987a. Essential and non-essential domains in the symbiotic nitrogen fixation regulatory protein *nifA* of *Bradyrhizobium japonicum*, *Gesellschaft für Biologische Chemie* 368:1040.

Fischer, H. -M., and Hennecke, H. 1987b. Direct response of *Bradyrhizobium japonicum nifA*-mediated *nif* gene regulation to cellular oxygen status, *Mol. Gen. Genet.* 209:621–626.

Fischer, H. -M., Alvarez-Morales, A., and Hennecke, H. 1986. The pleiotropic nature of symbiotic regulatory mutants: *Bradyrhizobium japonicum nifA* gene is involved in control of *nif* gene expression and formation of determinate symbiosis, *EMBO J.* 5:1165–1173.

Fischer, H. -M., Bruderer, T., and Hennecke, H. 1988. Essential and non-essential domains in the *Bradyrhizobium japonicum* NifA protein: identification of indispensable cysteine residues potentially involved in redox reactivity and/or metal binding, *Nucleic Acids Res.* 16:2207–2224.

Fogher, C., Dusha, I., Barbot, P., and Elmerich, C. 1985. Heterologous hybridization of *Azospirillum* DNA to *Rhizobium nod* and *fix* genes, *FEMS Microbiol. Lett.* 30:245–249.

Foor, F., Reuveny, Z., and Magasanik, B. 1980. Regulation of the synthesis of glutamine synthetase by the P_{II} protein in *K. aerogenes*, *Proc. Natl. Acad. Sci. USA* 77:2636–2640.

Forrai, T., Vincze, E., Vanfalvi, S., Kiss, G. B., Rhandhawa, G. S., and Kondorosi, A. 1983. Localization of symbiotic mutations in *Rhizobium meliloti*, *J. Bacteriol.* 153:635–643.

Fuhrmann, M., Fischer, H. -M., and Hennecke, H. 1985. Mapping of *Rhizobium japonicum nifB-, fixBC-* and *fixA*-like genes and identification of the *fixA* promoter, *Mol. Gen. Genet.* 199:315–322.

Gober, J. W., and Kashket, E. R. 1987. Potassium regulates bacteroid-associated functions of *Bradyrhizobium*, *Proc. Natl. Acad. Sci. USA* 84:4650–4654.

Golden, J. W. 1987. Rearrangement of nitrogen fixation genes in *Anabaena*, *Plant Physiol.* 83 (Suppl.):25.

Golden, J. W., Mulligan, M. E., and Haselkorn, R. 1987. Different recombination site specificity of two developmentally regulated genome rearrangements, *Nature* 327:526–529.

Golden, J. W., Robinson, S. J., and Haselkorn, R. 1985. Rearrangement of nitrogen fixation genes during heterocyst differentiation in the cyanobacterium *Anabaena*, *Nature* 314:419–423.

Grönger, P., Manian, S. S., Reiländer, H., O'Connell, M., Priefer, U. B., and Pühler, A. 1987. Organization and partial sequence of a DNA region of the *Rhizobium leguminosarum* symbiotic plasmid pRL6JI containing the genes *fixABC*, *nifA*, *nifB* and a novel open reading frame, *Nucleic Acids Res.* 15:31–49.

Gubler, M., and Hennecke, H. 1986. FixA, B and C genes are essential for symbiotic and free-living, microaerobic nitrogen fixation, *FEBS Lett.* 200:186–192.

Gubler, M., and Hennecke, H. 1988. Regulation of the *fixA* gene and *fixBC* operon in *Bradyrhizobium japonicum*, *J. Bacteriol.* 170:1205–1214.

Guerinot, M. L., and Chelm, B. K. 1987. Molecular aspects of the physiology of symbiotic nitrogen fixation in legumes. In *Plant-Microbe Interactions, Molecular and Genetic Perspectives*, T. Kosuge and E. W. Nester (eds.), vol. 2. Macmillan, New York, pp. 103–146.

Gussin, G. N., Ronson, C. W., and Ausubel, F. M. 1986. Regulation of nitrogen fixation genes, *Annu. Rev. Genet.* 20:567–591.

Hardy, R. W. F., and Havelka, U. D. 1975. Nitrogen fixation research: a key to world food?, *Science* 188:633–643.

Hageman, R. V., and Burris, R. H. 1980. Electrochemistry of nitrogenase and the role of ATP. In *Current Topics in Bioenergetics*, D. Rao Sanadi (ed.), vol. 10. Academic, New York, pp. 279–292.

Hageman, R. V., Orme-Johnson, W. H., and Burris, R. H. 1980. Role of magnesium adenosine 5'-triphosphate in the hydrogen evolution reaction catalyzed by nitrogenase from *Azotobacter vinelandii*, *Biochemistry* 19:2333–2342.

Haselkorn, R. 1986. Organization of the genes for nitrogen fixation in photosynthetic bacteria and cyanobacteria, *Annu. Rev. Microbiol.* 40:525–547.

Haselkorn, R., Rice, D., Curtis, S. E., and Robinson, J. 1983. Organization and transcription of genes important in *Anabaena* heterocyst differentiation, *Ann. Inst. Pasteur Microbiol.* 1384:181–193.

Hennecke, H., Fischer, H. -M., Ebeling, S., Gubler, M., Thöny, B., Göttfert, M., Lamb, J., Hahn, M., Ramseier, T., Regensberger, B., Alvarez-Morales, A., and Studer, D. 1987. *nif*, *fix*, and *nod* gene clusters in *Bradyrhizobium japonicum*, and NifA-mediated control of symbiotic nitrogen fixation. In *Molecular Genetics of Plant-Microbe Interactions*, D. P. S. Verma and N. Brisson (eds.). Martinus Nijhoff, Dordrecht, The Netherlands, pp. 191–196.

Hill, S., and Kavanaugh, E. P. 1980. Roles of *nifF* and *nifJ* gene products in electron transport to nitrogenase in *Klebsiella pneumoniae*, *J. Bacteriol.* 141:470–475.

Hill, S., Kennedy, C., Kavanaugh, D., Goldberg, R. B., and Hanau, R. 1981. Nitrogen fixation gene (*nifL*) involved in oxygen regulation of nitrogenase synthesis in *Klebsiella pneumoniae*, *Nature* 290:424–426.

Hirschman, J., Wong, P. -K., Sei, K., Keener, J., and Kustu, S. 1985. Products of nitrogen regulatory genes *ntrA* and *ntrC* of enteric bacteria activate *glnA* transcription in vitro: evidence that the *ntrA* product is a sigma factor, *Proc. Natl. Acad. Sci. USA* 82:7525–7529.

Hoover, T. R., Imperial, J., Ludden, P. W., and Shah, V. K. 1988. Homocitrate cures the *nifV⁻* phenotype in *Klebsiella pneumoniae*, *J. Bacteriol.* 170:1978–1979.

Hoover, T. R., Robertson, A. D., Cerny, R. L., Hayes, R. N., Imperial, J., Shah, V. K., and Ludden, P. W. 1987. Identification of the V factor needed for synthesis of the iron-molybdenum cofactor of nitrogenase as homocitrate, *Nature* 329:855–857.

Houmard, J., Bogusz, D., Bigault, R., and Elmerich, C. 1980. Characterization and kinetics of the biosynthesis of some nitrogen fixation (*nif*) gene products in *Klebsiella pneumoniae*, *Biochemie* 62:267–275.

Hunt, T. P., and Magasanik, B. 1985. Transcription of *glnA* by purified *Escherichia coli* components: core RNA polymerase and the products of *glnF*, *glnG*, and *glnL*, *Proc. Natl. Acad. Sci. USA* 82:8453–8457.

Hynes, M. F., Simon, R., Muller, P. R., Niehaus, K., Labes, M., and Pühler, A. 1986. The two megaplasmids of *Rhizobium meliloti* are involved in the effective nodulation of alfalfa, *Mol. Gen. Genet.* 202:355–362.

Iismaa, S. E., and Watson, J. M. 1987. A gene upstream of the *Rhizobium trifolii nifA* gene encodes a ferredoxin-like protein, *Nucleic Acids Res.* 15:3180.

Imperial, J., Ugalde, R. A., Shah, V. K., and Brill, W. J. 1984. Role of the *nifQ* gene product in the incorporation of molybdenum into nitrogenase in *Klebsiella pneumoniae, J. Bacteriol.* 158:187–194.

Israel, D. W., Howard, R. L., Evans, H. J., and Russell, S. A. 1974. Purification and characterization of the molybdenum-iron protein component of nitrogenase from soybean nodule bacteriods, *J. Biol. Chem.* 249:500–508.

Jacobson, M. R., Premakumar, R., and Bishop, P. E. 1986. Transcriptional regulation of nitrogen fixation by molybdenum in *Azotobacter vinelandii, J. Bacteriol.* 167:480–486.

Kahn, D., Batut, J., Boistard, P., Daveran, M. L., David, M., Domergue, O., Garnerone, A. M., Ghai, J., Hertig, C., Infante, D., and Renalier, M. H. 1987. Molecular analysis of a *fix* cluster from *Rhizobium meliloti*. In *Molecular Genetics of Plant-Microbe Interactions*, D. P. S. Verma and N. Brisson (eds.). Martinus Nijhoff, Dordrecht, The Netherlands, pp. 258–263.

Kaluza, K., and Hennecke, H. 1984. Fine structure analysis of the *nifDK* operon encoding the α and β subunits of dinitrogenase from *Rhizobium japonicum, Mol. Gen. Genet.* 196:35–42.

Kaluza, K., Fuhrmann, M., Hahn, M., Regensburger, B., and Hennecke, H. 1983. In *Rhizobium japonicum* the nitrogenase genes *nifH* and *nifDK* are separated, *J. Bacteriol.* 155:915–918.

Kennedy. C., Cannon, F., Cannon, M., Dixon, R., Hill, S., Jensen, J., Kumar, S., McLean, P., Merrick, M., Robson, R., and Postgate, J. 1981. Recent advances in the genetics and regulation of nitrogen fixation. In *Current Perspectives in Nitrogen Fixation*, A. Gibson and W. Newton (eds.). Elsevier/North-Holland, Amsterdam, pp. 146–156.

Kim, C. -H., Ditta, G., and Helinski, D. R. 1985. Transcriptional organization of a *Rhizobium meliloti nif/fix* gene cluster downstream from P2-overlapping transcription. In: *Nitrogen Fixation Research Progress*, H. J. Evans, P. J. Bottomley, and W. E. Newton (eds.). Martinus Nijhoff, Dordrecht, The Netherlands, p. 186.

Kim, C. -H., Helinski, D. R., and Ditta, G. 1987. Overlapping transcription of the *nifA* regulatory gene in *Rhizobium meliloti, Gene* 50:141–148.

Klugkist, J., Haaker. H., Wassink, H., and Veeger, C. 1985. The catalytic activity of nitrogenase in intact *Azotobacter vinelandii* cells, *Eur. J. Biochem.* 146:509–515.

Kong, Q. T., Wu, Q. L., Ma, Z. F., and Shen, S. C. 1986. Oxygen sensitivity of the *nifLA* promoter of *Klebsiella pneumoniae, J. Bacteriol.* 166:353–356.

Krajewska-Grynkiewicz, K., and Kustu, S. 1983. Regulation of *glnA*, the structural gene encoding glutamine synthetase, in *glnA*::Mud1 (Ap^R, *lac*) fusion strains of *Salmonella typhimurium, Mol. Gen. Genet.* 192:187–197.

Kranz, R. G., and Haselkorn, R. 1986. Anaerobic regulation of nitrogen-fixation genes in *Rhodopseudomonas capsulata, Proc. Natl. Acad. Sci. USA* 83:6805–6809.

Kustu, S., Sei, K., and Keener, J. 1986. Nitrogen regulation in enteric bacteria. In *Regulation of Gene Expression—25 Years On*, I. Booth and C. F. Higgens (eds.). Cambridge University Press, Cambridge, U.K., pp. 139–154.

Lammers, P. J., Golden, J. W., and Haselkorn, R. 1986. Identification and sequence of a gene required for a developmentally regulated DNA excision in *Anabaena, Cell* 44:905–912.

Lechevalier, M. P. 1983. Cataloging *Frankia* strains, *Can. J. Bot.* 61:2964–2967.

Ligon, J. M. 1986. Isolation and characterization of the *nifD* and *nifK* genes from *Frankia* that code for the molybdenum-iron protein of nitrogenase. Ph.D. thesis, State University of New York, College of Environmental Science and Forestry, Syracuse, New York.

Ligon, J. M., and Nakas, J. P. 1987. Isolation and characterization of *Frankia* sp. strain FaC1 genes involved in nitrogen fixation, *Appl. Environ. Microbiol.* 53:2321–2327.

Ligon, J. M., and Nakas, J. P. 1988. Nucleotide and deduced amino acid sequences of *nifK* encoding the β-subunit of dinitrogenase and partial sequences of *nifD* encoding the α-subunit of *Frankia* sp. strain FaC1, *Nucleic Acids Res.* 16:11843.

Lowe, D. J., Smith, B. E., and Eady, R. R. 1980. The structure and mechanism of nitrogenase. In *Recent Advances in Biological Nitrogen Fixation*, N. S. Subba Rao (ed.). Oxford and IBH, New Delhi, pp. 34–87.

Magasanik, B. 1982. Genetic control of nitrogen assimilation in bacteria, *Annu. Rev. Genet.* 16:135–168.

McClure, W. R. 1985. Mechanisms and control of transcription initiation in prokaryotes, *Annu. Rev. Biochem.* 54:171–204.

McLean, P. A. and Dixon, R. A. 1981. Requirement of *nifV* gene for production of wild-type nitrogenase enzyme in *Klebsiella pneumoniae*, *Nature* 292:655–656.

Merrick, M. 1983. Nitrogen control of the *nif* regulon in *Klebsiella pneumoniae*: involvement of the *ntrA* gene and analogies between *ntrC* and *nifA*, *EMBO J.* 2:39–44.

Merrick, M. J., and Gibbins, J. R. 1985. The nucleotide sequence of the nitrogen regulation gene *ntrA* of *Klebsiella pneumoniae* and comparison with conserved features in bacterial RNA polymerase sigma factors, *Nucleic Acids Res.* 13:7607–7620.

Merrick, M. S., and Stewart, W. D. P. 1985. Studies on the regulation and function of the *Klebsiella pneumoniae ntrA* gene, *Gene* 35:297–303.

Merrick, M., Gibbins, J., and Toukdarian, A. 1987. The nucleotide sequence of the sigma factor gene *ntrA* (*rpoN*) of *Azotobacter vinelandii*: analysis of conserved sequences in NtrA proteins, *Mol. Gen. Genet.* 210:323–330.

Merrick, M., Hill, S., Hennecke, H., Hahn, M., Dixon, R., and Kennedy, C. 1982. Repressor properties of the *nifL* gene product in *Klebsiella pneumoniae*, *Mol. Gen. Genet.* 185:75–81.

Mortenson, L. E., and Thorneley, R. N. F. 1979. Structure and function of nitrogenase, *Annu. Rev. Biochem.* 48:387–418.

Nakos, G., and Mortenson, L. E. 1971. Molecular weight and subunit structure of molybdoferridoxin from *Clostridium pasteurianum* W5, *Biochim. Biophys. Acta.* 229:431–436.

Nieva-Gomez, D., Roberts, G. P., Klevickis, S., and Brill, W. J. 1980. Electron transport to nitrogenase in *Klebsiella pneumoniae*, *Proc. Natl. Acad. Sci. USA* 77:2555–2558.

Ninfa, A. J., and Magasanik, B. 1986. Covalent modification of the *glnG* product, NR_I, by the *glnL* product, NR_{II}, regulates the transcription of the *glnALG* operon in *Escherichia coli*, *Proc. Natl. Acad. Sci. USA* 83:5909–5913.

Nixon, B. T., Ronson, C. W., and Ausubel, F. M. 1986. Two component regulatory systems responsive to environmental stimuli share strongly conserved domains with the nitrogen assimilation genes *ntrB* and *ntrC*, *Proc. Natl. Acad. Sci. USA* 83:7850–7854.

Norel, F., and Elmerich, C. 1987. Nucleotide sequence and functional analysis of the two *nifH* copies of *Rhizobium* ORS571, *J. Gen. Microbiol.* 133:1563–1576.

Normand, P., and Lalonde, M. 1986. The genetics of actinorhizal *Frankia*: a review, *Plant and Soil* 90:429–453.

Normand, P., Simonet, P., and Bardin, R. 1988. Conservation of *nif* sequences in *Frankia*, *Mol. Gen. Genet.* 213:238–246.

Noti, J. D., Folkerts, O., Turken, A. N., and Szalay, A. A. 1986. Nitrogenase promoter-*lacZ* fusion studies of essential nitrogen fixation genes in *Bradyrhizobium japonicum* I-110, *J. Bacteriol.* 167:784–791.

Novak, P. D., and Maier, R. J. 1987. Inhibition of hydrogenase synthesis by DNA gyrase inhibitors in *Bradyrhizobium japonicum*, *J. Bacteriol.* 169:2708–2712.

Orme-Johnson, W. H. 1985. Molecular basis of biological nitrogen fixation, *Annu. Rev. Biophys. Chem.* 14:419–459.

Orme-Johnson, W. H., and Munck, E. 1980. The prosthetic groups of nitrogenase. In *Molybdenum and Molybdenum Containing Enzymes*, M. P. Coughlin (ed.). Pergamon, Oxford, pp. 427–438.

Ow, D. W., and Ausubel, F. M. 1983. Regulation of nitrogen metabolism by *nifA* gene product in *Klebsiella pneumoniae*, *Nature* 301:307–313.

Ow, D. W., Sundaresan, V., Rothstein, D., Brown, S. E., and Ausubel, F. M. 1983. Promoters regulated by the *glnG* (*ntrC*) and *nifA* gene products share a heptameric consensus sequence in the -15 region, *Proc. Natl. Acad. Sci. USA* 80:2524–2528.

Ow, D. W., Xiong, Y., Gu, Q., and Shen, S. -C. 1985. Mutational analysis of the *Klebsiella pneumoniae* nitrogenase promoter: sequences essential for positive control by *nifA* and *ntrC* (*glnG*) products, *J. Bacteriol.* 161:868–874.

Powlowski, K., Ratet, P., Schell, J., and de Bruijn, F. J. 1987. Cloning and characterization of *nifA* and *ntrC* genes of the stem nodulating bacterium ORS-571, the nitrogen fixing symbiont of *Sesbania rostrata*: regulation of nitrogen fixation *nif* genes in the free living versus symbiotic state, *Mol. Gen. Genet.* 206:207–219.

Prakash, R. K., and Cummings, B. 1988. Creation of novel nitrogen-fixing actinomycetes by protoplast fusion of *Frankia* with streptomyces, *Plant Mol. Biol.* 10:281–289.

Pühler, A., Aguilar, M. O., Hynes, M., Müller, P., Klipp, W., Priefer, U., Simon, R., and Weber, G. 1984. Advances in the genetics of free-living and symbiotic nitrogen fixing bacteria. In: *Advances in Nitrogen Fixation Research*, C. Veeger and W. E. Newton (eds.). Nijhoff/Junk, The Hague, pp. 609–619.

Quinto, C., de la Vega, H., Flores, M., Leemans, J., Cevallos, M. A., Pardo, M. A., Azpiroz, R., de Lourdes, G. M., Calva, E., and Palacios, R. 1985. Nitrogenase reductase: a functional multigene family in *Rhizobium phaseoli*, *Proc. Natl. Acad. Sci. USA* 82:1170–1174.

Regensberger, B., Meyer, L., Filser, M., Weber, J., Studer, D., Lamb, J. W., Fischer, H. -M., Hahn, M., and Hennecke, H. 1986. *Bradyrhizobium japonicum* mutants defective in root-nodule bacteroid development and nitrogen fixation, *Arch. Microbiol.* 144:355–366.

Reitzer, L. J., and Magasanik, B. 1983. Isolation of the nitrogen assimilation regulator NR$_I$, the product of the *glnG* gene of *Escherichia coli*, *Proc. Natl. Acad. Sci. USA* 80:5554–5558.

Reitzer, L. J., and Magasanik, B. 1985. Expression of *glnA* in *Escherichia coli* is regulated at tandem promoters, *Proc. Natl. Acad. Sci. USA* 82:1979–1983.

Reitzer, L. J., and Magasanik, B. 1986. Transcription of *glnA* in *Escherichia coli* is stimulated by activator bound to sites far from the promoter, *Cell* 45:785–792.

Reuveny, Z., Foor, F., and Magasanik, B. 1981. Regulation of glutamine synthetase by regulatory protein P$_{II}$ in *Klebsiella aerogenes* mutants lacking adenylyltransferase, *J. Bacteriol.* 146:740–745.

Rice, D., Mazur, B. J., and Haselkorn, R. 1982. Isolation and physical mapping of nitrogen fixation genes from the cyanobacterium *Anabaena* 7120, *J. Biol. Chem.* 257:13157–13163.

Riedel, G. E., Brown, S. E., and Ausubel, F. M. 1983. Nitrogen fixation in *Klebsiella pneumoniae* is inhibited by certain multicopy hybrid *nif* plasmids, *J. Bacteriol.* 153:45–56.

Roberts, G. P., and Brill, W. J. 1980. Gene product relationships of the *nif* regulon of *Klebsiella pneumoniae*, *J. Bacteriol.* 144:210–216.

Roberts, G. P., MacNeil, T., MacNeil, D., and Brill, W. J. 1978. Regulation and characterization of protein products coded by the *nif* genes of *Klebsiella pneumoniae*, *J. Bacteriol.* 136:267–279.

Robson, R., Kennedy, C., and Postgate, J. 1983. Progress in comparative genetics of nitrogen fixation, *Can. J. Microbiol.* 29:954–967.

Ronson, C. W., and Astwood, P. M. 1985. Genes involved in the carbon metabolism of bacteroids. In *Nitrogen Fixation Research Progress*, H. J. Evans, P. J. Bottomley, and W. E. Newton (eds.). Martinus Nijhoff, Dordrecht, The Netherlands, pp. 201–207.

Ronson, C. W., Lyttleton, P., and Robertson, J. G. 1981. C$_4$-dicarboxylate transport mutants of *Rhizobium trifolii* form ineffective nodules on *Trifolium repens*, *Proc. Natl. Acad. Sci. USA* 78:4284–4288.

Ronson, C. W., Nixon, B. T., Albright, L. M., and Ausubel, F. M. 1987. *Rhizobium meliloti ntrA* (*rpoN*) gene is required for diverse metabolic functions, *J. Bacteriol.* 169:2424–2431.

Ruvkun, G. P., and Ausubel, F. M. 1980. Interspecies homology of nitrogenase genes, *Proc. Natl. Acad. Sci. USA* 77:191–195.

Ruvkun, G. B., Sundaresan, V., and Ausubel, F. M. 1982. Directed transposon Tn5 mutagenesis and complementation analysis of *Rhizobium meliloti* symbiotic nitrogen fixation genes, *Cell* 29:551–559.

Ryle, G. J. A., Powell, C. E., and Gordon, J. A. 1979. The respiratory costs of nitrogen fixation in soybean, cowpea and white clover. I. Nitrogen fixation and the respiration of the nodulated root, *J. Exp. Bot.* 30:135–144.

Schetgens, R. M. P., Hontelez, J. G. J., van den Bos, R. C., and van Kammen, A. 1985. Identification and phenotypical characterization of a cluster of *fix* genes, including a *nif* regulatory gene, from *Rhizobium leguminosarum* PRE, *Mol. Gen. Genet.* 200:368–374.

Shah, V. K., and Brill, W. J. 1977. Isolation of an iron-molybdenum cofactor from nitrogenase, *Proc. Natl. Acad. Sci. USA* 74:3249–3253.

Shah, V. K., Stacey, G., and Brill, W. J. 1983. Electron transport to nitrogenase. Purification and characterization of pyruvate flavodoxin oxidoreductase, the *nifJ* gene product, *J. Biol. Chem.* 258:12064–12068.

Shah, V. K., Imperial, J., Ugalde, R. A., Ludden, P. W., and Brill, W. J. 1986. In vitro synthesis of the iron-molybdenum cofactor of nitrogenase, *Proc. Natl. Acad. Sci. USA* 83:1636–1640.

Sibold, L., and Elmerich, C. 1982. Constitutive expression of nitrogen fixation (*nif*) genes of *Klebsiella pneumoniae* due to a DNA duplication, *EMBO J.* 1:1551–1558.

Simonet, P., Haurat, J., Normand, P., Bardin, R., and Moiroud, A. 1986. Localization of *nif* genes on a large plasmid in *Frankia* sp. strain ULQ0132105009, *Mol. Gen. Genet.* 204:492–495.

Smith, B. E. 1977. The structure and function of nitrogenase: a review of the evidence for the role of molybdenum, *J. Less. Common Met.* 54:465–475.

Smith, B. E. 1983. Reactions and physiochemical properties of the nitrogenase MoFe proteins. In *Nitrogen Fixation: The Chemical-Biochemical-Genetic Interface*, A. Müller and W. E. Newton (eds.). Plenum, New York.

St. John, R. T., Johnson, H. M., Seidman, C., Garfinkel, D., Gordon, J. K., Shah, V. K., and Brill, W. J. 1975. Biochemistry and genetics of *Klebsiella pneumoniae*, *J. Bacteriol.* 119:266–269.

Sundaresan, V., Jones, J. D. G., and Ausubel, F. M. 1983. Regulation of *nif* genes in *Klebsiella pneumoniae* and *Rhizobium meliloti*. In *Gene Expression*, UCLA Symposium in Molecular and Cellular Biology, D. Hammer and M. Rosenberg (eds.), vol. 8. Liss, New York, pp. 175–185.

Szeto, W. W., and Ausubel, F. M. 1987. An *ntrC* homologue in *B. japonicum*. In *Molecular Genetics of Plant-Microbe Interactions*, D. P. S. Verma and N. Brisson (eds.). Martinus Nijhoff, Dordrecht, The Netherlands, pp. 250–254.

Szeto, W. W., Zimmerman, J. L., Sundaresan, V., and Ausubel, F. M. 1984. A *Rhizobium meliloti* symbiotic regulatory gene, *Cell* 36:535–543.

Szeto, W. W., Nixon, B. T., Ronson, C. W., and Ausubel, F. M. 1987. Identification and characterization of *Rhizobium meliloti ntrC* gene: *R. meliloti* has separate regulatory pathways for activation of nitrogen fixation genes in free-living and symbiotic cells, *J. Bacteriol.* 169:1423–1432.

Thöny, B., Fischer, H.-M., Anthamatten, D., Bruderer, T., and Hennecke, H. 1987. The symbiotic nitrogen fixation regulatory operon (*fixRnifA*) of *Bradyrhizobium japonicum* is expressed aerobically and is subject to a novel, *nifA*-independent type of activation, *Nucleic Acids Res.* 15:8479–8499.

Thorneley, R. N. F., and Lowe, D. J. 1983. Nitrogenase of *Klebsiella pneumoniae*. Kinetics of the dissociation of oxidized iron protein from molybdenum-iron protein: identification of the rate limiting step for substrate reduction, *Biochem. J.* 215:393–403.

Thorneley, R. N. F., and Lowe, D. J. 1984. The mechanism of *Klebsiella pneumoniae* nitrogenase action: Stimulation of the dependence of hydrogen evolution rate on component protein concentration and ratio and sodium dithionite concentration, *Biochem. J.* 224:903–910.

Tjepkema, J. D., Schwintzer, C. R., and Benson, D. R. 1986. Physiology of actinorhizal nodules, *Annu. Rev. Plant Physiol.* 37:209–232.

Tso, M. Y. W. 1974. Some properties of the nitrogenase proteins from *Clostridium pasteurianum*. Molecular weight, subunit structure, isoelectric point and EPR spectra, *Arch. Microbiol.* 99:71–80.

Tumer, H. E., Robinson, S. J., and Haselkorn, R. 1983. Different promoters for the *Anabaena* glutamine synthetase gene during growth using molecular or fixed nitrogen, *Nature* 306:337–342.

Ueno-Nishio, S., Mango, S., Reitzer, L. J., and Magasanik, B. 1984. Identification and regulation of the *glnL* operator-promoter of the complex *glnALG* operon of *Escherichia coli, J. Bacteriol.* 160:379–384.

Upchurch, R. G., and Mortenson, L. E. 1980. In vivo energetics and control of nitrogen fixation: changes in the adenylate energy charge and adenosine 5'-diphosphate/adenosine 5'-triphosphate ratio of cells during growth on dinitrogen versus growth on ammonia, *J. Bacteriol.* 143:274–284.

Watson, R. J., Chan, Y.-K., Wheatcroft, R., Yang, A.-F., and Han, S. 1988. *Rhizobium meliloti* genes required for C_4-dicarboxylate transport and symbiotic nitrogen fixation are located on a megaplasmid, *J. Bacteriol.* 170:927–934.

Weber, G., Aguilar, O. M., Gronemeier, B., Reiländer, H., and Pühler, A. 1985. Genetic analysis of symbiotic genes of *Rhizobium meliloti*: mapping and regulation of nitrogen fixation genes. In *Advances in the Molecular Genetics of the Bacteria-Plant Interaction*, A. A. Szalay and R. P. Legocki (eds.). Cornell University Publishers, Ithaca, New York, pp. 13–15.

Wittenberg, J. B., Bergersen, F. J., Appleby, C. A., and Turner, G. L. 1974. Facilitated oxygen diffusion: the role of leghemoglobin in nitrogen fixation by bacteroids isolated from soybean root nodules, *J. Biol. Chem.* 249:4057–4066.

Microbial Colonization of Plant Roots

D. A. Klein

*Departments of Microbiology and Environmental
Health
Colorado State University
Fort Collins, Colorado 80523*

J. L. Salzwedel

*Department of Microbiology and Public Health
Michigan State University
East Lansing, Michigan 48824*

F. B. Dazzo

*Department of Microbiology and Public Health
Michigan State University
East Lansing, Michigan 48824*

Introduction

Studies of the rhizosphere, and of the colonization of plant roots, have developed into a diversity of research areas ranging from the molecular to the community levels of biological complexity. These newly evolving areas include studies of microbial populations in the rhizosphere, metabolites mediating microbial-plant communication, and plant-microbe factors influencing soil structure and aggregation. In addition, plant strategies to manage nutrient fluxes and nutrient acquisition are being given greater consideration, especially during ecological succession.

Molecular aspects of plant root-microbe interactions are developing into exciting and challenging subdisciplines of rhizosphere ecology. These include studies of signal exchange, as found with the *Rhizobium*-legume symbiosis, and molecular aspects of microbial competition and survival in the root region, which have important implications for plant pathology and biological control. Studies of soil protozoology and nematology are also developing rapidly, and these also have largely become independent subdisciplines. Since the classic work of Hiltner (1904), the complexities of our understanding of the root environment have only expanded. With each new level of analytical refinement (molecular, immunological techniques, etc.), the problems of maintaining a balanced perspective in this area have only increased. Unfortunately, by emphasizing a specific interest, the ability to fully assess the biotechnological implications of any finding can become deemphasized and less apparent. To attempt to bridge these concerns, we have developed a more integrated view of root colonization in this chapter to better allow the plant biotechnologist-microbiologist to assess the implications and importance of findings about specific aspects of the rhizosphere for the more general process of colonization.

The Rhizosphere and Colonization

Many valuable general treatments of rhizosphere biology are available, and these provide an excellent foundation for the present effort. Curl and Truelove (1986) have provided a general review of the rhizosphere; however, as a point of interest, the term "colonization" is not included in their subject index. Vancura and Kunc (1988) have provided additional information related to colonization in a recent symposium volume.

"Plant root colonization" has been discussed on a general level by many workers (Krasilnikov 1958, Macura and Vancura 1963), and the term has been used in a more general sense to include colonization by microbes found in the rhizosphere (Samtsevich 1965). During the period of these latter studies, there was still controversy as to whether plant roots themselves were actually colonized. Many workers suggested that microbes only penetrate necrotic tissues and cells when their resistance might fall. In relation to the ecology of soilborne pathogens, earlier symposia (Baker and Snyder 1965, Toussoun et al. 1970) do not consider colonization or root colonization as a specific topic. Much of this earlier work did not distinguish between plant root and rhizosphere colonization.

A large number of more general treatments of root and rhizosphere biology have been completed over the last two or three decades, and these provide a valuable background for considering colonization processes. These include reviews by Bowen (1980), Brown (1975), Parkinson (1967), Rovira (1969), Rovira and Davey (1974), and Woldendorp (1978).

Bowen (1980) made three major points that have important implications for this chapter:

1. The plant is the main driving force for rhizosphere biodynamics.

2. Distribution of roots and microbial movement to them are critical.

3. Microbial presence and growth can affect photosynthate losses from the root.

The extent to which roots alter the soil environment depends on the type of plant community. For permanent plant communities, root turnover is the norm, and most soil in the upper horizon of the root zone will be rhizosphere soil. Newman (1985) has suggested that in permanent grasslands, bacteria can migrate and fungi can ramify from one root to another. Thus, the greater portion of the soil may be permanently rhizosphere soil under these conditions. In managed systems, however, the volume of rhizosphere soil will increase during the period of active plant growth.

Djordjevic et al. (1987) discussed the gradation of relationships between a particular plant and associated microbes (both nonpathogenic and pathogenic), and emphasized this as a multiaspect process. Nonpathogenic microbes can have major effects on plant root physiology, and this leads to increased exudation rates. However, if the exudation rate is increased to a "macroscopically appreciable" extent, the microbe may be functioning as a pathogen. The transition from a commensalistic to a parasitic relationship may occur when "nonwaste" nutrition is taken from the host. This transition is best characterized by necrotrophic bacteria, which destroy plant tissue before colonization. Important examples in this area include the soft-rot erwinias and pseudomonads.

More advanced colonization interactions involve a series of strategies to "avoid recognition" (Djordjevic et al. 1987). This results in a lack of disease symptoms, except in times of plant stress. Direct contact between bacterium and host plant can occur, including specific genetic interactions, as with *Xanthomonas* spp., which, to take an important example, interact with products of BS1 resistance locus in pepper plants. More in-depth treatment of these areas was provided by contributions to a volume edited by Kosuge and Nester (1986).

As discussed by Mansfield and Brown (1985), most phytopathogenic bacteria depend on colonization of the host plant for survival and overwintering. These bacteria vary in their effectiveness as saprophytes and usually do not survive in the soil in the absence of a plant. This has been established for *Agrobacterium tumefaciens, Pseudomonas solanacerum*, and perhaps *Erwinia* spp. Others survive as components of the rhizoplane and phyllosphere microflora, including *Pseudomanas syringae* pv. *syringae* (Blakeman and Fokkema 1982). These workers note the lack of information on host specificity for colonization, in terms of pathogen growth on plant surfaces. Successful disease establishment involves entry by a variety of mechanisms and four stages of interaction between the microbe and the plant:

1. Avoidance of resistance processes

2. Establishment of nutritional relationships

3. Colonization of tissue

4. Development of disease symptoms

Disease may or may not result from the establishment of a nutritional relationship (Billings 1982), and these factors interact during microbial colonization of plant roots.

Plant Root Architecture and Colonization

Plant root colonization is first a problem of physical boundary definition, as the root itself is a diffuse and ill-defined structure, especially on a microbial scale. This will have major effects on the colonization process. The rooting strategies of mono- and dicotyledons are quite different (Curl and Truelove 1986). In most dicots the primary root bears lateral branches, which also progressively branch. Monocotyledons develop a more fibrous root system tending to expand more horizontally and be shallow. The establishment of root hairs behind the root tip as unicellular extensions of the epidermal cells increases the surface area and potential interactions with the soil. The structure of the apical meristem, and of the overlying root-cap cells is influenced by many factors, and abrasion will result in lysis of these cells and accumulation of part of the mucigel as the root penetrates into the soil (Foster et al. 1983).

As discussed by Schmidt (1979), the penetration of a root into a soil volume creates a unique set of changes. Where the soil may originally have been dominated by microbes utilizing the native soil organic matter or partially stabilized and decomposed residues, the entrance of the plant root into an environment creates a new zone for microbial exploitation, leading to the development of a markedly different balance of microbial physiological types and possibly a change in the ratio of bacteria to fungi. Much of this response will depend on the substrate characteristics and resources available for microbial use (Lynch 1987c).

A critical, but not fully appreciated, aspect of the plant root surface is the accumulation of mucigel-like materials (Foster 1982). As discussed by Foster (1986), histochemical techniques allow a more realistic assessment of root-surface characteristics. Based on this level of study, several important observations were made: (1) the low recovery of viable populations and (2) the location of microbes *in* the mucigel materials. As summarized by these workers, electron microscopic studies suggest that three models of root-surface ultrastructure may be applicable: the "cuticle"; the "microfibril"; and, most generally accepted, the "mucigel" model. The important concept is that the gel has *no firm boundary* (Jenny and Grossenbacher 1963), and that the delineation of the outer root zone requires the use of chemical procedures, such as treatment with collodial iron particles or ruthenium red (Greaves and Darbyshire 1972).

Scott et al. (1958) emphasized that this gel can become inhabited by microorganisms, some of which can also synthesize and secrete polysaccharide glycocalyx. At this point, root and microbial gels become intermixed, and the term "mucigel" should be used to describe this complex mixture. The term "rhizodermis" has a different meaning, and this structure can differentiate and mature. Six stages occur in this process of root-surface maturation (Foster and Bowen 1982). As a confirmation of this observation, Webley et al. (1965) noted that the root surfaces of grasses have a greater proportion of bacterially produced capsules and/or slime layers than are found in either rhizosphere or nonrhizosphere soil. The gels on the root surface also vary in density. The denser gels, more closely associated with the plant root surface, were suggested to be morphologically equivalent to the rhizoplane (Foster 1986), as summarized in Table 1. The process of

TABLE 1   The Physical and Microbiological Characteristics of the Major Plant Root Zones

Zone	Characteristics
Root cap	External gel secretion Cell sloughing Low microbial population
Meristemic zone	Cell junctions visible Few bacteria, some local sites of dense colonies
Cell extension zone	Growth at greater than 1,000 μm/h Columnar cell formation Surfaces more acidic Lower microbial population Microfibril reorganization
Root-hair zone	Secondary wall formed Dense root-hair mucilage secreted; rhizosphere soil zone to 3–5 mm, gel impregnated; zone may exceed diameter of the root Possible rhizosheath formation Microbes within gel are not $CHCl_3$-sensitive Microbes at funnel-shaped root-hair base Rapid microbial growth and root-surface gel Attachment is end-on (polar), as exemplified by *Rhizobium*
Mature rhizodermis cell zone	Close physical barrier Mucilagenous wall, often cuticlelike Microbes may be under this structure
Cortical zone	Cell death in the absence of microorganisms, which is root-age-related Microbial cells grow between rhizodermal cells Shift to unique microbial community = PHB and P accumulation; possible anaerobic conditions Within the rhizodermis, cell growth in lumena, continued microbial invasion of this area

SOURCE: From Foster (1986).

change in ultrastructure also has been documented in other studies, as discussed by Marshall (1976).

The structure of roots can also be modified by the microbial community and by grazing organisms. For instance, certain rhizosphere microbes, e.g., *Rhizobium* and *Azospirillum*, can induce prolific root-hair development in certain plants. As discussed by Huang (1986), microbes penetrate plants using a variety of strategies. This can occur by entering wounds, by entering openings of the epidermis created by lateral root emergence, as a result of root desquamation during penetration of the soil, or by active penetration of epidermal cell walls. For example, fungi can cause major root-surface changes, as observed with

Septoria nodorum (Hargreaves and Kean 1986). Attempted penetration resulted in major changes in the structures of cells at the plant root surface.

Another major factor that can influence root morphology and turnover will be the presence and activities of nematodes (Freckman and Caswell 1985) and soil insects. Although most (70%) nematodes do not feed directly on plant roots, they can significantly modify the structure of a root system. Complex feedback loops provide indirect interactions among these nematode trophic levels, which can influence plant morphology and nutrient availability to the microbial community.

These types of interactions suggest that plant root structure, as the environment for colonization, can be influenced by a series of factors important for predicting the longer-term fate and effects of microorganisms. It is also evident that the terms "endorhizosphere," "ectendorhizosphere," and "rhizoplane" (Lynch, 1987b) often will not adequately describe the actual environment in which colonization is occurring. The term "rhizosphere," in comparison, does imply an environment with some degree of physical separation from the root surface and its accumulated gel-like materials. Rhizosphere colonization may not be related to colonization of the root itself. These rhizosphere organisms may be less fit to colonize the actual plant root, and may represent a distinct and separate population dependent on a totally different set of nutrient resources and physical requirements.

Root Exudates and Their Characteristics

Root exudates vary in composition, dependent on plant stress, soil nutrient availability, and other factors affecting the normal physiology of the plant. A variety of factors influence exudation rates. As summarized by Rovira (1969) and Vancura (1964), these factors include plant physiological and pathological responses, the physical environment (light, temperature, water, plant type, plant age, nutrition, etc.), and the associated rhizosphere microflora. Stress, including herbage removal, animal grazing, and foliar sprays also can have distinct effects. The effect of plant age on exudation rates is more difficult to assess, especially if a growing plant has root turnover. Hamlen et al. (1972) found that rates of carbohydrate release from alfalfa roots decreased with age. Plants of different seral stages appear to be able to modify the characteristics of their exudates, especially in response to nutrient availability (Klein et al. 1989). This response to competition during succession (Bazzaz 1979) suggests that plants, through their physiological processes, exert close control of exudates.

The general classes of exudate materials have been summarized by

Curl and Truelove (1986) based on earlier sources (Table 2). These materials range from large polymeric components to transient volatile species. The major concept in the literature is the stimulatory effect of microbial presence on the rates of root exudation (Barber and Martin 1976, Prikryl and Vancura 1980). Lynch (1987a) has estimated that up to 40% of the total photosynthate produced by the plant can be allocated to belowground maintenance, primarily of the soil microbial community. The significance of this value is that rhizosphere microorganisms exert a major impact on plant productivity by acting as a sink for photosynthate.

A wide range of specific materials can be released from plant roots, as summarized in Table 3. The quantities of these components in root exudates also vary widely (Curl and Truelove 1986, Newman 1978). The soluble exudates, in general, equal 1 to 10% of the plant root weight, but can approach 25%. Klein et al. (1988), studying range grasses, found that residual soluble amino acids, sugars, and organic acids were in the range of 28 to 46 milligrams per gram of root under sterile conditions, and 14 to 28 milligrams per gram of root under nonsterile conditions. These values were similar to those (6 to 250 milligrams per gram of root) found by others for a wider variety of plants, as noted by Newman and Watson (1977).

Information on assumed fluxes was used in the development of models of microbial populations maintained at different distances from the plant root. When these materials are utilized by the responding mi-

TABLE 2   Classification of Root Exudates

Root exudates	Chemicals and elaborated metabolites released to the surface of the root or released into the root environment
Root exudations	The process of exudate release, involving several pathways and biochemical mechanisms
Leakages	Compounds of low molecular weight that diffuse into the apoplast and, via the apoplast, move to the root surface, or leak directly from epidermal or cortical cells
Secretions	Compounds that cross membrane barriers as a result of expenditure of metabolic energy
Mucilages	Four sources contributing to organic materials in the rhizosphere; these include materials derived from root-cap cells, polysaccharide hydrolysate of primary cell walls, epidermal and root-hair secretions from cells only with primary walls, and products of bacterial degradation of old dead cells
Lysates	From autolysis of older sloughed cells that become heavily colonized or from released microbial metabolites

SOURCE: Developed from Hale et al. (1981).

TABLE 3   Substances Detected in Plant Root Exudates

Kind of compound	Exudate components
Sugars	Glucose, fructose, sucrose, maltose, galactose, rhamnose, ribose, xylose, arabinose, raffinose, oligosaccharide, various other polysaccharides
Amino compounds	Asparagine, α-alanine, glycine, cystine/cysteine, methionine, phenylalanine, tyrosine, threonine, lysine, proline, tryptophane, β-alanine, arginine, homoserine, cystathionine
Organic acids	Tartaric, oxalic, citric, malic, acetic, propionic, butyric, succinic, fumaric, glycolic, valeric, malonic
Fatty acids and sterols	Palmitic, stearic, oleic, linoleic acids: cholesterol, campesterol, stigmasterol, sitosterol
Growth factors	Biotin, thiamine, niacin, pantothenate, choline, inositol, pyridoxine, p-aminobenzoic acid, n-methyl nicotinic acid
Nucleotides, flavonones, and enzymes	Flavonones, adenine, guanine, uridine, cytidine, phosphatase, invertase, amylase, protease, polygalacturonase
Miscellaneous compounds	Auxins, scopoletin, fluorescent substances, hydrocyanic acid, glycosides, saponin (glucosides), organic phosphorus compounds, nematode cyst or egg-hatching factors, nematode attractants, fungal mycelium growth stimulants, mycelium growth inhibitors, zoospore attractants, spore and sclerotium germination stimulants and inhibitors, bacterial stimulants, and parasitic weed germination stimulators

SOURCE: Curl and Truelove (1986).

crobial community, the level of soluble substrate will remain low, in the range of 0.26 micrograms of carbon per cubic centimeter of soil (Newman 1978). Several important estimates were made from these calculations: 55% of exudate carbon would be used for maintenance in the first 10 days and 74% within 20 days.

Major differences can be expected to occur between plants involved in natural successional processes, where available nutrients will be limited, as compared with intensive agricultural crops, where mineral nitrogen and other nutrients are added. Major shifts in plant and soil resource allocation can be expected to occur depending on the availability of fertilizer-derived nutrients.

A major concept related to exudation is that excreted metabolites will diffuse at different rates from the root source, and hence be found at different distances (Foster 1986, McDougall and Rovira 1970). Soluble materials were able to move 1000 μm from the root, while volatile components could move 400 to 3,000 μm to the outer rhizosphere. These different materials, which move various distances, should stimulate different components of the microbial communities.

A major concern is the calculation of the dimensions of the rhizosphere accurately in relation to root structure. Drury et al. (1983) used a mathematical approach to estimate the rhizosphere zone for *Fusarium oxysporum* chlamydospore germination. Values in the range of 1 mm or less, based on exudate-dependent germination, or 1.73 mm based on calculations, were made. In the Newman-Watson model (Newman and Watson 1977) diffusion through moist soil was calculated to be as great as 4 mm from the root surface (Bowen 1980). Newman also suggested that volatile components may be effective at greater distances, depending on physical conditions. Other studies (Woldendorp 1978) suggest that a width approaching 18 mm exists based on the relative increase in viable populations. This width may also reflect the development of root hairs and of secondary root-branching processes. Papavizas and Davey (1961) have shown expected increases in populations of bacteria, streptomycetes, and fungi at varied distances from the root (Table 4).

A critical variable influencing the characteristics of exudates is nitrogen availability, although this has not been considered specifically in models of rhizosphere processes (Newman 1978). Griffin et al. (1976) found that the C/N ratios of peanut plant exudate varied with mineral nitrogen availability. The ratio was 9:1 for plants grown in ¼-strength Hoaglands, and was 18.9:1 for plants grown in the same medium containing no nitrogen.

Exudates, if they have varied C/N ratios, can provide the energetic basis for establishment of different groups of rhizosphere microbes

TABLE 4 **Microbial Populations at Various Distances from the Root Surface of Lupin Seedlings Grown in Unamended Elsinboro Sandy Loam**

Distance from root surface (mm)	Bacteria[a]	Streptomycetes[a]	Fungi[a]
0[b]	159,000	46,700	355
0–3	49,700	15,500	176
3–6	38,000	11,400	170
9–12	37,400	11,800	130
15–18	34,170	10,100	117
80[c]	27,300	9,100	91
R/S ratio[d]	5.8	5.1	3.9

[a]Figures in thousands of colony-forming units per gram of oven-dried soil.
[b]Rhizoplane.
[c]Nonrhizosphere soil (control).
[d]R/S ratio = No. of microorganisms 0 mm from root surface/No. of microorganisms 80 mm from root surface.
SOURCE: Papavizas and Davey (1961).

and can be expected to have major effects on the microbial populations in the rhizosphere. With exudates having higher C/N ratios, microbial mineralization of N from soil organic matter can be stimulated (Clarholm 1985) or diazotrophic microbes can assume a greater role (Döbereiner et al. 1988). With general increases in exudation, an increased demand for oxygen by microbes will result. This, in combination with root respiration, will decrease the oxidation-reduction status of the smaller pores, where the autochthonous microbial population is primarily located. This in turn would decrease the activity of this population, which utilizes more refractile organic substrate under oligotrophic conditions. These conditions have been suggested to be primarily aerobic due to the lower energy yield from these substrates (Poindexter 1981).

More important in terms of colonization and control of this process are substances not only serving as substrates but also interacting (stimulation, inhibition) with the microbial community (Krasilnikov 1958). These compounds may also provide information concerning mechanisms of plant-microbe interactions, which is critical in terms of biotechnological applications. For instance, *Rhizobium meliloti* contains a plasmid which selectively utilizes calystegin compounds leading to competitive success in the *Artam* rhizosphere, the degradation of which is plasmid-encoded (Tepfer et al. 1988).

A variety of additional volatile components have been identified in exudates, which may have important implications for microbial colonization and functioning in the plant root zone (French 1985). These include compounds that stimulate germination, including *n*-nonanal (Table 5), benzonitrile, benzylcyanide, heptatol, and substituted benzaldehydes such as vanillin. This group of compounds may also have allelopathic properties (Rice 1984).

As noted by Vancura and Jandera (1986), biologically active com-

TABLE 5   The Most Effective Germination Stimulators and the Natural Products in Which They Have Been Found

Stimulator	Product
Nonanal	In urediniospores: wheat stem, stripe, rust; bean, sunflower, corn, oat crown rust; rose aroma, oil of cinnamon, citrus peel oils, turpentines (*Pinus* sp.)
6-Methyl-5-hepten-2-one	In urediniospores: wheat stem, stripe, rust; fungi (*Ceratocystis* sp.)
β-Ionone	In raspberry, green tea, corn kernels, carrot roots, alfalfa leaves, algal bloom, *Cyanidium caldarium*
Benzaldehyde	In flavor volatiles of peach, plum, other stone fruits; widely distributed in cyanogenic glucosides

SOURCE: French (1985).

pounds in root exduate can include phenolics of various types, e.g., pyrocatechols. Scheffer et al. (1962) suggested that these active factors provided nematode resistance for the grass *Eragrostis survula*. Bokhari et al. (1979) found that polyphenolics were present at higher concentrations in the rhizosphere of range grasses than in nonrhizosphere soils.

Studies of flaxes (Timonin 1940) suggested that HCN could be released from roots as a degradation product of the cyanogenic β-glucoside laminarin, which might provide the basis for the genetic modification of plants to increase the release of such materials. HCN production requires iron, and deprivation of this enzyme cofactor by microbial siderophores has been proposed as a mechanism of biological control by certain plant-growth-promoting rhizobacteria (Schippers 1988). These types of volatile compounds have also been discussed by Curl and Truelove (1986). These can include organic sulfides (methyl and propyl sulfides) and isothiocyanates. Biologically active compounds that are released from germinating seeds include ethanol, methanol, formaldehyde, acetaldehyde, propionaldehyde, formate, acetone, ethylene, and propylene (Vancura and Stotzky 1976). Catska (1979) found that the exudates of most plants that their group studied inhibited the germination of saprophytic fungal spores and stimulated the germination of phytopathogenic fungi. In these studies, differences in plant production of ethanol and methanol were suggested to be related to plant sensitivity to smut and fusariosis.

An additional important group of metabolites produced by plant roots in response to microorganisms are the phytoalexins (Ebel 1986). These compounds, defined as antimicrobial compounds of low molecular weight, are primarily phenylpropanoid, isoprenoid, and acetylenic in structure. They are induced not only by microbes of specific types, but also by elicitor molecules. Based on available information, over 150 different phytoalexins are produced.

A variety of elicitors have been described that induce phytoalexin production (Table 6). These include complex carbohydrates from fungal and plant cell walls, lipids, microbial enzymes, and polypeptides. In addition, abiotic elicitors include heavy metals, detergents, autoclaved ribonuclease enzyme, cold, and UV light. Clearly, the induction of phytoalexin synthesis is important for influencing the colonization of roots and rhizosphere, and is only beginning to be understood. Based on information available to date, elicitors appear to stimulate multiple responses in plant cells, some of which may be linked with disease resistance (Ebel 1986). The primary target site and the mode of action of phytoalexins have not been characterized. The presence of a vast number of factors that can elicit these plant responses highlights the need to consider whether there is a unifying concept for their activity.

TABLE 6   Selected Elicitors of Phytoalexin Synthesis and Necrosis in Plants

Source	Chemical nature
Fungi	
Phytophthora megasperma f. sp. *glycinea* culture filtrate and mycelial cell walls	Branched β-glucan with 3-, 6-, and 3, 6-linked glucosyl residues
P. infestans mycelium	Eicosapentaenoic acid Arachidonic acid
Cladosporium fulvum	Peptide
Plant elicitors	Pectin fragments
Ricinus communis, citrus pectin, polygalacturonic acid	Oligo-α-1,4-galacturonide
Glycine max cell walls, citrus pectin	Oligo-α-1,4-galacturonide

SOURCE: Ebel (1986).

More detailed studies of effects of phytoalexins and related isoflavonoids are available (Stössel 1987). The zoopathic yeast *Candida albicans* was found to be sensitive to a series of these materials. Delay of growth as well as inhibition was observed. With some of these compounds, detoxification was suggested as the means of finally initiating growth. Tolerance by the microbes was felt, in some cases, to involve nondegradative mechanisms. Lucy et al. (1988) showed that pisatin detoxification by pisatin demethylase was a pathogenicity factor for the fungus *Nectria haematococca*, and thus is an important line of evidence that phytoalexins are important contributors to plant defense against invasive microbes.

Another interesting group of compounds produced by plants are the agglutinins, which can serve as antimicrobial compounds (Mansfield and Brown 1985) and play a role in colonization of roots (Anderson et al. 1988). These were observed around bacteria following infection. Extracts from a variety of infected plants have been shown to possess agglutinating activity. Jasalavich and Anderson (1981) found that an agglutinin with glycoprotein characteristics was present in water extracts of leaves, roots, and stems of *Phaseolus*. These compounds were able to agglutinate saprophytic pseudomonads, but not plant pathogens or nonpseudomonad saprophytes. The mode of action of these compounds, in terms of suppression, is still being evaluated. *Pseudomonas putida* mutants altered in agglutination were less capable of colonizing bean roots, suggesting a role of the agglutinin in bacterial colonization (Anderson et al. 1988). The concept that plant agglutinins may contribute to the colonization success of rhizosphere bacteria is also supported by studies with pea rhizosphere bacteria (Chao et al. 1988). These studies showed that agglutinability of rhizosphere bacteria in pea root exudates was correlated with enhanced bacterial attachment to roots.

Specific lectin agglutinins released from legumes interact with rhizobial symbionts in the external root environment (Dazzo and Hubbell 1975, Dazzo and Hrabak 1981, Sherwood et al. 1986, Truchet et al. 1984, Halverson and Stacey 1986, Kijne et al. 1988). These mediate specific interactions with bacterial cell surface receptors involved in bacterial attachment and subsequent infection of the legume host's root hairs. Polysaccharide-degrading enzymes that degrade these bacterial surface receptors in the root environment are also released from legume roots (Dazzo et al. 1982).

Mixed plant communities, leading to interspecific competition, represent another critical factor that can influence the plant and its microbial community. Direct plant competition can influence root-exudate characteristics (Newman 1978), and toxic exudates from one plant species may influence the growth of microbes on another plant. Thus, the interactions that might occur are dependent on plant type and the composition of the plant community. As discussed by Parrish and Bazzaz (1982), interactions between plants of the same and different types can be markedly different, especially when plants from different successional stages are involved. Thus, interspecific interactions must be considered in order to understand and possibly predict microbial colonization.

Microbial Communities That Develop in the Rhizosphere

The microbial communities that develop in the rhizosphere have been intensely investigated over the last half century (Baker 1987, Foster 1986). Although this has been considered to be a distinct population, more recently it was proposed (Lynch 1987a) that the rhizosphere represents a transient population that develops in addition to the microbial communities normally present in the stabilized soil aggregates.

The nonrhizosphere soil microbial community is primarily an autochthonous one (Lynch 1984). Thus, the major nutrients used by such a microbial community are the more stabilized and clay-associated residual organic and humic materials having lower C/N ratios (Parnas 1976). In addition, such a microbial population will largely be in a starved, nongrowing state of imposed microbiostasis (Morita 1985), and responses to added nutrients (as derived from the rhizosphere) will be limited by microbiostatic factors [as described by Ho and Ko (1986)] in the absence of sufficient fluxes of growth-sustaining substrates.

Based on concepts of Hattori and Hattori (1977), the bacterial community in soil can be assumed to be primarily located in smaller soil pores, while the fungal component will tend to bridge between aggregates (Lynch and Bragg 1985) because of its ability to translocate nutrients from and between different regions in the soil (Paustian and Schnurer 1987, Holland and Coleman 1986). This difference in strategies of fungi and bacteria suggests that rhizosphere bacteria depend primarily on soluble exudates from the root, while many of the filamentous fungi obtain their substrates from outside the root zone (Newman 1985). Another reason why the bacterial component is maintained predominantly in the smaller pores may be the protection that this microenvironment provides against protozoan predation (Coleman 1985).

The increased availability of readily utilizable organic matter from root exudates will modify the location of microbial development in soil. These areas of newly developing populations will also be more accessible for predation by the soil protozoans. Thus, the rhizosphere will have a major impact on dynamics of microbial predator-prey relationships.

Another critical factor that will influence microbial responses to the rhizosphere will be the level and flux rate of mineral nitrogen and possibly phosphorus. In more intensively managed and fertilized soils, nitrogen availability will decrease the population of filamentous fungi, which have the functional advantage of being able to translocate nitrogen to the site of available organic carbon. This will increase the relative amount of carbon processed through the bacterial biomass (Turner 1983). This increased nitrogen availability also decreases the demand for native soil organic matter as a source of nitrogen for microbial growth.

Other factors that will shift the relative bacterial-to-fungal dominance in utilization of plant root exudate are the levels of various inhibitors—e.g., heavy metals, pesticides, herbicides, and fungicides—in this environment (Anderson et al. 1981, Duah-Yentumi and Johnson 1986, Jackson and Watson 1977). In general, fungi are more efficient than bacteria in their capacity to bioaccumulate heavy metals in soil (Zamani et al. 1985).

A large number of microscopic and physiological studies have been used to evaluate the microbial community structure in the rhizosphere. Among the most important classical studies are those of Lochhead and Chase (1943) and Katznelson et al. (1948). A major limitation has been the use of viable enumeration techniques, which have emphasized the bacterial and largely disregarded the fungal component of the microbial community. These viable enumeration procedures will markedly underestimate the populations of bacteria. By use of microscopic procedures, values in excess of 10^{10} bacteria per gram of rhizosphere soil have been documented (Torsvik 1980, Klein et al. 1989.) In a similar way, the populations of fungi in the rhizosphere, when based on viable propagule recovery on conventional growth media, will favor the isolation of abundantly sporulating genera such as *Penicillium, Aspergillus, Trichoderma*, and *Fusarium* (Curl and Truelove 1986). In addition, shifts in the components of the fungal community were observed in range plants, in comparison with nonrhizosphere soil (Fresquez and Lindemann 1982). In spite of these limitations, useful information was derived concerning the characteristics of microbes in the rhizosphere. To more effectively evaluate these populations and their activities, more refined microscopic procedures will be required, as discussed by Brock (1987).

Until such techniques are used on a more routine basis, most of the information on R/S ratios and physiological diversity, although providing interesting insights, will be incomplete, and possibly misleading.

The studies of Lochhead and Chase (1943) and Dahm (1984) on the nutritional groupings of bacteria revealed distinct differences in the requirements for amino acids, growth factors, and vitamins between rhizosphere and nonrhizosphere isolates. In addition, marked differences in respiration rates were observed between the two groups of organisms (Dahm 1984). Unfortunately, except in limited instances, these types of studies did not compare rhizospheres from different plant types and plants grown under different conditions. Studies with range grasses of different successional stages indicate that there may be marked shifts in the soluble amino acids, growth factors, and vitamins present in the root zone depending on the plant seral stage and nutrient resource availability (Table 7). These differences are related

TABLE 7   Soluble Sugars, Organic Acids, and Amino Acids in the Rhizospheres of *Agropyron smithii* (AGSM) and *Sitanion histrix* (SIHY), Piceance Basin, Colorado, 1986.

Treatment	Component[a]			
	Sugars	Organic acids	Amino acids	Total
Control:				
AGSM	96.5	54.3	54.0	204.8
SIHY	4.1	84.7	57.8	146.6
Fumigated:				
AGSM	20.9	27.6	49.4	97.9
SIHY	6.4	92.4	49.2	148.0
Fertilized:				
AGSM	4.1	136.0	115.5	255.6
SIHY	4.1	27.6	38.6	70.3

[a]Figures in micrograms of component per gram of soil.

to the nutritional requirements, amino acid growth factors, and vitamins of bacteria isolated from the rhizospheres of these varied plants (data not shown).

These considerations suggest that earlier work in this area has provided useful concepts relative to potential changes that can occur in the rhizosphere and rhizoplane microbial communities. It is also evident that the microbial populations are much more responsive and that more detailed analyses should make it possible to effectively assess and possibly control the nutritional responses of plants and their associated microbial communities to environmental change.

Microfauna are an often neglected group of organisms occupying an important trophic level in rhizospheres, and these organisms play a significant role in the dynamics of microbial populations. Soils have an abundant population of protozoa (ciliates, flagellates, and amoeboid forms), which develop strong predator-prey interactions with the bacterial and fungal communities (Clarholm 1981, Coleman 1985, Fenchel 1986, Stout 1980).

Especially with the possibility of protozoa preferentially feeding on more nutritionally attractive bacteria grown in the rhizosphere or especially with bacteria added as amendments to establish desired microbes in the rhizosphere (Goldstein et al. 1985), the protozoa should be given serious consideration in terms of their possible grazing on added bacteria. The use of bacterial amendments, in terms of modifying the soil and rhizosphere microbial community, should be planned with the consideration of possible preferential protozoan grazing.

The fungal population, in terms of nutritional functions, has not been given a similar emphasis, and physiological microscopic procedures must be developed, as used with fluorescein diaretate (FDA) hydrolysis (Ingham and Klein 1984). The recovery procedure markedly affects the fungal community available for analysis. Direct-plating studies have shown that *Fusarium* comprises a greater portion of the fungal community than would be expected based on viable enumeration procedures (Curl and Truelove 1986).

A variety of factors can influence the magnitude and characteristics of the saprophytic populations in the rhizosphere. Foster (1986) noted that mycorrhizal infection of roots resulted in 10-fold to 100-fold higher populations of bacteria in this "mycorrhizosphere"; however, even with these increases, only 7% of the root surface was covered with microbes. The mycorrhizal component was sensitive to the presence of pesticides, as summarized by Trappe et al. (1984).

These interactions can also involve tripartite associations in which an additional set of energetic costs are placed on the plant for maintenance of both mycorrhizal and nitrogen-fixing functions. Pacovsky et al. (1986) found a positive interaction between vesicular-arbuscular

(VA) mycorrhizae and *Rhizobium*, with nitrogen fixation (acetylene reduction) being significantly higher in the tripartite system. Of the increase, 80% was attributed to increased nodule mass and 20% to increased nodule activity. VA mycorrhizal fungal colonization was also significantly increased following nodulation. If water is not limiting the development of this three-membered symbiosis, the fourfold to fivefold increase in phytomass will lead to increased root exudation and possibly a changed potential for colonization. It was concluded that endophyte influences on the host were not limited to effects of N and P nutrients, and could involve plant hormone changes.

These varied concepts, in summary, provide a physical and physiological framework that will be important for understanding the physical association of colonizing microbes with the plant. Mansfield and Brown (1985) emphasized that most phytopathogenic bacteria, if not colonizing the plant, will be maintained as a part of the rhizosphere and rhizoplane microbial community. This zone is also important for the maintenance and proliferation of bacteria originally associated with the seed surface.

Root exudates can also play a major role in fungal colonization. These are more important in the utilization of insoluble, partially lignified polymeric root products. Lasik et al. (1979) found that polysaccharides produced by *Agrobacterium* spp. were used preferentially by *Gaeumannomyces* strains. Low concentrations of added polysaccharides increased wheat growth, while higher concentrations inhibited growth. The possibility of bacterial polysaccharides being preferred substrates could lead to an increased potential for disease control of *Gaeumannomyces graminae*. This interaction could be related to two possible factors that influence the plant: carbon source supply and vitamin production. Utilization of fungal substrates by other fungi has been used as a biocontrol strategy under more controlled conditions (Lynch 1987*b*).

Important but still poorly documented factors controlling rhizosphere populations are volatile fungistatic and bacteriostatic compounds. As discussed by Ho and Ko (1986) and Hora and Baker (1974), a variety of often poorly defined materials can lead to regulation of microbial populations in soils, especially in the absence of available nutrient resources. The ability of the rhizosphere environment to relieve such fungistasis (Drury et al. 1983) is a critical factor in the ability of the colonizing population not only to survive, but also, during the more active phases of plant root colonization, to grow.

Colonization will also be influenced by the presence of monocultures or mixed assemblages of plants. Bowen (1980) reviewed examples of assemblages: e.g., *Lolium* and *Plantago*, in which pairing *Lolium* decreased mycorrhizal infection of *Plantago* (Christie et al. 1978). These effects were suggested as being indirect, related to changes in plant nutrient status, assimilation, and root-exudation dynamics. The challenge at the next level of understanding will be to measure the impact of these plant-plant interactions on the C, N, and P content of the exudates, and if possible the microbial biomass and the organic matter in the rhizosphere, particularly in a successional context.

Newman (1978) has stressed that "intermingling plant species can influence the abundance of each other's root saprophytes and mycorrhizae." Direct competition, exudate modification, and toxic exudates may play a role in plant allelopathy. The exudates from one plant may support and stimulate growth of microbes in the rhizosphere of the other plant. In addition, saprophytic fungal hyphae and especially mycorrhizae (Whittingham and Read 1982) may be capable of transferring nutrients from one plant to another. The recurrent theme is that the terms "rhizosphere," "exudate," and "rhizosphere microbial community" must be used in a much more specific manner, as the microbes appear to be more responsive to particular plants, plant communities, and management than had previously been thought. Each of these factors will have to be considered in assessing the significance of colonization and interactions of plants with the microbial community.

Nutrients and Plant Community Productivity

The plant provides the major source of carbon for maintenance of the microbial community in the rhizosphere. The plant is also dependent on the ability of the microbial community to make required nutrients available, including nitrogen, phosphorous, and iron. Thus, the plant root and the rhizosphere microflora maintain a mildly mutualistic relationship. The rhizosphere microbial community also plays a critical role in the dynamics of accumulation and degradation of organic matter, providing a chemical energy base for storage of plant nutrients (O'Neill and Reichle 1980) resulting from microbial contributions to the formation of soil organic matter.

In a long-term teleonomic (Harold 1986) view, the expenditure of carbon for the development of organic matter, including its nutrient reserves, provides a more stable physiochemical environment for the continued development of the plant community. In a disturbance or secondary successional context, early successional and weedy plants, which expend minimum photosynthate for the development of organic matter, are able to proliferate briefly in the initial exploitation of disturbed sites where mineral nutrients accumulated by other plant types are available at higher levels.

Microbial utilization of accumulated plant residues and photosynthate results in a gradual decrease in C/N ratio as decomposition and stabilization of organic matter progresses (Table 8). The soil insects, animals, and microbial biomass have a low C/N ratio and a rapid turnover rate, and thus serve as an important source of nitrogen, phosphorous, and other minerals for the plants and other microbes (Table 9). The dynamics of microbial maintenance in the root zone, and the final processing of plant materials into the various organic-matter pools, are dependent on the levels of minerals available for plant use, as well as the mean temperature of the particular plant-soil system.

Phenolic plant decomposition residues can accumulate with continued growth of specific plants. This leads to the phytotoxic situation generally referred to as "soil sickness" (Vancura and Catska 1979). As discussed by Catska et al. (1982), the microbial community in the rhizosphere gradually changes with increasing plant age. Early studies on this subject are reviewed by McCalla and Haskins (1964). Increases in fungi and actinomycetes, and decreases in bacteria, occurred, together with a decrease in the soil pH in the rhizosphere environment. Phytotoxicity was felt to be associated with the presence of saprophytic fungi of the genera *Fusarium, Verticillium, Trichoderma*, and *Mucor* and especially of *Penicillium*. Griseofulvin, methylsalicylic acid, and patulin, produced in vitro, have been suggested as active agents (Vancura et al. 1983). The use of microbial inoculation to decrease the number of phytotoxic fungi developing under this situation was suggested.

In relation to the aquisition of nitrogen from soil organic matter or from diazotrophic nitrogen fixation, plant roots will tend to excrete soluble compounds having higher C/N ratios under nitrogen-limited

TABLE 8   Nutrient Resources and Estimated Half-Life Turnover Times for Major Plant Components in a Grassland Soil

Component	Nutrient resources[a]		C/N ratio	Approx. turnover time (years)
	Carbon	Nitrogen		
Aboveground, green	43.0	1.4	31	1–2
Aboveground, dead	64.0	2.5	26	2
Plant crowns	160.0	4.6	35	2
Live roots	88.0	2.3	38	3–4
Senescent roots	282.0	14.0	20	3–4
Detrital roots	149.0	9.0	17	4
Soil organic matter	1,327.0	127.0	10	100–1,000 and longer

[a]Figures in grams per square meter, 0 to 10 centimeters soil depth.
SOURCE: Developed from Hunt et al. (1987) and Woodmansee et al. (1978).

TABLE 9   The Biotic and Microbial Communities in a Grassland Soil: Resources and Annual Turnover

Trophic group	Nutrient resources[a]		C/N ratio	Approx. turnover time (years)
	Carbon	Nitrogen		
Bacteria and fungi:				
Bacteria	30.400	7.6	4	0.83
Saprophytic fungi	6.300	0.63	10	0.83
Vesicular-arbuscular fungi	0.700	0.070	10	0.50
Protozoans:				
Amoebae	0.378	0.054	7	0.17
Flagellates	0.0161	0.0023	7	0.17
Nematodes:				
Predaceous nematodes	0.108	0.0108	10	0.63
Omnivorous nematodes	0.065	0.0065	10	0.23
Bacterivorous nematodes	0.580	0.0580	10	0.37
Fungivorous nematodes	0.041	0.0041	10	0.32
Phytophagous nematodes	0.290	0.029	10	0.93
Insects:				
Collembola	0.00464	0.00058	8	0.54
Predaceous mites	0.0160	0.0020	8	0.54
Nematophagous mites	0.0160	0.0020	8	0.54
Fungivorous mites	0.304	0.019	8	0.68

[a]Figures in grams per square meter.
SOURCE: Adapted from Hunt et al. (1987).

conditions, as is observed with sugarcane (Döbereiner et al. 1988) and with unfertilized range plants (Klein et al. 1988). Under these conditions, the plant appears to invest in the synthesis of exudates with higher C/N ratios. This will allow nitrogen mobilization from the soil organic matter (Clarholm 1985) and tend to favor rhizosphere nitrogen fixation. Depending on the availability of nitrogen from soil organic matter, the presence of these higher-C/N-ratio soluble components in the rhizosphere will result in increased mineralization or nitrogen fixation.

Under conditions in which mineral nitrogen is available, the C/N ratio of exudates can be expected to be lower. Turner (1983) grew *Lolium perrene* in soil with or without added mineral nitrogen and/or phosphorous. The addition of nutrients resulted in decreased fungal dominance on the rhizoplane of these plants, whereas without nutrient amendments, the fungal component was a higher proportion of the microbial community (Table 10). These results again indicate the more distinct sensitivity of the fungal community to nutrient availability. Nutrient availability from soil organic matter also can be influenced by the mean soil temperature and water availability. The

TABLE 10   Ratio of Percentage of Root Surface Covered by Fungi to Percentage Covered by Bacteria

(Data from experiments of Turner in which *Lolium perenne* was grown in soil with various mineral nutrients supplied weekly)

Nutrients supplied to plants	Experiment		
	1	2	3
Complete[a]	0.62	0.61	0.06
Complete minus N[b]	3.0	3.6	1.9
Complete minus P[c]	5.4	nd[d]	2.3
None	1.25	3.6	1.8

[a]Complete modified Hoagland's solution
[b]N omitted
[c]P omitted
[d]nd = not determined
SOURCE: Turner (1983).

maintenance of a mean soil temperature of above approximately 25°C, or a greater degree of oxygenation, will result in decreased organic-matter retention (Mohr et al. 1972), which will result in decreased nutrient accumulation in the chemical energy reservoir for possible later use by microbes and plants. Under higher-temperature conditions, exudates can be used to support nitrogen fixation, as the organic-matter nitrogen pool is decreased due to the higher turnover and decomposition rate. In tropical soils, the plant residue and exudate available for microbial use can be high (Döbereiner et al. 1988), and decomposition will also be rapid.

Another important aspect of plant-microbe relationships affecting productivity is the synthesis of vitamins and growth hormones (auxins, cytokinins, etc.), which can influence plant phenology (Brown 1975, Hussain and Vancura 1969, Prikryl et al. 1985) and root growth. As an example, *Rhizobium* elaborates a low-molecular-weight, aromatic (nonheterocyclic) metabolite that, at concentrations as low as 10 nanograms per plant, can induce prolific development of root hairs (Dazzo et al. 1987). Baker (1987) has discussed the possible role of growth stimulants from *Trichoderma* in influencing plant growth. The field of "bacterization" or, more broadly, "microbialization" (Bruehl 1987) of plants is developing rapidly, and colonization plays an important role in these interactions, as discussed by Weller (1988), and in Chaps. 8 and 9 of this book. As recent examples, Ordentlich et al. (1987) found that rhizosphere colonization by *Serratia marcescens* led to control of *Sclerotium rolfsii* in beans under greenhouse conditions. Knight and Langston-Unkefer (1988) found that symbiotic dinitrogen fixation by alfalfa could be enhanced by colonization with a toxin-releasing plant pathogen. This effect was suggested to be caused by changes in the glutamine synthetase–catalyzed steps in ammonia assimilation by the plant.

Siderophores (see Chap. 9) are of increasing interest and can be important in the dynamics of iron utilization and competition between the plant and the microbial community. With the important role of iron in the development and proliferation of plant pathogens (Leong 1986), the rapid colonization of roots with fluorescent pseudomonads has been found to play a major role in increasing crop yields and decreasing the effects of some deleterious fungi and bacteria. The use of exudates to support colonization by these organisms can be of direct benefit to the plant. The role of iron competition in soil, in terms of physical, biological, and plant-microbe interactions, continues to be of intense interest (Baker et al. 1986, de Weger et al. 1986, Schippers et al. 1987) in terms of the development of colonization strategies.

Biotechnological Implications and Opportunities

Colonization is an essential step in the process of initiating microorganism interactions with plants. This process has many physical steps and biochemical interactions that can be exploited by biotechnology. Based on the varied modes and strategies of colonization, either to stimulate or retard this process, an understanding of these processes on a molecular level will lead to the most rapid advances.

General factors in colonization

With the need to rigorously define the root in terms of the structural aspects of the microenvironment, it is necessary to consider general factors influencing colonization, and more specific mechanisms of the plant for microbe recognition. General factors influencing colonization have been discussed by Baker (1987) and others. As discussed by Schippers et al. (1987), a variety of interactions can occur within the microbial community in the course of colonization. These can involve interactions of beneficial and deleterious microorganisms, and as noted by these authors, a major problem important in potential biotechnological applications is inadequate root colonization.

The physiological state of the organism is important in colonization. Iswandi et al. (1987) observed that the beneficial effects of rhizopseudomonad strains for growth of the barley cultivar Iban was a function of the storage time under laboratory conditions. After 6 months storage, most of the strains lost their beneficial effect on plant growth. Evidence was presented to suggest a relationship between the ability to colonize roots and expression of antifungal activity. This phenomenon has also been observed with *Rhizoctonia*, which after extended laboratory culture loses its ability to induce germination of orchid seeds (Caullery 1952).

Colonization can also involve interactions at the genetic level as discussed by Djordjevic et al. (1987). Increased understanding of the genetic basis for expression of chemical communication is beginning to suggest important biotechnical applications.

The area of fungal infection of plants has developed rapidly in the last several years, and the outcomes of infection involve such important phenomena as recognition and resistance (Pegg and Ayres 1988). Especially with mycorrhizal infection, a series of carefully timed steps occur, and genetic and immunological probes will be important in understanding these interactions.

Colonization processes

Random colonization.   This area, which has been discussed by Schmidt (1979), involves the movement and growth of the root through a volume of soil, and the changing of a low-nutrient-flux environment to one dominated by rhizosphere exudates and the presence of the root itself. Colonization, in this case, can simply involve the eventual random physical association of microbes in the plant root zone (passive retention) or an active increase in populations, once the microbes have begun to grow, to become a significant portion of the root-rhizosphere microbial community.

Seed-surface-related colonization.   Organisms originally present on the seed, or added intentionally as an agricultural inoculant, must maintain themselves as a part of the root-surface–rhizosphere community if colonization is to be successful. Extensive root colonization by seed-applied Gram-negative bacteria has been observed, as discussed by Baker and Sher (1987) and Suslow and Schroth (1982). This colonization process involves several steps, as the seed germinates and the root penetrates through the soil volume. From a biotechnological viewpoint, this approach to colonization provides a unique opportunity to establish desired microbes, and has been used extensively with *Rhizobium* (Burton 1976) and other potentially beneficial organisms such as *Pseudomonas putida* (Dupler and Baker 1984). It is important to note, however, that the seed coat of many legumes is an inhospitable environment for *Rhizobium*, and thus seed inoculation just before planting is recommended. This inhibition is believed to be due to tannins and phenolic materials in the seed coatings.

Chao et al. (1988) observed that the adsorption of seed-coated bacteria to roots involves specific pectic compounds and other acidic polysaccharides inducing the agglutination of various bacteria. These workers found a greater ability of such bacteria to colonize the pea rhizosphere, and these organisms were considered to follow the downward growth of the pea root through the soil profile. Exudates from different plants varied in their abilities to agglutinate these bacteria.

Facilitated colonization and infection. Colonization of the root surface, and possible breaching of the more resistant surface cells of the plant root, can be facilitated by injury (Huang 1986), and by grazing and feeding on roots by nematodes (Mai and Abawi 1982). This can result in the "unintentional" or nonspecific colonization by microbes, or it can be part of a specific plant-grazing–organism-infection cycle. *Agrobacterium tumefaciens*, as an example, is not independently infective, but requires the occurrence of root wounding to allow infection.

Exudates as attractants. At a physical distance from the plant root, specific compounds synthesized by the plant can result in microorganism migration to the root surface. This phenomenon has been observed with *Azospirillum* (Bashan 1986), and zoospores of *Phytophthora cinnamomi* exhibited chemotaxis to roots of avocado seedlings (Zentmyer 1961). Once the particular organism has moved into the root zone, more specific aspects of colonization can take place. As discussed by Baker and Cher (1987), bacterial chemotaxis toward plant exudates is an important mechanism of colonization by beneficial and harmful microbes. Chet et al. (1973) found that *Pseudomonas lacrymans* responded to amino acids in plant leachates, and Sher et al. (1984) found that saprophytic root- and seed-colonizing pseudomonads could migrate 2 cm through moist, nonsterile soil toward an exuding root. Agglutinating factors are released from the root and cause the agglutination of saprophytic pseudomonads (Anderson 1983). These agglutinated bacteria appear to have a greater ability to adhere to and colonize the root surface.

Agricultural chemicals and colonization. A variety of agricultural chemicals have been shown to have effects on microbial colonization of plant roots and the rhizosphere. These effects, although often considered to result from agricultural chemical use, may have important biotechnological implications and applications. Catska and Vrany (1976) found that foliar application of chlorocholine chloride (CCC), urea, and 4-chloro-2-methylphenoxyacetic acid led to changes in numbers and composition of the rhizosphere mycoflora. The most distinct effect was observed with CCC. Levesque et al. (1987) found that glyphosate could influence *Fusarium* spp. colonization of weed roots, propagule density, and crop emergence.

Signal exchange in the rhizosphere

The most advanced level of colonization involves a series of plant-microbe interactions that are subject to a wide variety of genetic and physiological controls. These areas have been summarized by Djordjevic et al. (1987) and by Halverson and Stacey (1986). These can include factors influencing plant disease resistance and species recognition mechanisms.

One of the most highly advanced levels of plant root colonization occurs with the *Rhizobium*-legume symbiosis. In this plant-microbe interaction, not only do the bacteria inhabit the rhizosphere and colonize the surface of plant roots, but *Rhizobium* (fast growers) and *Bradyrhizobium* (slow growers) species possess the unique ability to specifically infect and multiply within the roots of plants in the legume family to form a mutually beneficial association. Infection of the roots leads to the morphogenesis of a root nodule, which contains the bacteria. The nodule protects the bacteria and serves as a sink for plant photosynthate, which fuels the bacteria as they fix atmospheric nitrogen to ammonia. This process supplies the legumes with fixed nitrogen compounds, minimizing or eliminating the need for nitrogenous fertilizer.

This mutualistic symbiosis is highly evolved, as demonstrated by the characteristic host specificity of the interactions. Rhizobia nodulate many legumes, e.g., *Lotus, Vigna, Macroptillium*, but only *Rhizobium leguminosarum* can infect peas, only *Rhizobium meliloti* can infect alfalfa, and only *Rhizobium trifolii* can infect white clover. Not all rhizobia display this high degree of host specificity. Some broad-range, or "promiscuous," strains exist that nodulate more than one legume species. In one particular case, the *Rhizobium* isolate NGR234 nodulates the nonlegume *Parasponia*.

The genes governing host specificity in the fast-growing rhizobia have been well characterized (Djordjevic et al., 1986). For *R. trifolii*, *R. leguminosarum*, and *R. meliloti*, successful symbiotic associations culminating in an N-fixing nodule filled with bacteroids depend on the presence of one to several large bacterial plasmids, appropriately known as symbiotic plasmids (pSym). In *R. trifolii*, there is one Sym plasmid on which genes essential for nodulation are found. To date, the *nod* genes that have been named are the common *nod* genes *nodDABC*; the host-specific nodulation (*hsn*) *nodFE(R)LMN* genes; and two genes that affect infection thread development, *nodIJ*. "Common *nod* genes" are defined as genes having homologous DNA sequences among the rhizobia, whose loss by mutation or deletion renders the bacteria unable to infect their own or any host, and which when mutated or deleted, can be replaced by the equivalent gene from another *Rhizobium* species to restore nodulation ability. To be designated an *hsn* gene, either (1) the gene must alter the bacterial host range when that gene function is lost (e.g., *Tn5* insertion in *nodE* extends *R. trifolii* host range to nodulate peas), or (2) the gene must be transferred to a recipient strain of a different *Rhizobium* species in order to allow that recipient to efficiently nodulate the homologous host of the donor *Rhizobium* (e.g., *R. trifolii nodFELMN* transferred to *R. leguminosarum* enable the latter to efficiently nodulate white clover) (Djordjevic et al. 1986).

Successful infection is dependent upon a chemical communication by the two symbiotic partners at many different steps in the physiological cascade of nodulation (Halverson and Stacey 1986). The events that occur before and during infection have been described as "attachment" (Roa), "hair curling" (Hac), "infection thread formation" (Inf), "nodule initiation" (Noi), "nodule formation" (Nod), "bacterial release from infection threads" (Bar), "bacteroid development" (Bad), and finally "nitrogen fixation" (Fix) (Vincent 1980).

The chemical signals controlling this plant-microbe symbiosis are currently an important area of research, with the ultimate goal of understanding the various molecular events leading to formation of nitrogen-fixing nodules on specific legume hosts. Which signal exchange occurs first has yet to be determined, but a plausible scenario provides that plant root exudates serve as the attractant for free-living bacterial movement through the soil (Frazier and Fred 1922). One particular class of compounds in root exudates, the flavones, are pivotal signals leading to successful infection. These are hydroxylated aromatic compounds synthesized from a branch of the plant's phenylpropanoid pathway, which also gives rise to phytoalexins. Flavones stimulate *Rhizobium nod* gene expression in the nanomolar to micromolar range (Peters et al. 1986, Redmond et al. 1986). Some flavones are stimulatory, some are inhibitory, and some have no effect, depending on their pattern of hydroxylation and the *Rhizobium* species being examined. Some host specificity is dictated by the specific flavone to which a particular *Rhizobium* responds, and the basis of this flavone specificity lies in specific domains of the protein product of the pSym regulatory gene, *nodD* (Spaink et al. 1987). Those genes shown by transcriptional *lacZ* fusions to be expressed after exposure to flavones have a conserved upstream region known as a "nod box". Whether the flavones interact with this region of *nod* box DNA directly, or interact with the regulatory *nodD* protein product after it binds to the *nod* box has yet to be determined (Fisher et al. 1988).

Recent work with white clover suggests that flavone induction of the rhizobial common *nod* genes occurs at the root-hair tips (Dazzo et al. 1988*a*), which are precisely the site of bacterial penetration during infection.

After the bacteria reach the root-hair surface, an initial reversible attachment occurs that is not symbiont-specific (Dazzo et al. 1984*b*) and may involve a Ca^{2+}-dependent bacterial adhesin (Smit et al. 1986). Within an hour, this step is followed by a symbiont-specific, lectin-mediated step of bacterial aggregation (Dazzo et al. 1984*b*, Kijne et al. 1988). The plant lectin is viewed as a signal for infection during symbiont-specific aggregation at root-hair tips. *Rhizobium* in the rhizosphere of the legume host interacts directly with the lectin synthesized by the roots and secreted into the external environment (Sherwood et al. 1986, Truchet et al. 1986). Preliminary results of D. Gerhold and G. Stacey (personal communication) indicate that specific binding of the excreted lectin to *Rhizobium* in the rhizosphere triggers gene expression involved in efficient nodulation. The ability of the bacteria to interact with the host lectin varies with physiological growth phase (Bhuvaneswari et al. 1977, Dazzo et al. 1979) and nutrient limitations (Kijne et al. 1988). These factors will undoubtedly influence the interactions of the bacteria with the lectin in the rhizosphere. Successful infections are now considered to be restricted to sites where the bacteria express lectin receptors (Kijne et al. 1988). Extracellular microfibrils are elaborated where bacteria have attached to the roots, and these projections are believed to also participate in the aggregation of bacteria at root-hair tips (Smit et al. 1986) and to firmly anchor the bacteria to the root-hair surface (Dazzo et al. 1984*b*). A similar process of receptor-ligand interaction, followed by elaboration of extracellular microfibrils, occurs in the attachment of *Agrobacterium* to plant cells (Matthysse 1983).

Bacterial signals are currently under study. Van Brussel et al. (1986) described the thick-short root (Tsr) phenotype elicited on roots treated with bacterial culture supernatant. Production of Tsr factor requires *Rhizobium* pSym *nodABC* genes. Dazzo et al. (1987) have isolated a compound (BF-1) produced by *R. trifolii* that stimulates development and elongation of white clover root hairs. Other bacterial factors, which cause root-hair deformation and curling, and may impact on determining host specificity, have been isolated by Hollingsworth and colleagues (Dazzo et al. 1988*b*, Faucher et al. 1988).

Two other bacterial signals reside on the *Rhizobium* cell surface, the capsular polysaccharide (CPS) (Abe et al. 1984) and the lipopolysaccharide (LPS) (Dazzo et al. 1984*a*). Culture-age-dependent chemical changes in these surface molecules can be correlated with the degree of infection-related biological activity. Pretreatment of clover roots with these glycoconjugates increases the number of infections per clover seedling at low concentrations and inhibits infection at higher concentration. Carlson et al. (1984) and Noel et al. (1986) have evidence that certain rough *Rhizobium phaseoli* mutants whose LPS lacks the O-antigen can initiate but cannot sustain infection thread growth in host root hairs. Thus, LPS may also be an important signal to allow continued growth of the infection thread as an indication of compatible interactions between the two symbiotic partners. There must also be some signal exchange between symbionts to elicit cortical cell proliferation, since root cells begin dividing before contact with the inward-advancing infection thread induced by these *Rhizobium* mutants.

There is even more direct evidence that bacterial factors signal root-nodule induction. Puhler et al. (1988) have shown that root-nodule structures can be induced by excreted bacterial metabolites without physical contact of the root by the bacteria. In these experiments, a membrane filter separates the bacteria from the root, but presumably bacterial substances can diffuse across membranes. Other workers have used growth regulators and growth regulator transport inhibitors to induce rudimentary nodule formation (Hirsch et al. 1989).

Finally, there is evidence that *Azospirillum*, when grown in the presence of grass roots, elaborates molecules that signal morphogenic changes in root-hair development (Tien et al. 1979, Umali-Garcia et al. 1980, and others).

Epilogue

Although colonization is being used in a series of agricultural crops to modify the effects of pathogens, or to increase growth and yields in selected commercial plants, this process has developed on a somewhat empirical basis, and results, in many cases, have not been predictable. This is especially evident when more controlled and successful glasshouse-level experiments are extended to the actual field environment, where less positive results are often achieved.

It is evident that further advances will involve integration at a series of levels of molecular refinement in carefully designed experiments. As specific chemical elicitors and agglutinin factors are identified in key steps in these colonization processes, it should be possible to select genetic variants for use in more rigorous comparative experiments. As an example, the use of nitrate reductase–negative *Azospirillum* has allowed the contributions of nitrate reductase to the accumulation of nitrogen by tropical plants colonized by this organism to be examined on a rigorous basis (Boddey et al. 1986).

Improvements in our understanding of basic aspects of microbial ecology will also make important contributions to this area. With such a framework in place, it will be possible to more fully understand the implications of a particular organism, biochemical process, or active chemical for the colonization process.

We foresee important biotechnological applications to improvement of crop productivity of microbial metabolites affecting plant growth. For instance, application of these substances or their enhanced elaboration by rhizosphere microbes in situ may promote root-hair development, leading to more absorptive area and an efficient root system capable of sustaining plant health under water- or nutrient-limited conditions that prevail in much of the world. However, much more research (both basic and applied) is necessary to meet this major goal of the rhizosphere microbial ecologist.

Acknowledgments

Portions of the work reported here were supported by the U.S. Department of Energy under grant DE-FG02-87ER 60612 to D. A. Klein, by National Institutes of Health grant GM34331-04, and by the National Science Foundation–Michigan State University Center for Microbial Ecology, subproject 710 to F. B. Dazzo. J. L. Salzwedel was supported by a Michigan State University Nitrogen Availability Program and the MSU-REED program.

References

Abe, M., Sherwood, J. E., Hollingsworth, R. I., and Dazzo, F. B. 1984. Stimulation of clover root hair infection by lectin-binding oligosaccharides from the capsular and extracellular polysaccharides of *Rhizobium trifolii, J. Bacteriol.* 160:517–520.

Anderson, A. J. 1983. Isolation from root and shoot surfaces of agglutinins that show specificity for saprophytic pseudomonads, *Can. J. Bot.* 61:3438–3443.

Anderson, A. J., Habibzadegah-Tari, P., and Tepper, C. 1988. Molecular studies on the role of a root surface agglutinin in adherance and colonization by *Pseudomonas putida, Appl. Environ. Microbiol.* 54:375–380.

Anderson, J. P. E., Armstrong, R. A., and Smith, S. N. 1981. Methods to evaluate pesticide damage to the biomass of the soil microflora, *Soil Biol. Biochem.* 13:149–153.

Baker, K. F. 1987. Evolving concepts of biological control of plant pathogens, *Annu. Rev. Phytopathol.* 25:67–85.

Baker, R., Elad, Y., and Sneh, B. 1986. Physical, biological and host factors in iron competition in soils. In *Iron, Siderophores, and Plant Diseases*, T. R. Swinburne (ed.). Plenum, New York, pp. 78–84.

Baker, R., and Cher, F. M. 1987. Enhancing the activity of biological control agents. In *Innovative Approaches to Plant Disease Control*, I. Chet (ed.). Wiley, New York, pp. 1–17.

Baker, K. F., and Snyder, W. C. 1965. *Ecology of Soil-Borne Plant Pathogens*. University of California Press, Berkeley.

Barber, D. A., and Martin, J. K. 1976. The release of organic substances by cereal roots into soil, *New Phytol.* 76:69–80.

Bashan, Y. 1986. Migration of the rhizosphere bacteria *Azospirillum brasilense* and *Pseudomonas fluorescens* towards wheat roots in the soil, *J. Gen. Microbiol.* 132:3407–3414.

Bazzaz, F. A. 1979. The physiological ecology of plant succession, *Annu. Rev. Ecol. Syst.* 10:351–371.

Bhuvaneswari, T. V., Pueppke, S. G., and Bauer, W. D. 1977. Role of lectins in plant-microorganism interactions, I. Binding of soybean lectin to rhizobia, *Plant Physiol.* 60:488–491.

Billings, E. 1982. Entry and establishment of pathogenic bacteria in plant tissues. In *Bacteria and Plants*, M. E. Rhodes-Roberts and F. A. Skinner (eds.). Academic, New York, pp. 51–70.

Blakeman, J. P., and Fokkema, N. J. 1982. Potential for biological control of plant diseases on the phylloplane, *Annu. Rev. Phytopathol.* 20:167–192.

Boddey, R. M., Baldani, J. I., Baldani, V. L. D., and Döbereiner, J. 1986. Effect of inoculation of *Azospirillum* spp. on nitrogen accumulation by field-grown wheat, *Plant and Soil* 95:109–121.

Bokhari, U. G., Coleman, D. C., and Rubink, A. 1979. Chemistry of root exudates and rhizosphere soils of prairie plants, *Can. J. Bot.* 57:1473–1477.

Bowen, G. D. 1980. Misconceptions, concepts and approaches in rhizosphere biology. In *Contemporary Microbial Ecology*, (ed.). Academic, New York, pp. 283–304.

Brock, T. D. 1987. The study of microorganisms *in situ*: progress and problems. In *Ecology of Microbial Communities*, M. Fletcher, T. R. G. Gray, and J. G. Jones (eds.). Symp. 41, Society for General Microbiology, Cambridge University Press, Cambridge, U.K.

Brown, M. E. 1975. Rhizosphere microorganisms—opportunists, bandits or benefactors. In *Soil Microbiology. A Critical Review*, N. Walker (ed.). Halstead Wiley, New York, pp. 21–38.

Bruehl, G. W. 1987. *Soilborne Plant Pathogens*. Macmillan, New York.

Burton, J. C. 1976. Methods of innoculating seeds and their effects on the survival of rhizobia. In *Symbiotic Nitrogen fixation in Plants*, P.S. Nulman (ed.). Cambridge Univ. Press, Cambridge, pp. 175–189.

Carlson, R. W., Kalembasa, S., Turowski, D., Pachori, P., and Noel, K. D. 1987. Characterization of the lipopolysaccharide from a *Rhizobium phaseoli* mutant that is defective in infection thread development, *J. Bacteriol.* 169:4923–4928.

Catska, V. 1979. Volatile and gaseous metabolites of germinating seeds and their role in fungus plant relationships. In *Soil-Borne Plant Pathogens*, B. Schippers and W. Gams (eds.). Academic, London.

Catska, V., Hudska, G., Prikryl, Z., and Vancura, V. 1982. Rhizosphere microorganisms in relation to the apple replant problem, *Plant and Soil* 69:187–197.

Catska, V., and Vrany, J. 1976. Rhizosphere mycoflora of wheat after foliar application of chlorocholine chloride, urea and 4-chloro-2-methylphenoxyacetic acid, *Folia Microbiol.* 21:268–273.

Caullery, M. 1952. *Parasitism and Symbiosis*. Sedgwick and Jackson, London.

Chao, W.-L., Li, R-K., and Chang, W-T. 1988. Effect of root agglutinin on microbial activities in the rhizosphere, *Appl. Environ. Microbiol.* 54:1838–1841.

Chet, I., Ziberstein, Y., and Henis, Y. 1973. Chemotaxis of *Pseudomonas lachrymans* to plant extracts and to water droplets collected from the leaf surfaces of resistant and susceptible plants, *Physiol. Plant Pathol.* 3:474–479.

Christie, P., Newman, E. I., and Campbell, R. 1978. The influence of neighboring grassland plants on each others endomycorrhizas and root-surface microorganisms, *Soil Biol. Biochem.* 10:521–527.

Clarholm, M. 1981. Protozoan grazing of bacteria in soil: impact and importance, *Microb. Ecol.* 7:343–350.

Clarholm, M. 1985. Possible roles for roots, bacteria, protozoa and fungi in supplying nitrogen to plants. In *Ecological Interactions in Soil*, W. Fitter (ed.). Blackwell Scientific, London, 36:355–365.

Coleman, D. C. 1985. Through a ped darkly: an ecological assessment of root-soil-microbial-faunal interactions. In *Ecological Interactions in Soil. Plants, Microbes and Animals*, A. H. Fitter (ed.). British Ecological Society, Blackwell Scientific, Oxford, U.K., pp. 1–21.

Curl, E. A., and Truelove, B. (eds.). 1986. *The Rhizosphere*. Springer-Verlag, Berlin.

Dahm, B. 1984. Metabolic activity of bacteria isolated from soil, rhizosphere and mycorrhizosphere of pine, *Acta. Microbiol. Polon.* 33(2):157–162.

Dazzo, F., and Hrabak, E. 1981. Presence of trifoliin A, a *Rhizobium*-binding lectin, in clover root exudate, *J. Supramol. Struct. Cell. Biochem.* 16:133–138.

Dazzo, F., and Hubbell, D. 1975. Cross-reactive antigens and lectin as determinants of symbiotic specificity in the *Rhizobium*-clover association, *Appl. Microbiol.* 30:1017–1033.

Dazzo, F. B., Truchet, G. L., Sherwood, J. E., Hrabak, E. M., and Gardiol, A. E. 1982. Alteration of the trifoliin A-binding capsule of *Rhizobium trifolii* 0403 by enzymes released from clover roots, *Appl. Environ. Microbiol* 44:478–490.

Dazzo, F. B., Urbano, M. R., and Brill, W. J. 1979. Transient appearance of lectin receptors on *Rhizobium trifolii, Curr. Microbiol.* 2:15–20.

Dazzo, F. B., Truchet, G. L., and Hrabak, E. M. 1984a. Specific enhancement of clover root hair infections by trifoliin A-binding lipopolysaccharide from *Rhizobium trifolii.* In *Advances in Nitrogen Fixation Research,* C. Veeger and W. E. Newton (eds.). Nijhoff/Junk and Pudoc, The Hague and Wageningen, p. 413.

Dazzo, F. B., Truchet, G. L., Sherwood, J. E., Hrabak, E. M., Abe, M., and Pankratz, S. H. 1984b. Specific phases of root hair attachment in the *Rhizobium trifolii*-clover symbiosis, *Appl. Environ. Microbiol.* 48:1140–1150.

Dazzo, F. B., Hollingsworth, R. I., Philip, S., Smith, K. B., Welsch, M. A., Salzwedel, J., Morris, P., and McLaughlin, L. 1987. Involvement of pSYM nodulation genes in production of surface and extracellular components of *Rhizohium trifolii* which interact with white clover root hairs. In *Molecular Genetics of Plant-Microbe Interactions,* D. P. S. Verma and N. Brisson (eds.). Martinus Nijhoff, Dordrecht, The Netherlands, pp. 171–172.

Dazzo, F. B., Hollingsworth, R., Philip-Hollingsworth, S., Robeles, M., Olen, T., Salzwedel, J., Djordjevic, M., and Rolfe, B. 1988a. Recognition processes in the *Rhizobium trifolii*-white clover symbiosis In *Nitrogen Fixation: 100 Years After,* H. Bothe, F. de Bruijn, and W. Newton (eds.). Gustav Fischer Verlag, Stuttgart, pp. 431–435.

Dazzo, F. B., Hollingsworth, R., Salzwedel, J., Phillip-Hollingsworth, S., Robeles, M., Olen, T., Appenzeller, L., Wang, S., Toro, I., Squartini, A., Anderson, S., Chen, J., Chapman, K., Maya-Flores, J., Cargill, L., Djordjevic, M., and Rolfe, B. G. 1988b. Signal-recognition responses in the *Rhizobium trifolii*- white clover symbiosis. In *Symposium Proceedings,* 4th International Symposium on Molecular Genetics of Plant-Microbe Interactions, R. Palacios and D. Verma (eds.). American Phytopathological Society Press, St. Paul, Minnesota, pp. 35–40.

de Weger, L. A., van Boxtel, R., van der Burg, B., Gruters, R. A., Geels, F. P., Schippers, B., and Lugtenberg, B. 1986. Siderophores and outer membrane proteins of antagonistic, plant-growth-stimulating, root colonizing *Pseudomonas* spp. *J. Bacteriol.* 105:585–594.

Djordjevic, M. A., Innes, R. W., Wijffelmen, C. A., Schofield, P. R., and Rolfe, B. G. 1986. Nodulation is controlled by several distinct loci in *Rhizobium trifolii, Plant Mol. Biol.* 6:389–401.

Djordjevic, M. A., Gabriel, D. W., and Rolfe, B. G.. 1987. *Rhizobium*: the refined parasite of legumes, *Annu. Rev. Phytopathol.* 25:145.

Döbereiner, J., Reis, V. M., and Lazarini, A. C. 1988. New N_2 fixing bacteria in association with cereals and sugar cane. In *Nitrogen Fixation: 100 Years After,* C. H. Bothe, F. de Bruijn, and W. Newton (eds.). Gustav Fisher Verlag, Stuttgart, pp. 717–722.

Drury, R. E., Baker, R., and Griffin, G. J. 1983. Calculating dimensions of the rhizosphere, *Phytopathology.* 73(10):1351–1354.

Duah-Yentumi, S., and Johnson, D. B. 1986. Changes in soil microflora in response to repeated applications of some pesticides, *Soil Biol. Biohem.* 18(6):629–635.

Dupler, M., and Baker, R. 1984. Survival of *Pseudomonas putida*: a biological control agent in soil, *Phytopathology* 74:195–200.

Ebel, J. 1986. Phytoalexin synthesis: the biochemical analysis of the induction process, *Annu. Rev. Phytopathol.* 24:235–264.

Faucher, C., Maillet, F., Vasse, J., Rosenberg, C., van Brussel, A., Truchet, G., and Denarie, J. 1988. *Rhizobium meliloti* host range *nodH* gene determines production of an alfalfa-specific extracellular signal, *J. Bacteriol.* 170:5489–5499.

Fenchel, T. 1986. The ecology of heterotrophic flagellates. *Adv. Microb. Ecol.* 9:57–97.

Fisher, R. F., Egelhoff, T., Mulligan, J. T., Yelton, M., and Long, S. R. 1988. *Rhizobium meliloti* nodD binds to DNA sequences upstream of inducible nodulation genes. In *Nitrogen Fixation: 100 years After*, C. H. Bothe, F. de Bruijn, and W. Newton (eds.). Gustav Fisher Verlag, Stuttgart, pp. 391–398.

Foster, R. C. 1982. The fine structure of epidermal cell mucilages of roots, *New Phytol.* 91:727–740.

Foster, R. C. 1986. The ultrastructure of the rhizoplane and rhizosphere, *Annu. Rev. Phytopathol.* 24:211–234.

Foster, R. C., and Bowen, G. D. 1982. Plant surfaces and bacterial growth: the rhizosphere and rhizoplane. In *Phytopathogenic Procaryotes*, vol. 1. M. S. Mount and G. H. Lacy (eds.). Academic, New York, pp. 159–185.

Foster, R. C., Rovira, A. D., and Cock, T. W. 1983. *Ultrastructure of the Root-Soil Interface*, American Phytopathological Society, St. Paul, Minnesota.

Frazier, W. C., and Fred, E. B. 1922. Movement of legume bacteria in soil, *Soil Science* 14:29–35.

Freckman, D. W., and Caswell, E. P. 1985. The ecology of nematodes in agroecosystems, *Annu. Rev. Phytopathol.* 23:275–296.

French, R. C. 1985. The bioregulating action of flavor compounds on fungal spores and other properties, *Annu. Rev. Phytopath.* 23:173–199.

Fresquez, P. R., and Lindemann, W. C. 1982. Soil and rhizosphere microorganisms in amended coal mine spoils, *Soil Sci. Soc. Am. J.* 46:751–756.

Goldstein, R. M., Mallory, L. M., and Alexander, M. 1985. Reasons for possible failure of inoculation to enhance biodegradation, *Appl. Environ. Microbiol.* 50(4):977–983.

Greaves, M. P., and Darbyshire, J. F. 1972. The ultrastructure of the mucilaginous layer on plant roots, *Soil. Biol. Biochem.* 4:443–449.

Griffin, G. J., Hale, M. G., and Shay, J. 1976. Nature and quantity of sloughed organic matter produced by roots of axenic peanut plants, *Soil Biol. Biochem.* 8:29–32.

Hale, M. G., Moore, L. D., and Griffin, G. J. 1981. Factors affecting root exudation and significance for the rhizosphere ecosystems. In *Biological and Chemical Interactions in the Rhizosphere, Symposium*, Ecological Research Committee, Swedish National Research Council, Stockholm, pp. 43–71.

Halverson, L. J., and Stacey, G. 1988. Signal exchange in plant-microbe interactions, *Microbiol. Rev.* 50:193–225.

Hamlen, R. A., Bloom, J. R., and Lukezic, F. L. 1972. Influence of age and stage of development on the neutral carbohydrate components in root exudates from alfalfa plants grown in a gnotobiotic environment, *Can. J. Plant Sci.* 52:633–642.

Hargreaves and Kean 1986. In Biology and Molecular Biology of Plant-Pathogen Interactions, J. A. Bailey (ed.). NATO ASI Series, Series H Cell Biology, 1:135.

Harold, F. M. 1986. *The Vital Force: A Study of Bioenergetics*, W. H. Freeman, New York.

Hattori, T., and Hattori, R. 1977. The physical environment in soil microbiology: an attempt to extend principles of microbiology to soil microorganisms, CRC *Crit. Rev. Microbiol.* 4(4):423–461.

Hiltner, L. 1904. Uber neuere Erfahrungen und Probleme auf dem Gebiete der Bodenbacteriologie und unter Berüsuchtigung der Gründungung und Brache, Deutsche Landwirtschafts Gesellschaft, Berlin, Arbeiten der DLG, 98:59–78.

Hirsch, A., Bhuvaneswari, T., Torrey, J., and Bisseling, T. 1989. Early modulation genes are induced in alfalfa root outgrowths elicited by Auxin transport inhibitors. *Proc. Nat. Acad. Sci., U.S.A.* 8b: 1244–1248.

Ho, W. C., and Ko, W. H. 1986. Microbiostasis by nutrient deficiency shown in natural and synthetic soils, *J. Gen. Microbiol.* 132:2807–2815.

Holland, E. A., and Coleman, D. C. 1986. Litter placement effects on microbial and organic matter dynamics in an agroecosystem, *Ecology* 68(2):425–433.

Hora, T. S., and Baker, R. 1974. Influence of a volatile inhibitor in natural or limed soil on fungal spore and seed germination, *Soil Biol. Biochem.* 6:257–261.

Huang, J. S. 1986. Ultrastructure of bacterial penetration in plants, *Annu. Rev. Phytopathol.* 24:141–157.

Hunt, H. W., Coleman, D. C., Ingham, E. R., Ingham, R. E., Elliott, E. T., Moore, J. C.,

Rose, S. L., Reid, C. P. P., and Morley, C. R. 1987. The detrital food web in a shortgrass prairie, *Biol. Fertil. Soils* 3:57–68.

Hussain, A., and Vancura, V. 1969. *Formation of Biologically Active Substances by Rhizosphere Bacteria and Their Effect on Plant Growth.* Department of Soil Microbiology, Institute of Microbiology, Czechoslovak Academy of Sciences, Prague 4. 15:469–478.

Ingham, E. R., and Klein, D. A. 1984. Soil Fungi: Relationships between hyphal activity and staining with fluorescein diacetate. *Soil Biol. Biochem.* 16:279–280.

Iswandi, A., Bossier, P., Vandenabeele, J., and Verstraete, M. 1987. Deterioration and reactivation of beneficial rhizopseudomonads of barley (hordeum vulgare), *Biol. Fertil. Soils* 4:125–128.

Jackson, D. R., and Watson, A. P. 1977. Disruption of nutrient pools and transport of heavy metals in a forested watershed near a lead smelter, *J. Environ. Qual.* 6:331–338.

Jasalavich, C. A., and Anderson, D. J. 1981. Isolation from legume tissues of an agglutinin of saprophytic pseudomonads, *Can. J. Bot.* 59:264–271.

Jenny, H., and Grossenbacher, K. D. 1963. Root-soil boundary zone as seen in the electron microscope, *Soil Sci. Soc. Am. Proc.* 27:273–277.

Katznelson, H., Lochhead A. G., and Timonin, M. I. 1948. Soil microorganisms and the rhizosphere, *Bot. Rev.* 14:543–587.

Kijne, J. W., Smit, G., Diaz, C. L., and Lugtenberg, B. J. J. 1988. Lectin-enhanced accumulation of manganese limited *Rhizobium leguminosarum* cells on pea root hair tips, *J. Bacteriol.* 170:2994–3000.

Klein, D. A., Biondini, M., Frederick, B. A., and Trlica, M. J. 1988. Rhizosphere microorganism effects on soluble amino acids, sugars and organic acids in the root zone of *Agropyron cristatum, A. smithii* and *Bouteloua gracilis, Plant and Soil.* 110:19–25.

Klein, D. A., Frederick, B. A., and Redentre. 1989. Fertilizer effects on soil microbial communities and organic matter in the rhizosphere of *sitanion hystrix* and *Agropyron smithii, Arid soil Res. Rehab.* in press.

Knight, R. T. J., and Langston-Unkefer, P. J. 1988. Enhancement of symbiotic dinitrogen fixation by a toxin-releasing plant pathogen, *Science* 241:951–954.

Kosuge, T., and Nester, E. W. (eds.). 1986. *Plant-Microbe Interactions, Molecular and Genetic Perspectives.* Macmillan, New York, 2 vols.

Krasilnikov, N. A. 1958. *Soil Microorganisms and Higher Plants.* Academy of Sciences of the U.S.S.R. (Israel Program for Scientific Translations, 1961).

Lasik, J., Stanek, M., Vancura, V., and Wurst, M. 1979. Effect of bacterial polysaccharides on the growth of *Gaeumannomyces graminis* var. *tritici* and wheat roots, *Folia. Microbiol.* 24:262–268.

Leong, J. 1986. Siderophores: their biochemistry and possible role in the biocontrol of plant pathogens, *Annu. Rev. Phytopathol.* 24:187–209.

Levesque, C. A., Rahe, J. E., and Eaves, D. M. 1987. Effects of glyphosate on *Fusarium* spp.: its influence on root colonization of weeds, propagule density in the soil, and crop emergence, *Can. J. Microbiol.* 33:354–360.

Lochhead, A. G., and Chase, F. E. 1943. Qualitative studies of soil microorganisms. Nutritional requirements of the predominant bacterial flora, *Soil Sci.* 55:185–195.

Lucy, M., Matthews, P., and VanEtten, H. 1988. Metabolic detoxification of the phytoalexins maackiain and medicarpin by *Nectria haematococca* field isolates: relationship to virulence on chickpea, *Physiol. Mol. Plant Pathol.* 33:187–199.

Lynch, J. M. 1984. Interactions between biological processes, cultivation and soil structure, *Plant and Soil* 76:307–318.

Lynch, J. M. 1987a. Biological control within microbial communities of the rhizosphere, In: Ecology of Microbial communities (M. Fletcher, T. R. G. Gray and J. G. Jones, eds.) pp. 55–82, *Symp. 41 Soc. Gen. Microbiol.*, Cambridge Univ. Press, Cambridge.

Lynch, J. M. 1987b. *Microbial Interactions in the Rhizosphere,* AFRC Institute of Horticultural Research, Soil Microorganisms, publication no. 30. Soil Microbiological Society of Japan, pp. 33–41.

Lynch, J. M. 1987c. Soil biology: accomplishments and potential, *Soil Sci. Soc. Am. J.* 51(6):1409–1412.

Lynch, J. M., and Bragg, E. 1985. Microorganisms and soil aggregate stability, *Adv. Soil Sci.* 2:133–171.

Macura, J., and Vancura, V. (eds.). 1965. Plant Microbes Relationships, Czechoslovak Academy of Sciences, Prague.

Mai, W. F., and Abawi, G. S. 1982. Interaction among root-knot nematodes and fusarium wilt fungi on host plants, *Annu. Rev. Phytopathol.* 25:317–338.

Mansfield, J. W., and Brown, I. R. 1985. The biology of interactions between plants and bacteria, In *Biology and Molecular Biology of Plant-Pathogen Interactions*, J. A. Bailey (ed.). NATO ASI Series, Series H Cell Biology, pp. 71–98.

Marshall, K. C. 1976. Bacterial adhesion in natural environments, In *Microbial Adhesion to Surfaces*, R. C. W. Berkeley, J. M. Lynch, J. Melling, P. R. Rutter, and B. Vincent (eds.). Ellis Horwood, Chichester, U.K., pp. 187–196.

Matthysse, A. G. 1983. Role of bacterial cellulose fibrils in *Agrobacterium tumifaciens* infections, *J. Bacteriol.* 154:906–915.

McCalla, T. M., and F. Haskins. 1964. Phytotoxic substances from soil microorganisms and crop residues, *Bacteriol. Rev.* 28:181–201.

McDougall, B. M., and Rovira, A. D. 1970. Sites of exudation of ^{14}C-labelled compounds from wheat roots, *New Phytol.* 69:999–1003.

Mohr, E. C. J., van Baren, F. A., and van Schuylenborgh, J. 1972. *Tropical Soils. A Comprehensive Study of their Genesis*, 3d ed., Mounton-Baru-Van Hoeve, The Hague.

Morita, R. Y. 1985. Starvation and miniaturisation of heterotrophs, with special emphasis on maintenance of the starved viable state. In *Bacteria in the Natural Environments: The Effect of Nutrient Conditions*, M. Fletcher and G. Floodgate (eds.). Academic, New York, pp. 11–130.

Newman, E. I. 1978. Root microorganisms: their significance in the ecosystem, *Biol. Rev.* 53:511–554.

Newman, E. I. 1985. The rhizosphere: carbon sources and microbial populations. In *Ecological Interactions in Soil. Plants, Microbes and Animals*, A. H. Fitter (ed.). British Ecological Society, Blackwell Scientific, Oxford, U.K., pp. 107–121.

Newman, E. I., and Watson, A. 1977. Microbial abundance in the rhizosphere: a computor model, *Plant and Soil* 48:17–56.

Noel, K., Vandenbcsch, K., and Kulpaca, B. 1986. Mutations in *Rhizobium phaseoli* that lead to arrested development of infection threads, *J. Bacteriol.* 168:1392–1401.

O'Neill, R. V., and Reichle. D. E. 1980. Dimensions of ecosystem theory. In *Forests, Fresh Perspectives from Ecosystem Analysis*, Proceedings of the 40th Annual Biology Colloquium, R. H. Waring (ed.). Oregon State University Press, Corvallis, pp. 11–26.

Ordentlich, A., Elad, Y., and Chet, I. 1987. Rhizosphere colonization by *Serratia marcescens* for the control of *Sclerotium rolfsii*, *Soil Biol. Biochem.* 19:747–751.

Pacovsky, R. S., Fuller, G., and Stafford, A. E. 1986. Nutrient and growth interactions in soybeans colonized with *Glomus fasiculatum* and *Rhizobium japonicum*, *Plant and Soil* 92:37–45.

Papavizas, G. C., and Davey, C. B. 1961. Extent and nature of the rhizosphere of *Lupinus*, *Plant and Soil* 14:215–236.

Parkinson, D. 1967. Soil microorganisms and plant roots. In *Soil Biology*, A. Burges and F. Raw (eds.). Academic, New York, pp. 449–478.

Parnas, H. 1976. A theoretical explanation of the priming effect based on microbial growth with two limiting substrates, *Soil Biol. Biochem.* 8:139–144.

Parrish, J. A. D., and Bazzaz, F. A. 1982. Responses of plants from three successional communities to a nutrient gradient. *J. Ecol.* 70:233–248.

Paustian, K., and Schnurer, J. 1987. Fungal growth response to carbon and nitrogen limitation: application of a model to laboratory and field data, *Soil Biol. Biochem.* 19(5):621–629.

Pegg, G. F., and Ayres, P. G. 1988. Fungal infection of plants. British Mycological Society Symposium no. 13, Cambridge University Press, Cambridge, U.K.

Peters, N. K., Frost, J. W., and Long, S. R., 1986. A plant flavone, luteolin, induces expression of *Rhizobium meliloti* nodulation genes, *Science* 233:977–980.

Poindexter, J. S. 1981. Oligotrophy: feast and famine existence, *Adv. Microbiol. Ecol.* 5:63–89.

Prikryl, Z., and Vancura, V. 1980. Root exudates of plants. VI. Wheat root exudation as dependent on growth, concentration gradient of exudates and the presence of bacteria, *Plant and Soil* 57:69–83.

Prikryl, Z., Vancura, V., and Wurst, M. 1985. Auxin formation by rhizosphere bacteria as a factor of root growth, *Physiologia Plantarum* 27(2–3):159–163.

Puhler, A., Enenkel, B., Kapp, D., Keller, M., Mueller, P., Niehaus, K., Priefer, U., and Schmidt, C. 1988. *Rhizobium meliloti* and *Rhizobium leguminosarum* mutants defective in surface polysaccharide synthesis and root nodule development. In *Nitrogen Fixation: 100 Years After*, C. H. Bothe, F. de Bruijn, and W. Newton (eds.). Gustav Fischer Verlag, Stuttgart, pp. 423–430.

Redmond, J. W., Batley, M., Djordjevic, M. A., Innes, R. W., Kuempel, P. L., and Rolfe, B. G. 1986. Flavones induce expression of nodulation genes in *Rhizobium, Nature* 323:632–635.

Rice, E. L. (ed.). 1984. *Allelopathy*, 2d ed. Academic, New York.

Rovira, A. D. 1969. Plant root exudates, *Bot. Rev.* 35:35–57.

Rovira, A. D., and Davey, C. B. 1974. Biology of the rhizosphere. In *The Plant Root and Its Environment*, E. W. Carson (ed.). University Press of Virginia, Charlottesville, pp. 153–204.

Samstevich, S. A. 1965. Colonization of plant roots with epiphytic and soil microorganisms. In *Plant Microbes Relationships*, J. Macura and V. Vancura (eds.). Czechoslovak Academy of Sciences, Prague, pp. 48–53.

Scheffer, F., Kickuth, R., and Visser, J. H. 1962. Die Wurzelausscheidungen von *Eragrostis curvula* (Schrad.) Nees und ihr Einfluss auf Wurzelknoten-Nematoden, *Z. Pflanzenernähr. Dng. Bodenkund.* 98:144–120.

Schippers, B. 1988. Biological control of pathogens with rhizobacteria, *Phil. Trans. R. Soc. Lond.* 318:283–293.

Schippers, B., Balder, A. W., and Bakker, H. M. 1987. Interactions of deleterious and beneficial rhizosphere microoganisms and the effect of cropping practices, *Annu. Rev. Phytopathol.* 25:339–358.

Schmidt, E. L. 1979. Initiation of plant root-microbe interactions, *Annu. Rev. Microbiol.* 33:355–376.

Scott, F. M., Bake, E., Bowler, E., and Hammner, K. C. 1958. Electron microscope studies of the epidermis of *Allium cepa, Am. J. Bot.* 45:449–461.

Sher, F. M., Kloepper, J. W., and Singleton, C. A. 1984. Chemotaxis of *Pseudomonas putida* to soybean and seed exudates in vitro and in soil, *Can. J. Plant Pathol.* 6:267.

Sherwood, J., Truchet, G., and Dazzo, F. 1986. Effect of nitrate supply on in vivo synthesis and distribution of trifoliin A, a *Rhizobium trifolii* binding lectin, in *Trifolium repens* seedlings, *Planta* 162:540–547.

Smit, G., Kijne, J., and Lugtenberg, J. 1986. Both cellulose fibrils and a Ca^{2+}-dependent adhesin are involved in the attachment of *Rhizobium leguminosarum* to pea root hair tips, *J. Bacteriol.* 169:4294–4301.

Spaink, H. P., Wijffelman, C. A., Pees, E., Okker, R. J. H., and Lugtenberg, B. J. J. 1987. *Rhizobium* nodulation gene *nodD* as a determinant of host specificity, *Nature* 328:337–340.

Stössel, P. 1987. Effect of soybean and bean phytoalexins and related isoflavonoids on growth of *Candida albicans, Can. J. Microbiol.* 33:461–464.

Stout, J. D. 1980. The role of protozoa in nutrient cycling and energy flow, *Adv. Microb. Ecol.* 4:1–50.

Suslow, T. V., and Schroth, M. N. 1982. Rhizobacteria of sugar beets: effects of seed application and root colonization on yield, *Phytopathology* 72:199–206.

Tepfer, D., Goldmann, A., Pamboukdjiam, N., Maille, M., Lepingle, Chevalier, D., Denarie, J., and Rosenberg, C. 1988. A plasmid of *Rhizobium meliloti* 41 encodes catabolism of two compounds from root exudate of *Calystegium sepium, J. Bacteriol.* 170:1153–1161.

Tien, T. M., Gaskins, M. H., and Hubbell, D. H. 1979. Plant growth substances produced by *Azospirillum brasilense* and their effect on the growth of pearl millet (*Pennisetum americanum* L.), *Appl. Environ. Microbiol.* 37:1016–1024.

Timonin, M. I. 1940. The interaction of higher plants and soil microorganisms: microbial population of rhizosphere of seedlings of certain cultivated plants, *Can J. Res.* C18:307.

Torsvik, V. L. 1980. Isolation of bacterial DNA from soil, *Microb. Ecol.* 8:163–168.

Toussoun, T. A., Bega, R. V., and Nelson, P. E. 1970. *Root Diseases and Soil-Borne Pathogens*, University of California Press, Berkeley.

Trappe, J. M., Castellano, M., and Molina, R. 1984. Reaction of mycorrhizal fungi and mycorrhizal formation to pesticides, *Annu. Rev. Phytopathol.* 22:331–359.

Truchet, G., Sherwood, J., Pankratz, H. S., and Dazzo, F. 1984. Clover root exudate contains a particulate form of the lectin, trifoliin A, which binds *Rhizobium trifolii, Plant Physiol.* 66:575–582.

Turner, S. M. 1983. Effects of Nitrogen and Phosphorous on Microorganisms of the Rhizoplane. Ph.D. thesis, University of Bristol (cited in Newman, 1985, p. 109).

Umali-Garcia, M., Hubbell, D., Gaskins, M., and Dazzo, F. B. 1980. Association between *Azospirillum* and grass roots, *Appl. Environ. Microbiol.* 39:219–226.

Van Brussel, A. A. N., Zaat, S. A. J., Canter Cremers, H. C. J., Wijffelman, C. A., Pees, E., Tak, T., and Lugtenberg, B. J. J. 1986. Role of plant root exudate and Sym plasmid-localized nodulation genes in the synthesis by *Rhizobium leguminosarum* of Tsr factor, which causes thick and short roots on common vetch, *J. Bacteriol.* 165:517–522.

Vancura, V. 1964. Root exudates of plants. I. Analysis of root exudates of barley and wheat in their initial phases of growth, *Plant and Soil* 21(2):231–248.

Vancura, V, and Catska, V. 1979. Biological aspects of soil sickness, *Rostl. Vyroba.* 25:1191–1202.

Vancura, V., and Jandera, A. 1985. Formation of biologically active metabolites by rhizosphere microflora. In *Microbial Communities in Soil*, V. Jensen, A. Kjoller, and L. H. Sorensen (eds.), FEMS Symposium no. 33, Elsevier, Barking, Essex, pp. 73–87.

Vancura, V., and Kunc, F. (eds.). 1988. *Interrelationships between Microorganisms and Plants in Soil*, Elsevier, Amsterdam.

Vancura, V., and Stotzky, G. 1976. Gaseous and volatile exudates from germinating seeds and seedlings, *Can. J. Bot.* 54:518–532.

Vancura, V., Catska, V., Hudska, G., and Prikryl, Z. 1983. Soil sickness in apple orchards, *Zentralbl. Bakteriol. Mikrobiol. Hyg.* 138:531–539.

Vincent, J. M. 1980. Factors controlling the legume-*Rhizobium* symbiosis. In *Nitrogen Fixation*, W. E. Newton and W. H. Orme-Johnson (eds.), vol. II. University Park Press, Baltimore, Maryland, pp. 103–129.

Webley, D. M., Duff, R. B., Bacon, J. S. D., and Farmer, V. C. 1965. A study of the polysaccharide-producing organisms occurring in the root region of certain pasture grasses, *J. Soil Sci.* 16:149–157.

Weller, D. W. 1988. Biological control of soilborne plant pathogens in the rhizophere with bacteria, *Annu. Rev. Phytopathol.* 26:379–407.

Whittingham, J., and Read, D. J. 1982. Vesicular-arbuscular mycorrhiza in natural vegetation systems, III. Nutrient transfer between plants with mycorrhizal interconnections, *New Phytol.* 90:277–284.

Woldendorp, J. W. 1978. The rhizosphere as part of the plant-soil system, In *Structure and Function of Plant Populations*, Verhandelingen der Koninklike Nederlandse Akademie van Wetenschappen, Afdeling Naturkunde, Tweede Reeks, p. 70.

Woodmansee, R. G., Dodd, J. L., Bowman, R. A., Clark, F. E., and Dickinson, C. E. 1978. Nitrogen budget of a shortgrass prairie ecosystem, *Oecologia*, 34:363–376.

Zamani, B., Knezek, B., Benneke, E., Flegler, S., and Dazzo, F. 1985. Autoradiographic method to screen for soil microorganisms which accumulate zinc, *Appl. Environ. Microbiol.* 49:137–142.

Zentmyer, G. A. 1961. Chemotaxis of zoospores for root exudates, *Science*, 133:1595–1596.

Mycorrhizal Associations

Hugh E. Wilcox

Professor Emeritus
Department of Environmental and Forest Biology
State University of New York
College of Environmental Science and Forestry
Syracuse, New York 13210

Introduction

Mutualistic associations (known as "mycorrhizae") between plant roots and fungi are ubiquitous in terrestrial ecosystems throughout the world, occurring in 83% of dicotyledonous and 79% of monocotyledonous plants thus far investigated (Trappe 1987). Because of the vast multiplicity of hosts, fungi, and soils involved, mycorrhizae are bound to show diversity in structure and function and variations in the benefits that they confer on host plants.

Under natural conditions, where pedogenesis remains undisturbed, plant roots and their attendant mycorrhizae have evolved together to cope with whatever stresses are associated with the particular soil environment. Among the many environmental stresses ameliorated by various types of mycorrhizal association are those caused by scarcity of essential soil nutrients, drought, heavy-metal toxicities, unfavorable pH, organic acid toxicities, and root pathogens. Other beneficial effects sometimes observed include increases in longevity of feeder roots, improvements in soil texture through increased aggregation of soil particles, formation of mycelial channels for plant-to-plant transfer of nutrients, and alterations of root morphology that may increase surface areas for absorptive uptake.

In return for the many direct and indirect benefits provided by the fungal associant ("mycobiont"), the host ("phytobiont," "autobiont") provides the fungus with carbohydrates and possibly with growth substances. In the initial phases of mycorrhizal establishment, the fungus obtains these substances from root excretions, but later secures them through transport mediated by host and fungus plasmalemmae. The root excretions that promote mycorrhizal formation also stimulate the growth of other microorganisms in the root region (rhizosphere) as discussed in the preceding chapter. Upon the formation of mycorrhizae, the alteration in root physiology significantly changes the root exudates and causes new microbial equilibria to be established. Rhizospheres then become mycorrhizospheres (Perry et al. 1987), and activities of microorganisms in the mycorrhizosphere may have diverse and sundry effects on mycorrhizal functions (Linderman 1987, 1988).

Under natural conditions the mycorrhizae and associative microorganisms form a dynamic interacting community in which species may at times replace one another yet retain a harmonious composition that is adapted to prevailing edaphic conditions. The disturbances of deforestation, mining, and agriculture disrupt this equilibrium, often leading to "shortfalls" (Bowen 1978) in plant production.

To understand the development and role of mycorrhizal associations involves consideration of the complex interactions among host roots, indigenous mycorrhizal fungi, and the soil environment itself, comprising an heterogeneous combination of concurrent biotic and abiotic variable factors. The root systems themselves are protean in character, showing metamorphism of a combined genotypic and phenotypic nature aside from their mutative interactions with mycorrhizal fungi.

The complexity and elusiveness of these innumerable constitutive interactions indicate that a definitive account of mycorrhizal associations is likely to be a limitless task. The best that can be hoped for in the present account is to provide a glimpse of some principal features governing the variability and influencing the development of the more common types of mycorrhizae in natural and near-natural ecosystems.

The subjects selected for brief presentation here include (1) characteristics of the more common types of mycorrhizae, including taxonomy of hosts and fungi, (2) geographical distribution and ecosystem relationships, (3) interdependency of mycorrhizal fungi and hosts, (4) root-growth strategies and mycorrhizal establishment, and (5) utilization of mycorrhizal fungi in forestry and agriculture.

Classification of Mycorrhizae

Mycorrhizal associations are ordinarily placed in categories based on the recognition of "natural" groups or types (Harley 1969). All mycorrhizal organs belonging to a particular type are markedly similar in their structural, developmental, and physiological characteristics, notwithstanding that symbionts may be from very diverse taxa (Harley and Harley 1987). Currently the simplest, and most widely used, classification consists of three broad mycorrhizal groups based on the siting of fungal mycelium in relation to root structure. Two of these, "endomycorrhizae" and "ectendomycorrhizae," are variously subdivided into a number of recognizable subgroups based on additional unique features in fungal or host characteristics. Mycorrhizal organs of the third broad group, "ectomycorrhizae," cannot be usefully subdivided since they conform to a single recognizable type despite their variable appearances. A brief description of these broad categories and their subdivisions follows.

Ectomycorrhizae

"Ectomycorrhizae" have been ascribed mainly to roots of woody plants, although recently they have been reported in herbaceous annuals and perennials as well (Warcup 1980, Warcup and McGee 1983). They are ordinarily delineated by two characteristic features: an organized mantle of hyphae on the root surface and a plexus of hyphae that penetrates the root intercellularly from the mantle and surrounds cells of the cortex to form the so-called "Hartig net" (Figure 1b). These two structures are critical to the functioning of the symbiotic association. All nutrients absorbed from the soil must pass through the mantle, and its inner hyphae together with those in the

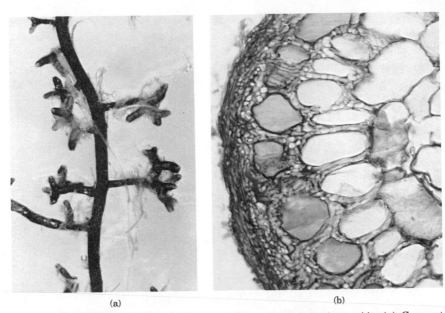

(a)   (b)

Figure 1 Ectomycorrhiza formed by *Suillus* sp. on *Pinus resinosa* Ait. (*a*) Compact mycorrhizal clusters showing external hyphae and rhizomorphs. Approx. 6*x*. (*b*) Transverse section showing a fungal mantle on the root surface and an intercellular Hartig net separating epidermal and cortical cells. Cortical cells are devoid of hyphae. Approx. 320*x*.

Hartig net are components of the interfaces between fungus and host through which nutrient exchanges occur.

Two features of the fungus mantle have been studied as means for identifying tree-fungus combinations. These are the organization of the mantle tissue as seen in plan view (from observations of cleared roots) and the structure of the associated rhizomorphs (Chilvers 1968). Success in achieving such an identification system has proven elusive. Reviews of the numerous attempts and methods are presented in a number of publications (Chilvers 1968, Zak 1973, Godbout and Fortin 1983, 1985, Agerer 1986*a,b*). Emphasis in the following pages is on the functional aspects of mycorrhizal structure rather than on taxonomy.

As previously mentioned, ectomycorrhizae are found predominantly on woody plants. Only 7 genera possessing herbaceous plants are reported in a tabulation of 140 genera (in 43 families of angiosperms and gymnosperms) of ectomycorrhizal plants (Harley and Smith 1983). These data cannot be considered as absolute because a relatively small number of plant species has been examined to determine their mycorrhizal status under natural conditions. The 2,930 species for which Newman and Reddell (1987) found relevant information represent about 1% of the total number of vascular plant species in the world. In those families with records of 20 or more species capable of forming some type of mycorrhizae, they found six in which the mycorrhizal status was predominantly ectomycorrhizal and another seven with decreasing percentages of ectomycorrhizae (and increasing percentages of endomycorrhizae). These families and their percentage of ectomycorrhizal species are as follows: Dipterocarpaceae (98%), Pinaceae (95%), Fagaceae (94%), Myrtaceae (90%), Salicaceae (83%), Betulaceae (70%), Leguminosae (16%), Rosaceae (12%), all other Gymnospermae (9%), Euphorbiaceae (7%), Scrophulariaceae (4%), Rubiaceae (3%), Cyperaceae (3%).

The known fungi involved in ectomycorrhizal formation are largely basidiomycetes, but an increasing number of ascomycetes, deuteromycetes, and zygomycetes have been added to the list. Both the deuteromycete *Cenococcum geophilum* Fr. (Trappe 1962) and the zygomycete *Endogone sensu stricto* (Warcup 1980) have been demonstrated to have remarkably wide host ranges.

Compilations of fungi that form ectomycorrhizal associations in plants have been presented periodically (Trappe 1962, 1971, Miller 1982*a,b*, Harley and Smith 1983).

Endomycorrhizae

In "endomycorrhizae" the fungi live within the root cortical cells. The fungus may also grow intercellularly but does not form a Hartig net. Neither does it form a mantle of mycelium around the root as do the ectomycorrhizae.

Several distinctly different types of endomycorrhizae have been recognized, the best known being the vesicular-arbuscular, ericoid, and orchidaceous. A number of other unclassified endophytic infections have also been encountered; those reported most frequently consist of dark, septate intracellular hyphae with numerous intercalary and terminal swellings. The same, or similar, dark-septate fungi have also been implicated in ericoid mycorrhizae and in various ectendomycorrhizal infections.

In orchidaceous mycorrhizae the mutualistic relationship is reversed, and the plant receives carbohydrate from the endophyte. Partly for this nonconformity, and partly in the interest of space, the orchidaceous mycorrhizae will be omitted from further consideration. This paper discusses only vesicular-arbuscular and ericoid endomycorrhizae.

Vesicular-arbuscular mycorrhizae. "Vesicular-arbuscular mycorrhizae" (VA mycorrhizae) are characterized by the formation of branched haustorial structures (arbuscles) (Figure 2*a*) within certain root corti-

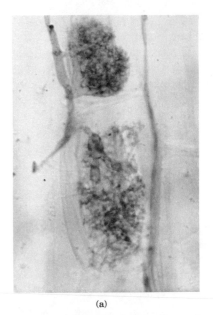
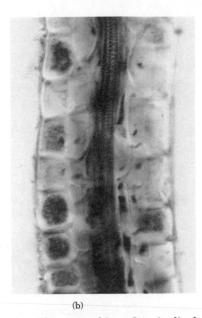

(a)   (b)

Figure 2 (a) Anatomical features of vesicular-arbuscular mycorrhizae. Longitudinal view from a freehand section of grape root showing fully developed arbuscles of an unknown vesicular-arbuscular fungus. Approx. 520x. (b) Anatomical features of ericoid mycorrhizae. Longitudinal view of ericoid mycorrhiza in a cleared root preparation of azalea showing an independent infection in each cortical cell. Fungus is unidentified. Approx. 140x.

cal cells, and by terminal and intercalary hyphal swellings (vesicles), which occur within or between the cells. An external mycelium forms a loose network in the soil surrounding the root. This mycelium is characterized by its great variation in diameter of the hyphae and the thickness of hyphal walls and often by the presence of hyphal vesicles, spores, and fruiting bodies (Nicolson 1967). Observations on the characteristics of the extramatrical phase of VA mycorrhizal fungi are scarce, but information on the intraradical phases is extensive. The anatomical features of VA mycorrhizae have been reviewed recently (Bonfante-Fasolo 1984) and will not be considered here.

VA mycorrhizae occur in plant families throughout the angiosperms, gymnosperms, pteridophytes, and bryophytes. They are more prevalent than ecto-, ectendo-, ericoid, and "dark-septate" mycorrhizae combined (Trappe 1987). In gymnosperms, VA mycorrhizae predominate in families other than the Pinaceae, and in Pinaceae they are found together with ectomycorrhizae in 2% of the species examined (Newman and Reddell 1987). In angiosperms with mycorrhizae, they predominate in all families except those five previously mentioned as being predominantly ectomycorrhizal. The predominance of VA mycorrhizae among all species of plants can be seen in the recent checklist of mycorrhizae in the British flora (Harley and Harley 1987).

With VA mycorrhizae so widespread, it is of interest to inquire whether there are plant families without any form of mycorrhizae whatsoever. In an early list, Gerdemann (1968) named 14 families as "possibly non-mycorrhizal or rarely mycorrhizal." The recent survey of Newman and Reddell (1987) indicates that VA mycorrhizae are found among at least 6 of the 14 families listed by Gerdemann, and they will very likely be found to some extent in the others as more species are examined.

The fungal species responsible for the formation of VA mycorrhizae have been placed in four genera (*Acaulospora, Gigaspora, Glomus*, and *Sclerocystis*) of the family Endogonaceae within the zygomycetes (Gerdemann and Trappe 1974, Hall 1984, Trappe 1982). According to Trappe (1982) the number of recognized zygomycetous mycorrhizal species is growing and could approach 200 by 1990.

Questions have been raised regarding the validity of the family Endogonaceae. Although there is an overall similarity in the morphology of spores and sporocarps produced by the species in the different genera, the family as now constituted consists of chlamydosporic, zygosporic, and sporangiasporic species as well as random chlamydospores found in soil sievings (Nicolson 1967, Hayman 1978, Hall 1984).

Ericoid mycorrhizae.   Mycorrhizae in the Ericales occur in two of the broad mycorrhizal groups, the endomycorrhizal "ericoid" type discussed here, and the two types of ectendomycorrhizae to be discussed later (arbutoid and monotropoid).

Ericoid mycorrhizae occur in the very fine roots ("hair roots") common in ericaceous plants. These roots possess only one to three cell layers of cortical cells surrounding a narrow central stele (Read 1983). The most striking feature of infected roots is the presence of characteristic coils of hyphae within the epidermal cells (Bonfante-Fasolo and Gianninazi-Pearson 1979). At maturity the entire volume of the cell appears to be occupied with fungal hyphae, and little or no vacuolar area is visible (Read 1983). Because infection occurs through the outer wall of these cells, each cell represents an individual infection unit and adjacent cells may have infections of different ages (Read 1983) (Figure 2*b*).

In contrast to the arbuscles of VA mycorrhizae, the intracellular hyphae of ericoid mycorrhizae retain their structural integrity within the host cell. The first sign of senescence is the degeneration of the contents of the host cell, not a degeneration of intracellular hyphae

(Harley and Smith 1983). Breakdown of the fungus begins only in the later stages of host degeneration.

Ericoid mycorrhizae are found in families of the Ericales, including the Epacridaceae, Empetraceae, and most Ericaceae except the subfamily Arbutoideae. Read (1983) has presented a table of taxonomic, geographical, and mycorrhizal relationships within the Ericales.

Both ascomycetes and basidiomycetes are involved in the formation of ericoid mycorrhizae (Harley and Smith 1983). Although dark-colored, slow-growing fungi have been periodically isolated from ericaceous roots and reinoculated into aseptically grown seedlings, the fungi mostly remained sterile and could not be identified. Read (1974) was first to stimulate a fungal isolate to fruit; this isolate has been named *Hymenoscyphus ericae* (Read) Korf & Kernan. Other dark-colored ascomycetous soil fungi are probably also involved in mycorrhiza formation. Dalpé (1986) was able to synthesize ericoid mycorrhizae in *Vaccinium angustifolia* Ait. using several species of the ubiquitous genus of soil fungi *Oidiodendron*. The perfect stages of this anamorphic genus belong to the ascomycetous genera *Arachniotus*, *Myxotrichum*, and *Toxotrichum*.

The identification of the basidiomycetous associants of ericoid mycorrhizae has proven even more intractable. The presence of dolipore septa in ericoid mycorrhizae indicates the presence of basidiomycetes, but none has been isolated (Bonfante-Fasolo and Gianninazi-Pearson 1979). Species of the genus *Clavaria* have been suggested from the presence of fruiting bodies around ericaceous plants. Immunological techniques and radioactive transport studies with ericoid mycorrhizae provide indirect support for this suggestion, but confirmation awaits isolation and reinoculation of an identified fungus in axenic culture (Harley and Smith 1983).

Ectendomycorrhizae

Several distinct groups of "ectendomycorrhizae" have been recognized. A group that is often found associated with nursery seedlings is formed by "E-strain fungi" and certain "black imperfect" fungi. Also there are the better-known arbutoid and monotropoid ectendomycorrhizae of certain ericalean families, formed by many fungi that are ectomycorrhizal in other plant families. The characteristics of all these are described below.

E-strain ectendomycorrhizae. Mikola (1965) first isolated and named "E-strain fungi" from coniferous nurseries in Finland. E-strain ectendomycorrhizae possess large-diameter intracellular hyphae, heavily ramified throughout the cortex of both long and short roots. A Hartig net of thick hyphae surrounds cortical cells of the small- and medium-diameter roots, but it is usually lacking in the larger roots in which the fungi remain scattered in the intercellular spaces or form a rudimentary net. Little or no fungus mantle occurs on the surfaces of most mycorrhizal roots (Figure 3a and b). A detailed description of a typical E-strain ectendomycorrhiza is provided by the more recent publication of Piché et al. (1986) on the mycorrhiza formed in *Pinus resinosa* Ait. by the E-strain fungus *Wilcoxina mikolae* var. *mikolae* (Yang and Korf 1985).

The experiments of Laiho (1965) performed with unidentified isolates of E-strain fungi showed that they formed mycorrhizae with all tree species that normally possess ectomycorrhizae. The association was ectendomycorrhizal in all species of the genera *Pinus* and *Larix*, whereas associations by the same fungus were only ectomycorrhizal in all species of *Picea, Abies, Tsuga, Pseudotsuga, Betula*, and *Populus*.

It now appears that some of Laiho's generalizations are too sweeping. Wilcox et al. (1983) reported that an E-strain fungus that was ectendomycorrhizal on *Pinus resinosa* Ait. was ectomycorrhizal on *Pinus elliottii* Engelm. In addition they reported that the type of asso-

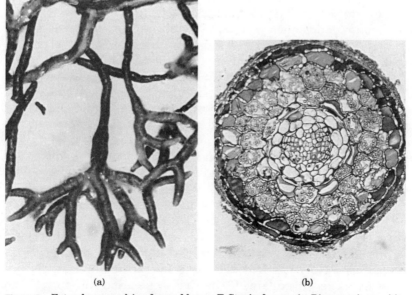

(a)   (b)

Figure 3   Ectendomycorrhiza formed by an E-Strain fungus in *Pinus resinosa* Ait. (a) Loose mycorrhizal clusters showing smooth surfaces with few extramatrical hyphae. Approx. 10x. (b) Transverse section showing thin hyphal mantle on the root surface, bulbous hyphae (Hartig net) between the cortical cells, and a dense intracellular infection. Approx. 140x.

ciation on the same pine species might vary too with the particular E-strain isolate. When tested on *Pinus resinosa*, some of their E-strain isolates produced ectendomycorrhizal infections and others only ectomycorrhizal. Obviously more work is necessary to clarify differences of specificity in nursery seedlings among host species, E-strain fungi, and the ecto- or ectendomycorrhizal associations that they form.

The first E-strain isolate to be identified was named as a new species of discomycete in the genus *Tricharina* (Yang and Wilcox 1984). Subsequently this genus was shown to be a mixture of saprobic and mycorrhizal species, and the mycorrhizal species were segregated into the new genus *Wilcoxina* by Yang and Korf (1985). At present *Wilcoxina* contains three species, one of which possesses two varieties. It is uncertain whether other E-strain species remain to be discovered.

Arbutoid ectendomycorrhizae.   "Arbutoid" ectendomycorrhizae possess a fungus mantle and morphology very much like that of ectomycorrhizae (Zak 1974, 1976). A Hartig net envelops the outer tier of cortical cells, which contain masses of branching hyphae. As a result the root-fungus association assumes characteristics of both ericoid and ectomycorrhizae. The anatomical and ultrastructural characteristics of arbutoid mycorrhizae have recently been described for six species of *Pyrola* (Robertson and Robertson 1985) and for *Arbutus unedo* (Fusconi and Bonfante-Fasolo 1984).

Both arbutoid ectendomycorrhizae and the previously described ericoid endomycorrhizae occur in families of the order Ericales. The arbutoid configuration is found in the Pyrolaceae (Robertson and Robertson 1985) and in the subfamily Arbutoideae of the Ericaceae, in which it is frequently reported in the genera *Arbutus* and *Arctostaphylos*. It is uncertain, however, whether arbutoid and ericoid mycorrhizae are strictly segregated among separate genera of the Ericaceae as the above statement would imply. Largent et al. (1980) reported on a field study in northern California that showed that occasional ericoid mycorrhizae occurred along with the normal arbutoid types in *Arctostaphylos patula*, and conversely arbutoid ectendomycorrhizae occurred together with the more frequent ericoid type in *Leucothoe davisiae*. The simultaneous occurrence of two sorts of mycorrhizae might indicate a dimorphic root system with ericoid mycorrhizae on the "hair" roots and the arbutoid on roots of larger diameter. Alternatively it might indicate some particular environmental condition as yet unexplained. The Ericaceae are a widespread and diverse family consisting of about 125 genera and 3,500 species (Cronquist 1981), so there is obvious scope for extensive investigation.

Zak (1974) was able to identify the fungal associant of an *Arbutus menziesii* ectendomycorrhiza by linking rhizomorphs from it to a sporophore of *Corticium zakii* occurring immediately above. This fungus was an ectomycorrhizal symbiont on Douglas fir [*Pseudotsuga menziesii* (Mirb.) Franco] and grand fir [*Abies grandis* (Dougl.) Lindl.] that were in the same stand with the *Arbutus*. Subsequently Zak (1976) synthesized arbutoid ectendomycorrhizae on *A. menziesii* using four other common ectomycorrhizal fungi and suggested that most arbutoid mycorrhizae are formed by the same fungi as are associated with ectomycorrhizae of forest trees.

Zak's suggestion was supported by a pure culture mycorrhizal synthesis experiment conducted by Molina and Trappe (1982) using 28 cultures of ectomycorrhizal fungi isolated from diverse hosts and habitat associations upon the ericaceous hosts *Arbutus menziesii* and *Arctostaphylos uva-ursi*. All but three fungi formed arbutoid ectendomycorrhizae with well-developed mantles that resembled the ones formed in ectomycorrhizal hosts. Except for *Cenococcum geophilum* (ascomycetous) all of the tested fungi were Basidiomycetes.

Not all arbutoid ectendomycorrhizae possess thick mantles formed by basidiomycetous ectomycorrhizal fungi. Fusconi and Bonfante-Fasolo (1984) described white-amber mycorrhizal roots of *Arbutus unedo* L. which had relatively inconspicuous mantles formed by an unknown ascomycetous fungus. The mantle formed an organized structure only at the base of the feeder roots, reminiscent of E-strain ectendomycorrhizae. It would be interesting to examine the possible relationship between E-strain and this type of arbutoid ectendomycorrhizae.

Monotropoid ectendomycorrhizae. "Monotropoid" ectendomycorrhizae are found in the Ericaceae in the achlorophyllous subfamily Monotropoideae (Duddridge and Read 1982, Robertson and Robertson 1982). These mycorrhizae possess a thick mantle and a Hartig net between the first layer of cortical cells. Fungal pegs penetrate from the mantle into the outer cortical cells, producing a structure resembling that of transfer cells (Gunning and Pate 1974). As in the achlorophyllous orchids, monotropoid plants can be considered epiparasites living off fungi, which in turn may be mycorrhizal on other higher plants. In most cases the fungi (largely basidiomycetes, Duddridge 1984) involved in monotropoid mycorrhizae are those associated with ectomycorrhizal plants.

Geographical Distribution and Ecosystem Relationships

Ectomycorrhizal associations

Ectomycorrhizal tree species play a dominant role in many important forest types in cool and temperate regions of the world, including boreal coniferous forests, subalpine forests, and temperate deciduous forests adjacent to boreal coniferous forests (Meyer 1973). Ectomycorrhizae also occur in evergreen sclerophyllous forests that are characterized by mild, rainy winters as well as hot, dry summers. Detailed information on the geographical distribution and characteristics of ectomycorrhizal associations in the various forest types is presented by Meyer.

Ectomycorrhizal associations were long considered to occur relatively infrequently in the tropics (Meyer 1973), mostly in low-diversity forests on poor soils (Malloch et al. 1980, Janos 1980, 1983). However, sufficient evidence has now accumulated to establish the dominance of ectomycorrhizae among the Dipterocarpaceae of the southeast Asian rain forest and certain legumes in the subfamily Caesalpinioideae, tribes Amherstieae and Detarieae, throughout Africa (Alexander 1987, Newbery et al. 1988). Ectomycorrhizal legumes are found in the rain forests of Ghana, Nigeria, Zaire, Senegal, and Cameroon in monospecific stands up to 500 m across in a matrix of diverse VA mycorrhizal species (Newbery et al. 1988). These legumes also extend into the drier Niombo woodlands of east and south central Africa. These woodlands cover large parts of Tanzania and extend 1600 km south through Zambia, Zaire, Malawi, and Mozambique, and into Zimbabwe, as well as west to Angola (Alexander 1987). In these locations the legume ectomycorrhizae mingle with ectomycorrhizal species in the Dipterocarpaceae, Euphorbiaceae, and Proteaceae to form a diversified ectomycorrhizal flora (Högberg 1982, Högberg and Piearce 1986).

The fungi in ectomycorrhizal associations and the predisposing habitat factors have common characteristics in both temperate and tropical forest regions. Their occurrence is generally associated with conditions of mineral nutrient stress (Grime 1977) in forest stands of limited species diversity. The uptake of phosphorous and/or nitrogen is promoted by the mycelial strands and rhizomorphs of the fungus. In temperate ecosystems with a pronounced seasonality, nutrient releases occur in flushes, and the fungal mantle provides a means of storage at such times for later release to the host. In addition, the mycelial fans spreading from infected roots act as inocula that infect the roots of adjacent plants, resulting in the formation of mycelial networks that connect both intraspecific and interspecific combinations of compatible host plants (Finlay and Read 1986). The hyphae of these networks spread the infection within and between plant root systems, function in nutrient uptake, and provide channels for interplant transfer of nutrients, including carbon (Read et al. 1985). The low species diversity of ectomycorrhizal forests is correlated with the relatively high specificity of the associated fungal symbionts, which have evolved in adaptation to conditions of mineral nutrient (and associated) stress (Grime 1977).

The underlying similarity of ectomycorrhizal associations does not mean that they adapt a common configuration in the soil. The particular manifestation of ectomycorrhizae is governed by a complex of interlinking physiological and ecological factors, which have been imperfectly elucidated (Meyer 1974). For example, the difference between pine ectomycorrhizae with their extensive mycelial fans and beech ectomycorrhizae with thick mantles and limited hyphal extension into the soil has been ascribed to differences in the mode of nutrient capture resulting from characteristics of the resource material (Read 1984). The characteristics of tropical ectomycorrhizae are yet to be investigated, but the extensive colonization of surface organic matter by hyphal strands combined with the seasonality of phosphorous uptake in groves of ectomycorrhizal rain forest legumes of Africa has been proposed as evidence of a virtually closed nutrient cycle dominated to a considerable extent by the ectomycorrhizal fungi (Newbery et al. 1988).

Endomycorrhizal associations

Vesicular-arbuscular mycorrhizae.   VA mycorrhizal associations occur throughout the world on both annual and perennial root systems. Arbuscles have been found in plant remains from the Early to Middle Triassic of Antarctica (Stubblefield et al. 1987), confirming a long history of mycotrophic symbioses and lending support to the proposal by Pirozynski and Malloch (1975) that vascular plants arose in symbiotic partnership with fungi. In further pursuit of this theme, Pirozynski (1981) has postulated that evolution of ectomycorrhizal symbiosis is relatively more recent and that endotrophism is showing signs of weakening, leading to an increase in facultative mycotrophism and the nonmycorrhizal condition.

In addition to their widespread distribution throughout the plant kingdom, VA mycorrhizae are also geographically ubiquitous and occur in plants growing in arctic, temperate, and tropical regions (Hayman 1978). A recent monograph on the ecophysiology of VA mycorrhizae summarizes their role in humid tropical ecosystems, grasslands and shrublands, field crop systems, and horticultural systems (Safir 1987).

There is widespread belief that VA mycorrhizae have a dominance in lowland humid tropical habitats that is analogous to that of ectomycorrhizae in northern coniferous forests. Janos (1983, 1987) has examined the reports on VA mycorrhizae from all three main tropical regions of the world (Asia, Africa, and neotropics). Although most studies were in accord with the ubiquity of VA infection in humid tropical forests, a few reports were found of nonmycorrhizal colonizing species dominating disturbed sites. VA mycorrhizae appear to be common in dry, lowland tropical areas, as in humid forests, but a few ectomycorrhizal species also occur (Högberg 1982, Janos 1987).

Most forage plants of arid and semiarid rangelands throughout the world are mycorrhizal, with most species forming associations with VA mycorrhizal fungi (Trappe 1981, Miller 1987) in climax and subclimax ecosystems. However, on disturbed sites weedy colonizing species are often nonmycorrhizal due to a reduction in numbers of mycorrhizal propagules (Reeves et al. 1979). The nonmycorrhizal species may hinder successional stages in ecosystem recovery because they do not provide an inoculum for subsequent species that require VA mycorrhizae for survival.

The worldwide occurrence of VA mycorrhizae in major crop plant families is well-established but is beyond the scope of this paper. Extensive information on this topic can be found in a number of excellent texts (Sanders et al. 1975, Harley and Smith 1983, Powell and Bagyaraj 1984, Safir 1987).

VA mycorrhizae, like ectomycorrhizae, improve the uptake of immobile mineral elements by extensive hyphal ramification of the soil. However, they appear to lack the saprophytic ability possessed by some ectomycorrhizal fungi (Janos 1983). Their greatest value is in increasing mineral uptake from infertile or nutritionally imbalanced soils. However, mycelial interconnections linking the root systems of host plants may also provide channels for direct interplant nutrient transfer as suggested by Francis et al. (1986); the occurrence of such interconnections is widespread. Since early stages of plant development take place in dense stands of established vegetation, it has been suggested that interplant transfer might aid in seedling establishment (Francis et al. 1986). Additional implications of mycorrhizal mycelia and nutrient cycling in plant communities have been reviewed by Read et al. (1985).

Ericoid mycorrhizae.   Ericoid mycorrhizal associations in the various families of the Ericales are distributed worldwide. Extensive investigations have been conducted on representatives of the Ericaceae, which occur as dominant plants over vast areas of heathland in the northern hemisphere. Preliminary investigations have revealed comparable ericoid mycorrhizae in African heath species despite differences in climatic and ecological features.

Read (1983) has reviewed the biology of mycorrhizae in the Ericales, giving a comprehensive characterization of ericoid, arbutoid, and monotropoid mycorrhizae. He characterizes soils in ericoid ecosystems as containing the bulk of their nutrients in recalcitrant organic forms, and vegetation as possessing only a small fraction of the nutrient fund. Under these circumstances the low nutrient availability leads to the production of tissues of low protein content and the consequent diversion of photosynthate into a range of secondary metabolites that accumulate in the needle-like leaves. The litter from these leaves has a high C/N ratio due to the precipitation of N in stable polyphenol-protein complexes. The ericoid mycorrhizal roots, which become concentrated in the upper humus layers of the soil, represent a special adaptation enabling the plants to obtain access to the slowly released nitrogen that is available under these circumstances. The ability of ericoid mycorrhizae to utilize various simple organic substances has been confirmed by Read and his coworkers.

Ectendomycorrhizal associations

The geographical distribution of arbutoid and monotropoid mycorrhizae is coincident with that of the few families and genera of Ericales to which they are restricted; their distribution and ecosystem relationships have not received special attention.

Little is known about the distribution and ecosystem relationships of E-strain, black-dematiaceous, and other types of ectendomycorrhizae. E-strain fungi were first reported in coniferous nursery seedlings (Mikola 1965, Wilcox 1971) but are now known to occur widely in natural ecosystems (Danielson 1982).

Ectendomycorrhizae produced by black dematiaceous fungi are ubiquitous but often appear to have been confused with *Cenococcum geophilum* mycorrhizae or with pseudomycorrhizal associations produced by parasitic fungi (Wang and Wilcox 1985). The frequent isolation of heterogeneous genera, species, and strains of Hyphomycetes from ectendomycorrhizae and pseudomycorrhizae on stressed sites points to the looming importance of these associations and the need to elucidate their taxonomic and ecosystem relationships.

Interdependence of Mycorrhizal Fungi and Hosts

The mycorrhizal associants of each ecosystem have evolved conjointly to meet the exigencies of the particular environment. The interdependence that has arisen is more often obligate for the fungus than for the host. However, differences in degree of dependency have been reported for both and are discussed here because of their relevance to ecological theory and to practical applications.

Mycobionts of ectomycorrhizae and ericoid mycorrhizae have been successfully cultured in vitro, but those from VA mycorrhizae have not. Thus, the latter have been presumed to possess an absolute dependency upon their hosts. Hepper (1984) has reviewed the efforts to grow VA mycorrhizal fungi in pure culture and discussed the problems involved.

The nutritional requirements of cultured ectomycorrhizal fungi indicate that they, too, are largely dependent upon their hosts for simple carbohydrates (Hacskaylo 1973) and can be considered obligate. The adaptive formation of cellulase in a few ectomycorrhizal fungi has been interpreted to indicate that they have been derived from those with litter-decomposing abilities by adaptation in successive degrees (Meyer 1974). Lewis (1973) envisaged an alternative hypothesis for the origin of biotrophy from necrotrophy but Harley and Smith (1983) eschew an evolutionary explanation for the ability of ectomycorrhizal fungi to produce exoenzymes that attack complex carbon compounds. They propose a kind of facultative mycorrhizal behavior that may exist at least temporarily in some ecological situations where mycorrhizal symbiosis does not occur.

The ericoid endophyte *Hymenoscyphus ericae* also appears to have an obligate dependency on its hosts for simple carbohydrates, but is able to use organic phosphates (Mitchell and Read 1981) and organic nitrogen sources in the form of amino acids (Pearson and Read 1975, Stribley and Read 1980). The situation for other ericoid endophytes is unknown.

The degree of dependency of hosts on their fungal associants is much less clear-cut. Woody plants exhibit greater dependency than herbaceous plants (Trappe and Fogel 1977, Trappe 1987). However within the woody plants appraisals of dependency are divergent. The fact that it is possible to grow mycotrophs in uninfected condition has been cited as evidence that they are not obligately associated with their fungi except in an ecological sense (Harley 1969). Similarly, the fact that numerous attempts to establish plantations of exotic pines in different parts of the world invariably failed until suitable mycorrhizal fungi had been imported is often cited as evidence that symbiosis is obligatory in *Pinus* and, in most cases, other ectomycorrhizal forest trees (Trappe and Strand 1969, Vozzo and Hacskaylo 1971, Mikola 1973). Recent studies on mutualistic symbioses in ectomycorrhizal forests have emphasized that phytobionts are capable of forming consortia with a wide range of mycobionts simultaneously and that phytobionts take up mycobionts selectively according to developmental phase and ecological conditions (Malloch et al. 1980). Some mycorrhizal fungi are characteristic of young forest stands, whereas others are associated with older ones (Dighton and Mason 1984). The factors that influence the species of mycobionts upon which the hosts depend are obscure.

VA mycorrhizae are ubiquitous in most habitats, and the nature of mycorrhizal dependency is correspondingly varied. The primary selective factors that produce mycorrhizal dependency in different habitats are most commonly mineral nutrient availability (especially phosphorous) and the probability of mycorrhizal infection (Janos 1983), but other factors may be involved.

Janos (1980) examined in pot culture 28 plant species of lowland tropical rain forests for their dependency on VA mycorrhizal fungi and found that 16 were completely dependent upon them for growth. The remainder were either nonmycorrhizal or facultatively mycorrhizal and utilized mycorrhizae in infertile but not in fertile soils. In subsequent reviews of tropical mycorrhizae, Janos (1983, 1987) discusses at greater length the distinctions between obligate, facultative, and nonmycorrhizal conditions. Most obligately mycorrhizal species in the tropics possess VA mycorrhizae; nevertheless, ectomycorrhizal associations do occur and are locally abundant in low-diversity forests on poor soils because of their specific, highly beneficial fungal associates.

Not only do ecological factors influence the type of mycorrhiza upon which a host depends, but different mycorrhizal associations, in turn, influence subsequent population and ecosystem dynamics. The roles of VA mycorrhizae in humid tropical ecosystems, grass and shrublands, field crop systems, and horticultural systems are reviewed in a recent monograph (Safir 1987) and will not be examined here.

Root-Growth Characteristics and Mycorrhizal Development

In addition to providing an accurate recognition of the symbionts, a meaningful investigation of mycorrhizal development must also scrutinize the normal growth and developmental characteristics of the root system involved. Too often, normal environmental responses of roots have been mistaken for the effects of mycorrhizal fungi; conversely, mycorrhizal effects have remained unrecognized because subtle morphological responses were overlooked. A few principles regarding root-growth activity and spatial and temporal distributions of fine root branches are discussed below in relation to the initiation and patterning of mycorrhizal development. The approaches followed in describing mycorrhizal development in the root systems of woody plants have been markedly different from those used in herbaceous plants, and each will be described separately below.

Woody root systems and mycorrhizal development

Most woody root systems show two root classes: short or small-diameter "feeder" roots, which are ephemeral, and long and more prominent skeletal roots, which form the permanent framework of the root system. The form and development of a root system is reflected in the characteristics and patterning of the two root classes, which in turn are determined partly by genetic variability and partly by modifications induced by environmental conditions (Wilcox 1983). The ability of developing root systems to make short-term and opportunistic adaptations to edaphic conditions has confounded many a mycorrhizal investigation.

Mycorrhizal development occurs in primary root tissues prior to their suberization. Consequently mycorrhizal investigations have focused on the distal root system in the vicinity of apical meristems. Originally attention was paid only to the feeder roots, which in ectomycorrhizal associations became conspicuously altered by mycorrhizal formation, but it is now recognized that mycorrhizal fungi may also colonize the distal portions of skeletal roots (Wilcox 1968).

Studies of mycorrhizal development in relation to root-system morphology have concentrated on the branching characteristics of the distal branches (Büsgen 1901, Wilcox 1983) and analyzed either the single root axis generated by one root meristem, or a "root subsystem" (Jenik 1976) represented by any defined part of a root system. This latter approach has been used in the sampling of hierarchical units for developmental studies of ecto- and ectendomycorrhizae (Wilcox 1982).

Lengths and diameters of primary roots show considerable inter- and intraspecific differences, and heterorhizic attributes are not always apparent in the distal branches. Some hardwood species, such as beech, hornbeam, hazel, and oak, possess numerous, strikingly thin root branches, the terminal portions usually being mycorrhizal. Primitive angiosperm roots, typified by those of the order Magnoliales, possess coarse rootlets generally from 0.5 to 1.5 mm, or more, in diameter (Baylis 1975, St. John 1980). Within the same root system, individual roots vary in diameter, length, rate of elongation, and longevity. The smaller root classes may elongate for only a few weeks, followed by suberization or frequent die-off (Coleman 1976, Fogel 1985). The larger root size classes, which are longer and longevous, become mother (or parent) roots with laterals that, in turn, may be ephemeral or become smaller mother roots.

Maturational and physiological changes along the length of the root, the distance from the apex to the first lateral, and the spacing of laterals are influenced by the activity of the root apical meristem. In the thinnest or shortest laterals the apical meristem generally has only a single period of growth activity, followed, sooner or later, by its death and replacement by other meristems. Those laterals which ultimately comprise the skeletal framework of the root system survive for more than a season. The apical meristems of these roots grow relatively rapidly over a long seasonal period and undergo annual, or periodic, dormancy.

Once established in the apical regions of roots, mycorrhizal infections tend to extend continuously and acrogenously with root growth and to progressively senesce proximally as cortical cells suberize. The qualitative and quantitative characteristics of the infections conform with the hierarchical variations in morphology and growth activity of the colonized root axis. Ectomycorrhizal fungi appear to maintain a particularly close affinity to the root apical meristem and to establish an organ whose meristematic activity is under joint control. These joint organs remain functional over a longer period of time commensurate with their role in exploitation of seasonal flushes of nutrient.

VA mycorrhizal associations in woody plants appear to be less intimately associated with apical meristems and to occur over longer axial increments of root growth. However, little definitive information is available on VA mycorrhizal development in relation to the architecture and seasonal growth activity of woody hosts.

Herbaceous root systems and mycorrhizal development

The length and complex branching of an intact herbaceous root system become extensive and difficult to describe after a few weeks growth (Russell 1977). The predominant mycorrhizal association in these root systems is formed by VA mycorrhizal fungi, which are extremely difficult to recognize and localize within the developing root system. These difficulties have led to sampling techniques in which root segments are cleared and checked for the presence of mycorrhizal infections. Such information has been used to estimate the proportion of the root system occupied by the fungi for correlation with various crop studies.

In most herbaceous species, VA mycorrhizal infections occur and are functional over a considerable portion of the total root length. Infected lengths of individual root axes are built up of discrete "infection units," which may or may not overlap longitudinally (Cox and Sanders 1974). The expansion and senescence of infection units is coordinated with the senescence of root axes, and in annual plants the mutualistic association terminates with the life cycle of the plant. In herbaceous perennials infections may remain systemic in older root systems. Techniques are being devised for the study of the architecture and mycorrhizal development of entire root systems from an ecological perspective with emphasis on genotypic and phenotypic plasticity in relation to environment and to possible influence on rate and pattern of nutrient uptake (Fitter 1987).

The cellular details of anatomical investigations have been described in a number of recent reviews (Carling and Brown 1982, Scannerini and Bonfante-Fasolo 1983, Bonfante-Fasolo 1984, Bonfante-Fasolo and Gianinazzi-Pearson, 1986, and Duddridge 1984) and will not be considered here.

Utilization of Mycorrhizal Fungi in Forestry and Agriculture

Ectomycorrhizae

Among practical applications of ectomycorrhizae have been large-scale afforestations in which exotic tree species are imported from far outside their natural range (Mikola 1970); afforestation of formerly treeless areas (grasslands) or of alpine areas above the present timberline (Mikola 1973); establishment of nurseries on former agricultural soils (Mikola 1973); and revegetation of adverse sites (Marx 1977, 1980). The possibilities for introducing ectomycorrhizae into existing forests are limited by competition from the natural populations of fungi that occur everywhere on undisturbed sites and are adapted for the ecosystems they occupy (Mikola 1973). However, the proportions of ectomycorrhizal types in natural forests shift after disturbance following harvest and site preparation, making reforestation increasingly difficult with time (Perry et al. 1987). In these circumstances seedling survival has been enhanced by unknown ectomycorrhizae reintroduced by transfer of small amounts of soil from natural forests to planting holes on the harvested area (Amaranthus and Perry 1987) or by specific ectomycorrhizae on tree seedlings from bare-root and container-seedling nurseries inoculated to improve reforestation efforts (Castellano and Trappe 1985, Marx and Cordell 1987).

Ectomycorrhizae were introduced early in the establishment of planted forests of exotic tree species by importation of litter, soil, or mycorrhizal seedlings from native habitats (Mikola 1970, 1973). Such natural inocula contain unidentified ectomycorrhizal fungi that may or may not be the most beneficial strains, and may also inadvertently contain deleterious organisms. Consequently efforts are now focused worldwide on technologies for utilization of pure cultures. These efforts involve recognition, selection, and culture of the most beneficial mycorrhizal fungi for particular host species (Trappe 1977), and large-scale commercial production and application of inoculum to bare-root and container nurseries or to direct-seeding operations (Marx 1980, Tommerup et al. 1987).

Both mycelia and basidiospores have been used as pure-culture inocula. Their early use in small-scale nursery and field application and the advantages and disadvantages of each have been comprehensively reviewed (Mikola 1970, 1973, Marx 1980, Marx and Kenney 1982, Tommerup et al. 1987). The following examples of ectomycorrhizal inoculation are chosen for discussion because of their commercial or near-commercial application in forestry practice.

The largest field application involving mycelial inoculation with a pure-culture ectomycorrhizal fungus to date is found in the pinelands of the southeastern U.S. Since the winter of 1973, a U.S. Forest Service research group at Athens, Georgia, under the leadership of Donald Marx has conducted extensive mycorrhizal research in forest tree nurseries, forest plantings, and on reclaimed mineland that has emphasized the practical use of *Pisolithus tinctorius* with southern pine species, predominantly on *Pinus taeda* L. (Marx et al. 1984, Cordell et al. 1987). Accomplishments include consistently effective vegetative inoculum, an effective vegetative inoculum applicator, and an extensive operational program organized cooperatively with several key nursery, forestry, and mineland reclamation agencies. Particularly noteworthy is the reforestation of severely disturbed sites, such as coal-mine and kaolin spoils, borrow pits, and severely eroded lands, enabled by the particular adaptation of *Pisolithus* ectomycorrhizae to high soil temperature and other adverse site factors. However, the ultimate success of this ectomycorrhizal fungus in routine forestation practice throughout the southeastern United States remains uncertain. Among the potential limiting factors are uncertainties regarding availability of an adequate inoculum supply, the added cost of inoculum, the reduced success of *Pisolithus* inoculations resulting from the use of the rust fungicide triadimefon in nurseries, and the variable benefits arising from occasional unfavorable nursery and planting-site factors. Recently Marx and Cordell (1988, in press) have reported advances in technology and additional sources of inoculum.

Another large-scale research effort involving a different set of forestry problems is centered in the Pacific Northwest. The U.S. Forest Service group at Corvallis, Oregon, together with scientists at Oregon State University, has conducted extensive investigations of ectomycorrhizae throughout this region (Castellano 1987, Linderman 1987, Perry et al. 1987). Here the research approaches to the ecology and management of ectomycorrhizal fungi have been influenced by the highly varied habitats and the great diversity of ectomycorrhizal hosts. Of particular concern are the delays and reductions of reforestation, varying with environment and tree species, which accompany clear-cutting practices. The input of spores and the survival of propagules decline with time, resulting in changes of mycorrhizal inoculation potential in relation to climate, soil, and slash-burning procedures (Amaranthus and Perry 1987, Perry et al. 1987). A reforestation technology has been developed using the genus *Rhizopogon*, after testing and rejection of a number of other promising ectomycorrhizal fungi (Castellano and Trappe 1985). Bare-root and container-grown seedlings of Douglas fir and ponderosa pine are being produced from commercially supplied inoculum, using techniques developed for basidiospore suspensions (Parke et al. 1983, Castellano and Trappe 1985, Castellano et al. 1985). Extensive reforestation trials conducted on *Rhizopogon vinicolor* Smith, which is host-specific to Douglas fir, showed that inoculated seedlings survived and grew better than uninoculated "nursery run" seedlings on difficult reforestation sites (Castellano and Trappe 1985) and were able to withstand and recover from drought (Parke et al. 1983). The genus *Rhizopogon* contains a mixture of about 150 host-specific and generalist species, several of which have been tested successfully with conifer hosts (Castellano 1987). It is likely that many of these will be used ultimately in wide-scale inoculation programs for conifers of the Pacific Northwest. One of the remaining intractable reforestation problems is to devise a successful ectomycorrhizal technology for the genus *Abies* on high-elevation sites (Castellano, personal communication).

The increasing importance of coniferous plantations throughout the world affords an increasing challenge for ectomycorrhizal research programs. Additional fungi must be screened for their ability to increase survival and yield of particular tree species. Suitable technologies must be developed for inoculum production and application. Synergistic effects between mycorrhizal fungi in mixed cultures and between ectomycorrhizal fungi and associative organisms in the mycorrhizosphere must be investigated. Continued research progress will require coordinated laboratory and field investigations.

Vesicular-arbuscular mycorrhizae

The vastly greater occurrence of VA mycorrhizae than ectomycorrhizae in plants and soils would presumably indicate a greater potential for practical application. Actually applications have been similar except for the emphasis on agricultural rather than forest plants. Also methods for production of mycorrhizal inoculum have differed, with large quantities of VA mycorrhizal inoculum needed for field inoculation being raised in pot cultures of mycorrhizal plants rather than in axenic culture. The procedures and safeguards for this method of inoculum production have been comprehensively reviewed (Menge 1983, 1984, Hayman 1987).

As in ectomycorrhizal technology, the benefits of VA mycorrhizal inoculation are influenced by indigenous populations. Much attention is being paid to the influence of agricultural practices and site-specific effects on the composition and efficiency of fungal populations as an alternative to VA mycorrhizal inoculation (Dehne 1987, Baltruschat and Dehne 1988). However, searches continue for highly efficient VA mycorrhizal isolates that are superior to native soil populations (Marx and Schenck 1983, Menge 1983). Recent field trials in the tropics have shown that many tropical crop plants are mycotrophic and that yields can be significantly increased by inoculation with highly efficient VA mycorrhizal isolates even under very rustic small farmers' conditions (Howeler et al. 1987). Elsewhere, commercial uses of VA mycorrhizae have been restricted largely to situations where the natural populations of VA mycorrhizal fungi have been destroyed or damaged, e.g., in fumigated soils or disturbed sites (Menge 1983, Johnson 1987). Large numbers of plants raised in fumigated nursery soils or in containerized culture are likely candidates for VA mycorrhizal inoculation. Examples include citrus seedlings, hardwood tree seedlings, and various fruit trees (Marx and Schenck 1983).

The utilization of VA mycorrhizae for the restoration of disturbed lands offers a potential that is still largely unexplored. The majority of studies of mine spoils have been conducted to determine the occurrence of mycorrhizae in the vegetation on the site rather than the effects of introducing VA mycorrhizae (Miller 1987). The restoration of arid shrub-grasslands following severe disturbance was promoted by planting shrub islands (patch planting) of VA mycorrhizal species to serve as foci for the dispersal of infection. This method was superior to and less costly than direct inoculation of the soil (Allen 1987). The growing knowledge of the ecophysiology of VA mycorrhizal plants suggests many future applications to natural ecosystem management.

References

Agerer, A. (1986a). Studies on ectomycorrhizae. II. Introducing remarks on characterization and identification. *Mycotaxon 26*:473–492.

Agerer, A. (1986b). Studies on ectomycorrhizae. III. Mycorrhizae formed by four fungi in the genera *Lactarius* and *Russula* on spruce. *Mycotaxon 27*:1–59.

Alexander, I. 1987. Ectomycorrhizas in indigenous lowland tropical forest and woodland. In *Mycorrhizae in the Next Decade. Practical Applications and Research Priorities*, Proceedings of the 7th North American Conference on Mycorrhizae, D. M. Sylvia, L. L. Hung, and J. D. Graham (eds.). Institute of Food and Agricultural Sciences, Gainesville, Florida, pp. 115–117.

Allen, M. F. (1987). Ecology of vesicular-arbuscular mycorrhizae in an ecosystem: use of natural processes promoting dispersal and establishment. In *Mycorrhizae in the Next Decade. Practical Applications and Research Priorities*, Proceedings of the 7th North American Conference on Mycorrhizae, D. M. Sylvia, L. L. Hung, and J. D. Graham (eds.). Institute of Food and Agric. Sci., Gainesville, Florida, pp. 133–135.

Amaranthus, M. P., and Perry, D. A. (1987). Effect of soil transfer on ectomycorrhiza formation and the survival and growth of conifer seedlings on old, nonforested clearcuts. *Can. J. For. Res. 17*:944–950.

Baltruschat, H., and Dehne, H. W. (1988). The occurrence of vesicular-arbuscular mycorrhiza in agro-ecosystems. I. Influence of nitrogen fertilization and green manure in continuous monoculture and in crop rotation on the inoculum potential of winter wheat. *Plant and Soil 107*:279–284.

Baylis, G. T. S. (1975). The magnolioid mycorrhiza and mycotrophy in root systems derived from it. In *Endomycorrhizas*, F. E. Sanders, B. Mosse, and P. B. Tinker (eds.). Academic, London, pp. 373–389.

Bonfante-Fasolo, P. (1984). Anatomy and morphology of VA mycorrhizae. In *VA Mycorrhiza*, C. Ll. Powell and D. J. Bagyaraj (eds.). CRC Press, Boca Raton, Florida, pp. 6–33.

Bonfante-Fasolo, P., and Gianinazzi-Pearson, V. (1979). Ultrastructural aspects of endomycorrhiza in the Ericaceae. I. Naturally infected hair roots of *Calluna vulgaris* L. Hull. *New Phytol. 83*:739–744.

Bonfante-Fasolo, P., and Gianinazzi-Pearson, V. (1986). Wall and plasmalemma modifications in mycorrhizal symbiosis. In *Mycorrhizae: Physiology and Genetics*, Proceedings of the 1st European Symposium on Mycorrhizae, V. Gianinazzi-Pearson and S. Gianinazzi (eds.). Inst. Natl. de Recherche Agron., Paris, pp. 67–73.

Bowen, G. D. (1978). Dysfunction and shortfalls in symbiotic responses. In *Plant Diseases*, J. G. Horsfall and E. B. Cowling (eds.), vol. 3. Academic, New York, pp. 231–256.

Büsgen, M. (1901). Einiges über Gestalt und Wachstumweise der Baumwürzeln. *Allg. Forst-. Jagdzg. 77*:273–278, 305–309.

Carling, D. E., and Brown, M. F. (1982). Anatomy and physiology of vesicular-arbuscular and nonmycorrhizal roots. *Phytopathology 72*:1108–1114.

Castellano, M. A. (1987). Ectomycorrhizal inoculum production and utilization in Pacific Northwestern U. S.—a glimpse at the past, a look at the future. In *Mycorrhizae in the Next Decade. Practical Applications and Research Priorities*, Proceedings of the 7th North American Conference on Mycorrhizae, D. M. Sylvia, L. L. Hung, and J. D. Graham (eds.). Institute of Food and Agricultural Sciences, Gainesville, Florida, pp. 290–292.

Castellano, M. A., and Trappe, J. M. (1985). Ectomycorrhizal formation and plantation performance of Douglas-fir nursery stock inoculated with *Rhizopogon* spores. *Can. J. For. Res. 15*:613–617.

Castellano, M. A., Trappe, J. M., and Molina, R. (1985). Inoculation of container-grown Douglas-fir seedlings with basidiospores of *Rhizopogon vinicolor* and *R. colossus*: effects of fertility and spore application rate. *Can. J. For. Res. 15*:10–13.

Chilvers, G. A. (1968). Some distinctive types of eucalypt mycorrhiza. *Aust. J. Bot. 16*:49–70.

Coleman, D. C. (1976). A review of root production processes and their influence on soil biota in terrestrial ecosystems. In *The Role of Terrestrial and Aquatic Organisms in Decomposition Processes*, 17th Symposium of the British Ecological Society, J. M. Anderson and A. Macfadyen (eds.). Blackwell Scientific, Oxford, U.K., pp. 417–434.

Cordell, C. E., Marx, D. H., Maul, S. B., and Owen, J. H. (1987). Production and utilization of ectomycorrhizal fungal inoculum in the Eastern United States. In *Mycorrhizae in the Next Decade. Practical Applications and Research Priorities*, Proceedings of the 7th North American Conference on Mycorrhizae, D. M. Sylvia, L. L. Hung, and J. D. Graham (eds.). Institute of Food and Agricultural Sciences, Gainesville, Florida, pp. 287–289.

Cox, G. C., and Sanders, F. E. (1974). Ultrastructure of the host-fungus interface in a vesicular-arbuscular mycorrhiza. *New Phytol. 73*:901–912.

Cronquist, A. (1981). *An Integrated System of Classification of Flowering Plants*. Columbia University Press, New York.

Dalpé, Y. (1986). Axenic synthesis of ericoid mycorrhiza in *Vaccinium angustifolium* Ait. by *Oidiodendron* species. *New Phytol. 103*:391–396.

Danielson, R. M. (1982). Taxonomic affinities and criteria for the identification of the common ectendomycorrhizal symbiont of pines. *Can. J. Bot. 60*:7–18.

Dehne, H. W. (1987). Management of VA mycorrhizae in temperate crops. In *Mycorrhizae in the Next Decade. Practical Applications and Research Priorities*, Proceedings of the 7th North American Conference on Mycorrhizae, D. M. Sylvia, L. L. Hung, and J. D. Graham (eds.). Institute of Food and Agricultural Sciences, Gainesville, Florida, pp. 5–6.

Dighton, J., and Mason, P. A. (1984). Mycorrhizal dynamics during forest tree development. In *Developmental Biology of Higher Fungi*, Symposium of the British Mycological Society, D. Moore, L. A. Casselton, D. A. Wood, and J. C. Frankland (eds.). Cambridge University Press, Cambridge, U.K., pp. 117–139.

Duddridge, J. A. (1984). A comprehensive ultrastructural analysis of the host-fungus interface in mycorrhizal and parasitic associations. In *Developmental Biology of Higher Fungi*, Symposium of the British Mycological Society, D. Moore, L. A. Casselton, D. A. Wood, and J. C. Frankland (eds.). Cambridge University Press, Cambridge, U.K., pp. 141–173.

Duddridge, J. A., and Read, D. J. (1982). An ultrastructural analysis of the development of mycorrhizae in *Monotropa hypopitys* L. *New Phytol. 92*:203–214.

Finlay, R. D., and Read, D. J. (1986). The structure and function of the vegetative mycelium of ectomycorrhizal plants. II. Uptake and distribution of mycelial strands interconnecting host plants. *New Phytol. 103*:157–165.

Fitter, A. H. (1987). An architectural approach to the comparative ecology of plant root systems. *New Phytol. 106*(suppl.):61–77.

Fogel, R. (1985). Roots as primary producers in below-ground ecosystems. In *Ecological Interactions in Soil: Plants, Microbes, and Animals*. Special Publ. no. 4, British Ecological Society, A. H. Fitter, D. H. Atkinson, D. J. Read, and M. B. Usher (eds.). Blackwell Scientific, Oxford, U.K., pp. 23–35.

Francis, R., Finlay, R. D., and Read, D. J. (1986). Vesicular-arbuscular mycorrhizas in natural vegetation systems. IV. Transfer of nutrients in inter- and intra-specific combinations of host plants. *New Phytol. 102*:103–111.

Fusconi, A., and Bonfante-Fasolo, P. (1984). Ultrastructural aspects of host-endophyte relationships in *Arbutus unedo* L. mycorrhizas. *New Phytol. 96*:397–410.

Gerdemann, J. W. (1968). Vesicular-arbuscular mycorrhiza and plant growth. *Annu. Rev. Phytopathol. 6*:397–418.

Gerdemann, J. W., and Trappe, J. M. (1974). The Endogonaceae in the Pacific Northwest. *Mycologia Memoir* no. 5, New York Botanical Garden, Bronx, New York, 76 pp.

Godbout, C., and Fortin, J. A. (1983). Morphological features of synthesized ectomycorrhizae of *Alnus crispa* and *A. rugosa*. *New Phytol. 94*:249–262.

Godbout, C., and Fortin, J. A. (1985). Synthesized ectomycorrhizae of aspen: fungal genus level of structural characterization. *Can. J. Bot. 63*:252–262.

Grime, J. P. (1977). Evidence for the existence of three primary strategies in plants and its relevance to ecological and evolutionary theory. *American Naturalist 111*:1169–1194.

Gunning, B. E. S., and Pate, J. S. (1974). Transfer cells. In *Dynamic Aspects of Plant Ultrastructure*, A. W. Robards (ed.). McGraw-Hill, New York, pp. 441–480.

Hacskaylo, E. (1973). Carbohydrate physiology of ectomycorrhizae. In *Ectomycorrhizae: Their Ecology and Physiology*, G. C. Marks and T. T. Kozlowski (eds.). Academic, New York, pp. 207–230.

Hall, I. R. (1984). Taxonomy of mycorrhizal fungi. In *VA Mycorrhiza*, C. Ll. Powell and D. J. Bagyaraj (eds.). CRC Press, Boca Raton, Florida, pp. 57–94.

Harley, J. L. (1969). *The Biology of Mycorrhiza*, 2d. ed. Leonard Hill, London, 334 pp.

Harley, J. L., and Harley, E. L. (1987). A check-list of mycorrhiza in the British flora. *New Phytol. 105*(suppl.):1–102.

Harley, J. L., and Smith, S. E. (1983). *Mycorrhizal Symbiosis*. Academic, London, 483 pp.

Hayman, D. S. (1978). Endomycorrhizae. In *Interations between Non-Pathogenic Soil Microorganisms and Plants*, Y. R. Dommergues and S. V. Krupa (eds.). Elsevier Scientific, Amsterdam, pp. 401–442.

Hayman, D. S. (1987). VA mycorrhizas in field crop systems. In *Ecophysiology of VA Mycorrhizal Plants*, G. R. Safir (ed.). CRC Press, Boca Raton, Florida, pp. 171–192.

Hepper, C. M. (1984). Isolation and culture of VA mycorrhizal (VAM) fungi. In *VA Mycorrhiza*, C. Ll. Powell and D. J. Bagyaraj (eds.). CRC Press, Boca Raton, Florida, pp. 95–112.

Högberg, P. (1982). Mycorrhizal associations in some woodland and forest trees and shrubs in Tanzania. *New Phytol. 92*:407–415.

Högberg, P., and Piearce, G. D. (1986). Mycorrhizas in Zambian trees in relation to host taxonomy, vegetation type, and successional patterns. *J. Ecol. 74*:775–785.

Howeler, R. H., Sieverding, E., and Saif, S. (1987). Practical aspects of mycorrhizal technology in some tropical crops and pastures. *Plant and Soil 100*:249–283.

Janos, D. P. (1980). Mycorrhizae influence tropical succession. *Biotropica 12*(suppl.):56–64.

Janos, D. P. (1983). Tropical mycorrhizas, nutrient cycles and plant growth. In *Tropical Rain Forest: Ecology and Management*. Special Publ. no. 2, British Ecological Society, S. L. Sutton, T. C. Whitmore, and A. C. Chadwick (eds.). Blackwell Scientific, Oxford, U.K., pp. 327–345.

Janos, D. P. (1987). VA mycorrhizas in humid tropical ecosystems. In *Ecophysiology of VA Mycorrhizal Plants*, G. R. Safir (ed.). CRC Press, Boca Raton, Florida, pp. 107–134.

Jenik, J. (1976). Roots and root systems in tropical trees: morphologic and ecologic aspects. In *Tropical Trees as Living Systems*, Proceedings of the 4th Cabot Symposium, P. B. Tomlinson and M. Zimmermann (eds.). Harvard Forest, Petersham, Massachusetts, pp. 323–349.

Johnson, C. (1987). Utilization of vesicular-arbuscular mycorrhizal fungi in greenhouse production of transplanted crops. In *Mycorrhizae in the Next Decade. Practical Applications and Research Priorities*, Proceedings of the 7th North American Conference on Mycorrhizae, D. M. Sylvia, L. L. Hung, and J. D. Graham (eds.). Institute of Food and Agricultural Sciences, Gainesville, Florida, pp. 275–277.

Largent, D. L., Sugihara, N., and Wishner, C. (1980). Occurrence of mycorrhizae on ericaceous and pyrolaceous plants in northern California. *Can. J. Bot. 58*:2274–2279.

Lewis, D. H. (1973). The relevance of symbiosis to taxonomy and ecology, with particular reference to mutualistic symbioses and the exploitation of marginal habitats. In *Taxonomy and Ecology*, Systematics Association Special Volume no. 5, Academic, New York, pp. 151–172.

Laiho, O. (1965). Further studies on the ectendotrophic mycorrhiza. *Acta For. Fenn. 79*:1–35.

Linderman, R. G. (1987). Perspectives on ectomycorrhizae research in the Northwest. In *Mycorrhizae in the Next Decade. Practical Applications and Research Priorities*, Proceedings of the 7th North American Conference on Mycorrhizae, D. M. Sylvia, L. L. Hung, and J. D. Graham (eds.). Institute of Food and Agricultural Sciences, Gainesville, Florida, pp. 72–74.

Linderman, R. G. (1988). Mycorrhizal interactions with the rhizosphere microflora: the mycorrhizosphere effect. *Phytopathology 78*:366–370.

Malloch, D. W., Pirozynski, K. A., and Raven, P. H. (1980). Ecological and evolutionary significance of mycorrhizal symbiosis in vascular plants (a review). *Proc. Natl. Acad. Sci. USA 77*:2113–2118.

Marx, D. H. (1977). The role of mycorrhizae in forest production. *Technical Association of the Pulp and Paper Industry Conference Papers, Annual Meeting*, Atlanta, Georgia, pp. 151–161.

Marx, D. H. (1980). Ectomycorrhizal fungus inoculation: a tool for improving forestation practices. In *Tropical Mycorrhiza Research*, P. Mikola (ed.). Clarendon, Oxford, U.K., pp. 13–71.

Marx, D. H., and Cordell, C. E. (1987). Ecology and management of ectomycorrhizal fungi in regenerating forests in the Eastern United States. In *Mycorrhizae in the Next Decade. Practical Applications and Research Priorities*, Proceedings of the 7th North American Conference on Mycorrhizae, D. M. Sylvia, L. L. Hung, and J. D. Graham (eds.). Institute of Food and Agricultural Sciences, Gainesville, Florida, pp. 69–71.

Marx, D. H., and Cordell, C. E. (1988). The use of specific fungi to improve artificial forestation practices. In *Biotechnology of Fungi for Improving Plant Growth*, Symposium Proceedings, Sussex University, Little Hampton, U.K. (in press).

Marx, D. H., and Kenney, D. S. (1982). Production of ectomycorrhizal fungus inoculum. In *Methods and Principles of Mycorrhizal Research*, N. C. Schenck (ed.). American Phytopathological Society, St. Paul, Minnesota, pp. 131–146.

Marx, D. H., and Schenck, N. C. (1983). Potential of mycorrhizal symbiosis in agricultural and forest production. In *Challenging Problems in Plant Health*, T. Kommendahl and P. H. Williams (eds.). 75th Anniv. Publ., American Phytopathological Society, St. Paul, Minnesota, pp. 334–347.

Marx, D. H., Cordell, C. E., Kenney, D. S., Mexal, J. G., Artman, J. D., Riffle, J. W., and Molina, R. J. (1984). Commercial vegetative inoculum of *Pisolithus tinctorius* and inoculation techniques for development of ectomycorrhizae on bare-root tree seedlings. Forest Science Monogr. 25. *For. Sci. 30*(suppl.):1–101.

Menge, J. A. (1983). Utilization of vesicular-arbuscular mycorrhizal fungi in agriculture. *Can. J. Bot. 61*:1015–1024.

Menge, J. A. (1984). Inoculum production. In *VA Mycorrhiza*, C. Ll. Powell and D. J. Bagyaraj (eds.). CRC Press, Boca Raton, Florida, pp. 187–203.

Meyer, F. H. (1973). Distribution of ectomycorrhizae in native and man-made forests. In *Ectomycorrhizae: Their Ecology and Physiology*, G. C. Marks and T. T. Kozlowski (eds.). Academic, New York, pp. 79–105.

Meyer, F. H. (1974). Physiology of mycorrhiza. *Annu. Rev. Plant Physiol. 25*:567–586.

Mikola, P. (1965). Studies on the ectendotrophic mycorrhiza of pine. *Acta For. Fenn. 75*:1–56.

Mikola, P. (1970). Mycorrhizal inoculation in afforestation. *Int. Rev. For. Res. 3*:123–196.

Mikola, P. (1973). Application of mycorrhizal symbiosis in forestry practice. In *Ectomycorrhizae: Their Ecology and Physiology*, G. C. Marks and T. T. Kozlowski (eds.). Academic, New York, pp. 383–411.

Miller, O. K., Jr. (1982a). Ectomycorrhizae in the Agaricales and Gasteromycetes. *Can. J. Bot. 61*:909–916.

Miller, O. K., Jr. (1982b). Taxonomy of ecto- and ectendomycorrhizal fungi. In *Methods and Principles of Mycorrhizal Research*, N. C. Schenck (ed.). American Phytopathological Society, St. Paul, Minnesota, pp. 91–101.

Miller, R. M. (1987). The management of VA mycorrhizae in semiarid environments. In *Mycorrhizae in the Next Decade. Practical Applications and Research Priorities*, Proceedings of the 7th North American Conference on Mycorrhizae, D. M. Sylvia, L. L. Hung, and J. D. Graham (eds.). Inst. Food and Agric. Sci., Gainesville, Florida, pp. 139–141.

Mitchell, D. T., and Read, D. J. (1981). Utilization of inorganic and organic phosphates by the mycorrhizal endophytes of *Vaccinium macrocarpon* and *Rhodendron ponticum*. *Trans. Br. Mycol. Soc. 76*:255–260.

Molina, R., and Trappe, J. M. (1982). Lack of mycorrhizal specificity by the ericaceous hosts *Arbutus menziesii* and *Arctostaphylos uva-ursi*. *New Phytol. 90*:495–509.

Newbery, D. M., Alexander, I. J., Thomas, D. W., and Gartlan, J. S. (1988). Ectomycorrhizal rain-forest legumes and soil phosphorous in Korup National Park, Cameroon. *New Phytol. 109*:433–450.

Newman, E. I., and Reddell, P. (1987). The distribution of mycorrhizas among families of vascular plants. *New Phytol. 106*:745–751.

Nicolson, T. H. (1967). Vesicular-arbuscular mycorrhiza: A universal plant symbioses. *Sci. Progr.* (Oxford) *55*:561–568.

Parke, J. L., Linderman, R. G., and Black, C. H. (1983). The role of ectomycorrhizas in drought tolerance of Douglas-fir seedlings. *New Phytol. 95*:83–95.

Pearson, V., and Read, D. J. (1975). The physiology of the mycorrhizal endophyte of *Calluna vulgaris*. *Trans. Br. Mycol. Soc. 64*:1–7.

Perry, D. A., Molina, R., and Amaranthus, M. P. (1987). Mycorrhizae, mycorrhizospheres, and reforestation: current knowledge and research needs. *Can. J. For. Res. 17*:929–940.

Piché, Y., Ackerley, C. A., and Peterson, R. L. (1986). Structural characteristics of ectendomycorrhizas synthesized between roots of *Pinus resinosa* and the E-strain fungus, *Wilcoxina mikolae* var. *mikolae*. *New Phytol. 104*:447–452.

Pirozynski, K. A. (1981). Interactions between fungi and plants through the ages. *Can. J. Bot. 59*:1824–1827.

Pirozynski, K. A., and Malloch, D. W. (1975). The origin of land plants: a matter of mycotrophism. *Biosystems 6*:153–164.

Powell, C. Ll., and Bagyaraj, D. J. (eds.). (1984). *VA Mycorrhiza*. CRC Press, Boca Raton, Florida.

Read, D. J. (1974). *Pezizella ericae* Sp. Nov., the perfect stage of a typical mycorrhizal endophyte of Ericaceae. *Trans. Br. Mycol. Soc. 63*:381–383.

Read, D. J. (1983). The biology of mycorrhizae in the Ericales. *Can. J. Bot. 61*:985–1004.
Read, D. J. (1984) The structure and function of the vegetative mycelium of mycorrhizal roots. In *The Ecology and Physiology of the Fungal Mycelium*, D. J. Jennings and A. D. M. Rayner (eds.), *Symposium Br. Mycol. Soc.*, Cambridge Univ. Press, Cambridge, U.K., pp. 215–240.
Read, D. J., Francis, R., and Finlay, R. D. (1985). Mycorrhizal mycelia and nutrient cycling in plant communities. In *Ecological Interactions in Soil: Plants, Microbes, and Animals*. Special Publ. no. 4, British Ecological Society, A. H. Fitter, D. A. Atkinson, D. J. Read, and M. B. Usher (eds.). Blackwell Scientific, Oxford, U.K., pp. 193–217.
Reeves, F. B., Wagner, D., Moorman, T., and Kiel, J. (1979). The role of endomycorrhizae in revegetation practices in the semiarid west. I. A comparison of incidence of mycorrhizae in severely disturbed vs. natural environments. *Am. J. Bot. 66*:6–13.
Robertson, D. C., and Robertson, J. A. (1982). Ultrastructure of *Pterospora andromedea* Nuttall and *Sarcodes sanguinea* Torrey mycorrhizas. *New Phytol. 92*:539–551.
Robertson, D. C., and Robertson, J. A. (1985). Ultrastructural aspects of *Pyrola* mycorrhizae. *Can. J. Bot. 63*:1089–1098.
Russell, R. S. (1977). *Plant Root Systems: Their Function and Interaction with Soil*. McGraw-Hill, London.
Safir, G. R. (ed.). (1987). *Ecophysiology of VA Mycorrhizal Plants*. CRC Press, Boca Raton, Florida.
St. John, T. V. (1980). Root size, root hairs and mycorrhizal infection: a re-examination of Baylis's hypothesis with tropical trees. *New Phytol. 84*:483–487.
Sanders, F. E., Mosse, B., and Tinker, P. B. (eds.) (1975). *Endomycorrhizas*. Academic, London.
Scannerini, S., and Bonfante-Fasolo, P. (1983). Comparative ultrastructural analysis of mycorrhizal associations. *Can. J. Bot. 61*:917–943.
Stribley, D. P., and Read, D. J. (1980). The biology of mycorrhiza in the Ericaceae. VII. The relationship between mycorrhizal infection and the capacity to utilize simple and complex organic nitrogen sources. *New Phytol. 86*:365–371.
Stubblefield, S. P., Taylor, T. N., and Trappe, J. M. (1987). Fossil mycorrhizae: a case for symbiosis. *Science 237*:59–60.
Tommerup, I. C., Kuek, C., and Malajczuk, N. (1987). Ectomycorrhizal inoculum production and utilization in Australia. In *Mycorrhizae in the Next Decade. Practical Applications and Research Priorities*, Proceedings of the 7th North American Conference on Mycorrhizae, D. M. Sylvia, L. L. Hung, and J. D. Graham (eds.). Institute of Food and Agricultural Sciences, Gainesville, Florida, pp. 293–295.
Trappe, J. M. (1962). Fungus associates of ectotrophic mycorrhizae. *Bot. Rev. 28*:538–606.
Trappe, J. M. (1971). Mycorrhiza-forming Ascomycetes. In *Mycorrhizae*, E. Hacskaylo (ed.). USDA Forest Service Misc. Publ. 1189, Washington, D.C., pp. 19–37.
Trappe, J. M. (1977). Selection of fungi for ectomycorrhizal inoculation in nurseries. *Annu. Rev. Phytopathol. 15*:203–222.
Trappe, J. M. (1981). Mycorrhizae and productivity of arid and semiarid rangelands. In *Advances in Food-Producing Systems for Arid and Semiarid Lands*, J. T. Manassah and E. J. Briskey (eds.). Academic, New York, pp. 581–599.
Trappe, J. M. (1982). Synoptic keys to the genera and species of Zygomycetous fungi. *Phytopathology 72*:1102–1108.
Trappe, J. M. (1987). Phylogenetic and ecologic aspects of mycotrophy in the angiosperms from an evolutionary standpoint. In *Ecophysiology of VA Mycorrhizal Plants*, G. R. Safir (ed.). CRC Press, Boca Raton, Florida, pp. 5–25.
Trappe, J. M., and Fogel, R. D. (1977). Ecosystematic functions of mycorrhizae. In *The Belowground Ecosystem: A Synthesis of Plant-Associated Processes*, J. K. Marshall (ed.). Range Science Dept. Sci. Ser. no. 26, Colorado State University, Fort Collins, Colorado.
Trappe, J. M., and Strand, R. F. (1969). Mycorrhizal deficiency in a Douglas-fir nursery. *For. Sci. 15*:381–389.

Vozzo, J. A., and Hacskaylo, E. (1971). Inoculation of *Pinus caribaea* with ectomycorrhizal fungi in Puerto Rico. *For. Sci. 17*:239–245.

Wang, C. J. K., and Wilcox, H. E. (1985). New species of ectendomycorrhizal and pseudomycorrhizal fungi: *Phialophora finlandia, Chloridium paucisporum,* and *Phialocephala fortinii. Mycologia 77*:951–958.

Warcup, J. H. (1980). Ectomycorrhizal associations of Australian indigenous plants. *New Phytol. 85*:531–535.

Warcup, J. H., and McGee, P. A. (1983). The mycorrhizal associations of some Australian Asteraceae. *New Phytol. 95*:667–672.

Wilcox, H. E. (1968). Morphological studies of the roots of red pine, *Pinus resinosa*. II. Fungal colonization of roots and the development of mycorrhizae. *Am. J. Bot. 55*:688–700.

Wilcox, H. E. (1971). Morphology of ectendomycorrhizae in *Pinus resinosa*. In *Mycorrhizae*, E. Hacskaylo (ed.). USDA Forest Service Misc. Publ. 1189, Washington, D.C., pp. 54–68.

Wilcox, H. E. (1982). Morphology and development of ecto- and ectendomycorrhizae. In *Methods and Principles of Mycorrhizal Research*, N. C. Schenck (ed.). American Phytopathological Society, St. Paul, Minnesota, pp. 103–113.

Wilcox, H. E. (1983). Fungal parasitism of woody plant roots from mycorrhizal relationships to plant disease. *Annu. Rev. Phytopathol. 21*:221–242.

Wilcox, H. E., Yang, C. S., and LoBuglio, K. F. (1983). Responses of pine roots to E-strain ectendomycorrhizal fungi. In *Tree Root Systems and Their Mycorrhizas*, D. L. Atkinson et al. (eds.). Martinus Nijhoff/Dr. W. Junk, The Hague, pp. 293–297.

Yang, C. S., and Korf, R. P. (1985). A monograph of the genus *Tricharina* and of a new segregate genus, *Wilcoxina* (Pezizales). *Mycotaxon 24*:467–513.

Yang, C. S., and Wilcox, H. E. (1984). An E-strain ectendomycorrhiza formed by a new species, *Tricharina mikolae. Mycologia 76*:675–684.

Zak, B. (1973). Classification of ectomycorrhizae. In *Ectomycorrhizae: Their Ecology and Physiology*, G. C. Marks and T. T. Kozlowski (eds.). Academic, New York, pp. 43–78.

Zak, B. (1974). Ectendomycorrhiza of Pacific madrone (*Arbutus Menziesii*). *Trans. Br. Mycol. Soc. 62*:202–205.

Zak, B. (1976). Pure culture synthesis of Pacific madrone ectendomycorrhizae. *Mycologia 68*:362–369.

Fungi as Biological Control Agents

Charles R. Howell

U.S. Department of Agriculture
Agricultural Research Service
Southern Crops Research Laboratory
Rt. 5, Box 805
College Station, Texas 77840

Introduction

The use of fungi to control pathogens that incite plant disease is a concept that has been in existence for some time, and the literature is replete with reports by scientists on the discovery of fungi that are antagonistic to pathogens and on the use of these fungi to control disease. Almost without exception, however, fungal biocontrol agents fail to consistently control diseases under the highly variable environmental conditions that occur in the field. These failures are due, in part, to a dearth of knowledge about the mechanisms that are employed by fungal antagonists during the biocontrol process, and of the effects of the biotic and abiotic environment on the biocontrol agent and/or the mechanisms it employs. Research up to this point has made one thing abundantly clear; the requirements of nature and of currently accepted agricultural practices are not likely to be fully met by the biocontrol agents that are isolated from nature. Presumably, recently developed techniques in biotechnology such as gene cloning and transfer will be required to construct fungal biocontrol agents with the traits necessary to successfully control plant diseases in the field.

The purpose of this chapter is to highlight those mechanisms that are believed to function in the biocontrol of plant disease by fungi, and to discuss the effects of environment on these mechanisms and on the biocontrol agents themselves. It is beyond the scope of this chapter to mention all instances of biocontrol of plant disease by fungi, so only a few examples will be mentioned to illustrate each point.

Mycoparasitism in Fungal Biocontrol Agents

Mycoparasitism is probably one of the most outstanding features of this class of biocontrol agents. Most were isolated from the resting structures or actively growing mycelia of their hosts. Two kinds of mycoparasites are recognized, "biotrophs" that do little harm to the host in the initial stages and "necrotrophs" that kill in advance or immediately upon contact with the host (Barnett and Binder 1973). Mycoparasites are considered to function in the biocontrol process by reducing the pathogen inoculum prior to host exposure and by preventing the pathogen from infecting the host at the infection court (Cook and Baker 1983).

The dematiaceous hyphomycetes *Sporidesmium sclerotivorum* and *Tetrasperma oligocladum* are two excellent examples of mycoparasitic biocontrol agents that function mainly by reducing the pathogen inoculum in the soil (Adams and Ayers 1980, 1981, Ayers and Adams 1981). Both are parasitic on the sclerotia of *Sclerotinia* spp. (Uecker et al. 1978, 1980). Sclerotial infection is initiated from macroconidia in close proximity in the soil, and subsequent infections come from conidia that are produced on threadlike hyphae emanating from infected sclerotia.

Infected sclerotia do not germinate and ultimately are degraded. Neither mycoparasite appears to invade the hyphae of *Sclerotinia* spp., and, although not classified as obligate parasites, both fungi are dependent on the sclerotia of host species for their existence in soil (Ayers and Adams 1979, 1981). Ayers and Adams (1985) have described an interesting phenomenon concerning these two mycoparasites. A third dematiaceous hyphomycte, *Laterispora brevirama*, is markedly similar in morphology to *S. sclerotivorum* and *T.*

oligocladum, and is often found in close association with them on infected sclerotia. It is unable to parasitize sclerotia on its own, and occurs only in the presence of one or the other of the primary
mycoparasites. It apparently functions either as a secondary parasite
of Sclerotinia spp., or as a direct parasite of *S. sclerotivorum* and *T.
oligocladum*.

Another mycoparasitic biocontrol agent that appears to function by
parasitizing and thus reducing the levels of pathogen propagules is
the coelomycete *Coniothyrium minitans*. It too attacks sclerotia in a
manner similar to that of the previous examples. However, it has a
much wider host range and can also attack the mycelia of its hosts
(Huang and Hoes 1976). In addition, it is easily cultured on ordinary
laboratory media. Its hosts include *Sclerotinia* spp. (Turner and Tribe
1976), *Sclerotium* spp. (Ahmed and Tribe 1977), *Botrytis* spp., and
Claviceps purpurea (Turner and Tribe 1976). Its activities are apparently not confined to the soil, as it produces pigmented conidia in
melanized pycnidia that would probably survive on aerial plant parts
(Trutman et al. 1982).

A group of closely related moniliaceous hyphomycetes that has fig-
ured prominently in research on biocontrol with mycoparasites comprises the genera *Verticillium, Trichoderma*, and *Gliocladium*. All are
necrotrophic mycoparasites and good saprophytes that grow well on
ordinary laboratory media. *Verticillium biguttatum* is often found in
association with the sclerotia of *Rhizoctonia solani* (Jager and Velvis
1983). It has been used to treat seed potatoes in Holland (Velvis and
Jager 1983). Treatment with *V. biguttatum* did not alleviate disease
symptoms on the plant, but it did substantially reduce the numbers of
pathogen sclerotia on new tubers. Of untreated tubers, 53% had surface sclerotia on them, while only 10% of the treated tubers were infested. This fungus also protects potato sprouts from infection by *R.
solani* (Jager and Velvis 1984).

The genus *Trichoderma* contains many species that are noted for
their capacity to parasitize other fungi and to act as biocontrol agents.
Trichoderma harzianum is the species most frequently reported to be
an effective disease antagonist. Its host range includes *Sclerotium
rolfsii* (Wells et al. 1972), *R. solani* (Hadar et al. 1978), *Pythium
aphanidermatum* (Sivan et al. 1984), *Macrophomina phaseolina* (Elad
et al. 1986), *Alternaria raphani* and *Alternaria brassicicola* (Vannacci
and Harman 1987), and *Rosellinia necatrix* (Freeman et al. 1986). *T.
harzianum* parasitizes its hosts by coiling around and penetrating the
hyphae and resting structures. Entrance is gained through holes
formed in host hyphae by the production of β-(1,3)-glucanase and
chitinase by the parasite (Elad et al. 1982). In *P. aphanidermatum*,
production of high levels of cellulase by the mycoparasite has been

associated with disease control (Elad et al. 1982). Growth of *P. aphanidermatum* is also inhibited by extracellular filtrate from *T. harzianum* (Sivan et al. 1984). Lectin binding is also believed to function in the host-parasite interaction (Elad et al. 1983, Barak et al. 1985). Recognition is provided by the binding of a lectin in the host's hyphae to galactose residues on the *Trichoderma* cell walls (Elad et al. 1983). *T. harzianum* has been reported to control diseases incited by many of the pathogens listed above, and it effectively controls diseases of cotton, wheat, and muskmellon incited by *Fusarium oxysporum* f. sp. *vasinfectum, Fusarium roseum* 'Culmorum,' and *F. oxysporum* f. sp. *melonis*, respectively (Sivan and Chet 1986). It has also been reported to control potato diseases incited by *Fusarium solani* and *F. oxysporum* (Dwived 1984). The biocontrol mechanism operating in the case of the *Fusarium* spp. is unknown. Other species of *Trichoderma*, such as *Trichoderma hamatum* (Chet et al. 1981), *Trichoderma koningii* (Dos Santos and Dhingra 1980), *Trichoderma viride* (Grosclaude et al. 1973), and *Trichoderma longibrachiatum* and *Trichoderma polysporum* (Hashioka and Fukita 1969) have also been reported to act as mycoparasites of plant pathogens. They function in the same manner as *T. harzianum*, and some, including strains of *T. harzianum*, have been shown to produce antifungal antibiotics as well (Claydon et al. 1987, Dennis and Webster 1971a,b, Okuda et al. 1981).

The genus *Gliocladium* contains many mycoparasitic species that have numerous features in common with those from *Trichoderma*. In fact, the former have on occasion been mistaken for the latter. Both are acid-loving, fast-growing saprophytes, whose fructifications are usually various shades of green. Both also produce chlamydospores in culture. They have similar host ranges with respect to mycoparasitism, and they tend to parasitize the host in a similar fashion. They differ in that the conidia of *Gliocladium* spp. are borne on the sporophore in slime balls, while those from *Trichoderma* are dry (Gillman 1966). In addition, *Gliocladium* spp. are noted for the production of secondary metabolites that act as antifungal agents, antibacterial agents, or phytotoxins. Among the better known of this group is *Gliocladium roseum*. It is an aggressive mycoparasite with a wide host range, and it differs from most others in this group by usually occurring in neutral to alkaline soils (Domsch et al. 1980). Also in contrast to the others, it parasitizes the oospores of *Phytophthora erythroseptica* (Wynn and Epton 1979), and directly penetrates the chlamydospores and sporangia of *Phytophthora palmivora* without parasitizing the host hyphae (Lim and Chan 1986). In its parasitism of *Botrytis allii*, *G. roseum* produces the cell-wall-degrading enzymes β-(1–3)-glucanase and chitinase, and elaborates low-molecular-weight toxins that are effective over a short distance (Pachenari and Dix 1980).

Another well-studied member of the genus *Gliocladium* is *Gliocladium virens*. This species has only been reported to parasitize *Rhizoctonia solani* (Tu and Vaartaja 1981, Howell 1982) and *Sclerotinia sclerotiorum* (Tu 1980), but it produces a number of antibiotic compounds that broaden its host range considerably with respect to biocontrol activity. *G. virens* was first reported to produce the antifungal compounds gliotoxin (Brian and Hemming 1945) and viridin (Brian and McGowan 1945), but was misidentified as *Trichoderma lignorum* (Weindling 1936) or *Trichoderma viride*. This difficulty was not cleared up until many years later when Webster and Lomas (1964) showed that these compounds were produced not by *Trichoderma* spp., but by *G. virens*. Aluko and Hering (1970) have subsequently shown that *G. virens* kills the sclerotia of *R. solani* on the surface of potato tubers, and they hypothesized that sclerotial death was due to antibiosis, not to parasitism. This was later confirmed by Howell (1986), who showed that mutants of *G. virens* deficient for mycoparasitic activity retained the capacity to act as biocontrol agents of seedling disease incited by *R. solani*, and to kill the sclerotia of the pathogen in natural soil. Other members of this group may parasitize exclusively in this manner. Huang (1978) has shown that *Gliocladium catenulatum* parasitized the hyphae and sclerotia of *Sclerotinia sclerotiorum*, and the hyphae and macroconidia of several *Fusarium* spp. by hyphal contact only. Penetration of the host cells, however, was never observed, and the chlamydospores of the *Fusarium* spp. were not infected. This method of parasitism is consistent with that observed by Howell (1982) in the parasitic interaction between *G. virens* and *Pythium ultimum*. Close proximity of parasite hyphae to those of the host resulted in granulation of the host cytoplasm and dissolution of host hyphae. In addition, *G. virens* produced an antibiotic that was strongly inhibitory to the growth of *P. ultimum*. This antibiotic, a previously unreported diketopiperazine (gliovirin), was highly active against members of the Oomycetes, but had little effect on other fungi or bacteria (Howell and Stipanovic 1983). Mutants of *G. virens* deficient for gliovirin production were overgrown in culture by *P. ultimum*, and they were unable to control cotton-seedling disease incited by this pathogen in soil. Mutants with enhanced gliovirin activity were equal to the parental isolate in biocontrol efficacy, even though their growth rate was reduced.

A representative of the Penicillia is also known to act as a biocontrol agent through a combination of mycoparasitism and antibiotic production. *Penicillium vermiculatum*, more commonly known now by its teleomorph *Talaromyces flavus* (Domsch et al. 1980), was first reported by Boosalis (1956) to vigorously parasitize the hyphae of *R. solani* and prevent seedling disease of peas in sterile soil. In nonsterile soil, the mycoparasite was much less effective, either as a biocontrol agent or mycoparasite. In green-manure-treated plots containing the mycoparasite, 82% of isolated *R. solani* propagules were not parasitized. However, only 20% of them survived after 8 weeks in soil, leading the author to speculate that antibiosis might be involved. *P. vermiculatum* (*T. flavus*) does indeed produce antibiotics. Talaron (Mizumo et al. 1974), vermiculine (Fuska et al. 1972), vermistatin (Fuska et al. 1979a), and vermicillin (Fuska et al. 1979b) have been reported from this fungus, and production of glucose oxidase by *T. flavus* has been implicated in its biocontrol of *Verticillium dahliae*, probably because of the reaction product hydrogen peroxide (Kim et al. 1988). *T. flavus* has been reported to be an effective biocontrol agent for verticillium wilt of eggplant (Marois et al. 1982), and for sclerotinia wilt of sunflower in the field (McLaren et al. 1985). *T. flavus* parasitizes the hyphae of *Sclerotinia sclerotiorum* by direct penetration, resulting in granulation of host cytoplasm and collapse of cell walls (McLaren et al. 1986).

A close relative of *P. vermiculatum* is *Penicillium frequentans*; it parasitizes the sclerotia of a number of fungi (Karhuvaara 1960, Makkonen 1960, Moubasher 1970). Seed treatment of mustard (Wright 1956) or beet (Liu 1965) with this fungus suppressed the development of *P. ultimum*–incited seedling disease on these crops.

Another well studied mycoparasite, *Laetisaria arvalis*, is a basidiomycete. It parasitizes *R. solani* (Odvody et al. 1980) and antagonizes *P. ultimum* through antibiosis (Hoch and Abawi 1979). *L. arvalis* was originally reported as a *Corticium* sp., but was subsequently assigned to *L. arvalis* (Burdsall et al. 1980). When added to seed or soil as sclerotia or colonized carrier, it is an effective biocontrol agent of *R. solani*–incited diseases of beans (Odvody et al. 1977), sugar beets (Odvody et al. 1980), and cucumber (Lewis and Papavizas 1980), as well as *P. ultimum*–incited root rot of table beets (Hoch and Abawi 1979).

The Oomycetes also contain members that are mycoparasites on other Oomycetes and on fungi from different classes. *Pythium oligandrum* is an aggressive mycoparasite that can attack such pathogens as *Gaeumanomyces graminis* var. *tritici* and *Fusarium* spp. (Deacon 1976), and *P. ultimum* and *R. solani* (Lutchmeah and Cooke 1984). Isolates within species of the hosts appear to vary greatly in resistance to attack by *P. oligandrum* (Foley and Deacon 1985). Some host species are penetrated by the mycoparasite, but it was never observed to penetrate the hyphae of *P. ultimum*. The mechanism in this case appears to be hyphal interference, a phenomenon in which the hyphae of the anatagonist make contact with those of the host and inhibit and distort the host's growth (Lutchmeah and Cooke 1984). The authors suggested that this process was primarily a mechanism for resource capture from other species rather than benefit arising from direct exploitation of the host. *P. oligandrum* is a good biocontrol agent of *P. ultimum*–induced seedling diseases of cress and sugarbeets, and of *Mycocentrospora acerina*–induced seedling disease of carrot (Lutchmeah and Cooke 1985). It has also been implicated in the suppression by soils of *P. ultimum* (Martin and Hancock 1984), and in the control of sugar beet damping-off incited by *P. ultimum* (Martin and Hancock 1987).

Also implicated in soil suppression of plant-parasitic *Pythium* spp. is *Pythium nunn* (Lifshitz et al. 1984*b*). This fungus is a mycoparasite of *Pythium* spp., *Phytopthora* spp., and *R. solani* (Lifshitz et al. 1984*a*). The host range of this mycoparasite may be limited by the outer cell wall components of the host. The presence of surface mucilage on the host may determine whether cell-wall-degrading enzymes are produced by the parasite (Elad et al. 1985). In addition to its hyperparasitic activity, *P. nunn* also produces a soluble, filterable factor that inhibits the growth of *Pythium* spp. and *R. solani* in vitro, and the germination of *Pythium* sporangia in soil (Elad et al. 1985).

Finally, there is a group of mycoparasites belonging to the Dematiaceae that parasitizes the rusts, powdery mildews, or other leaf-spot fungi. *Scytalidium uredinicola* parasitizes the aecia of several rust species belonging to the genera *Cronartium* and *Endocronartium* (Kuhlman et al. 1976, Tsuneda et al. 1980). It reduces the amount of aeciospore inoculum available to infect the alternate host. *Sphaerellopsis* (formerly *Darluca*) *filum* is another rust parasite. It attacks the uredial sori of *Puccinia recondita* on wheat (Swendsrud and Calpouzos 1972) and the telial sori of *Cronartium strobilinum* on oak (Kuhlman et al. 1978), thus reducing the inoculum potential for subsequent infections.

Ampelomyces quisqualis is a hyperparasite of powdery mildews. It is not host-specific and is commonly found on many species of this pathogen. As a biocontrol agent, *A. quisqualis* has been most useful in the control of powdery mildews of greenhouse cucumbers incited by *Erysiphe cichoracearum* and *Sphaerotheca fuliginea* (Sundheim 1982). It has also shown tolerance to triforine and quinomethionate, fungicides commonly used for powdery mildews, and has been used in combination treatments with them (Sundheim and Amundsen 1982). *A. quisqualis* has also been used to control other powdery mildew species on sugar beets, watermelon, carrot, apple, mulberry, pepper, and zinnia in the greenhouse (Sztejnberg 1979).

Dicyma (Hansfordia) pulvinata is a mycoparasite of several leaf-spot fungi. It has been reported to parasitize *Cercospora personatum* (Hughes 1951) and *Cercosporidium personatum* on peanut (Taber et al. 1981), *Cladosporium fulvum* on tomato (Peresse and Le Picard 1980), and *Aerisporium caricae* on papaya (Hepperly 1986). *D. pulvinata* quickly colonizes the leaf-spot lesions of its host and destroys the conidiophores and conidia of the pathogen. It does not prevent initial infection by the pathogen, but suppresses secondary inoculum formation and dispersal (Mitchell et al. 1987). The mycoparasite has also been shown to produce a fungitoxic sesquiterpene, 1, 3-desoxyphomenone, that is active against *Cladosporium* spp. (Tirilly et al. 1987).

Nonmycoparasitic Fungal Biocontrol Agents

Many of the fungal biocontrol agents used to control plant diseases are not mycoparasites of other fungi, or at least that phase of their life cycle is not involved in the biocontrol phenomenon. The mechanisms employed by these fungi are thought to be antibiosis, competition, a combination of antibiosis and competition, or stimulation of plant defenses prior to infection by the pathogen.

Athelia bombacina is a basidiomycete of the family Corticaceae that has been shown to antagonize the apple scab pathogen *Venturia inaequalis* (Heye and Andrews 1983). Antagonism was apparently effected through antibiosis and competition for nutrients. *A. bombacina* readily colonized fallen apple leaves in the orchard, which are the natural substrate of *V. inaequalis*, and prevented ascospore production by the pathogen.

Chaetomium globosum, an ascomycete, is a common soil saprophyte that has been reported to control both foliar and soilborne plant pathogens. Ascospore suspensions of this antagonist sprayed on apple leaf surfaces during the growing season significantly reduced the incidence of apple scab incited by *V. inaequalis* during 2 years of field testing (Cullen et al. 1984). Control of this disease by *C. globosum* has been ascribed to the production of the antibiotic chetonin by the antagonist (Cullen and Andrews 1984). Chetonin is also known to be a mycotoxin, and it has been associated with ovine ill-thrift in livestock (Brewer et al. 1972). *C. globosum* has been used to control *Fusarium nivale*–incited blight of oat seedlings (Tveit and Wood 1955), and corn root infection has been controlled by coating the seed with spores and biomass of the antagonist (Kommedahl and Chang-Mew 1975). In another study, *C. globosum* treatment of seed reduced the numbers of seed coat microflora, and an unidentified antibiotic was isolated from ascospore-treated seed (Hubbard et al. 1982).

A mixture of *Trichoderma* spp. and a *Scytalidium* sp. have been used to successfully control decay of creosoted poles and timbers caused by *Lentinus lepideus* (Bruce and King 1983). Disease control is still retained in wooden blocks in which the antagonists have been killed and the blocks leached to remove water-soluble materials. The inhibitory factor produced by *Trichoderma* is unknown, but the protective mechanism of *Scytalidium* in wood has been attributed to the production of water-soluble, water-insoluble, and heat-stable antibiotics, one of which is the antibacterial and antifungal compound scytalidin (Overeem and Mackor 1973, Stillwell et al. 1973).

A number of species of the genus *Penicillium* are known to be effective biocontrol agents without benefit of mycoparasitic activity. *Penicillium oxalicum* protected pea seeds from infection by a complex of *Fusarium, Rhizoctonia, Aphanomyces,* and *Pythium* when spores of the antagonist were dusted onto seed (Windels 1981). *Penicillium chrysogenum* protected tomato seedlings from *Verticillium albo-atrum* when the antagonist was applied as a root dip (Dutta 1981). *Penicillium* spp. also decrease crown gall incited by *Agrobacterium radiobacter* pv. *tumefaciens* on cherry seedlings (Moore 1981). Cowpea root diseases incited by *Fusarium solani* and *Macrophomina phaseolina* have reportedly been controlled by mixing cornmeal cultures of *Penicillium funiculosum* into infested soil before planting (Odunfa 1982).

There are some fungal biocontrol agents that do not function either as mycoparasites or as antibiotic producers. These fungi apparently stimulate the host plant to protect itself against subsequent infections by the pathogen. A *Phialophora* sp. with lobed hyphopodia, isolated from the roots of winter wheat, protects wheat from the take-all disease incited by *Gaeumannomyces graminis* var. *tritici* (Martyniuk and Myskow 1974). The antagonist colonizes the root cortex and stem bases of wheat plants, where it produces darkly pigmented cells and lobed hyphopodia. The exact means of making the root more resistant to infection is unknown, but induction of plant resistance by the antagonist may well be involved (Deacon 1976). When bean cultivars susceptible to *Colletotrichum lindemuthianum* were infected with the cucurbit pathogen *Colletotrichum lagenarium*, local and systemic protection against the bean pathogen resulted (Elliston et al. 1976). Infection of the first true leaf of cucumber, followed by a booster inoculation with *C. lagenarium* in 2 to 3 weeks, immunized cucumber against diseases caused by *C. lagenarium, Cladosporium cucumerinum,* and *Pseudomonas lachrymans* (Kuc and Preisig 1984). Immunity lasted through the fruiting period. Injection of the sporangia of *Peronospora tabacina,* the blue-mold pathogen, into the stems of tobacco plants systemically protects tobacco foliage from this disease

(Cohen and Kuc 1980). This protection is graft-transmissible (Tuzun and Kuc 1985), and resistance to blue mold can be maintained in to-bacco plants differentiated from callus or tissue culture of tobacco plants with induced resistance (Tuzun and Kuc 1987). Protection of bean seedlings from Rhizoctonia root rot by a binucleate *Rhizoctonia*-like fungus (BNR) was demonstrated by Cardoso and Echandi (1987). BNR did not inhibit *R. solani* in culture, but exudates from roots ex-tensively colonized by BNR were inhibitory to hyphal growth and sclerotial germination of *R. solani*. Treated bean roots that were surface-sterilized to remove BNR still retained the protective capabil-ity against *R. solani*. The authors concluded that the protective mech-anism was a BNR-induced metabolic response by bean seedlings that suppressed *R. solani* at the infection site.

Fungal Biocontrol Agents of Nematodes

As with fungal plant pathogens, nematodes that incite plant diseases may also be parasitized by fungi, and thus may be subject to biological control by fungal biocontrol agents. Fungi parasitic to nematodes may be conveniently separated into two groups:

1. Nematode-trapping fungi that parasitize motile larval stages

2. Nematode-destroying fungi that parasitize eggs and adults of nem-atodes

The former group contains nonspecific parasites that are not aggres-sive toward nematodes, and their control of plant disease has been marginal (Sayre 1986). The members of this group are moniliaceous hyphomycetes belonging to the genera *Arthrobotrys, Candelabrella, Dactylaria, Dactylella,* and *Genicularia*. These fungi employ adhesive networks, loops, and knobs, or constricting rings to ensnare their hosts. A lectin-to-carbohydrate binding phenomenon initiates the cap-ture process (Nordbring-Hertz and Mattiasson 1979, Nordbring-Hertz et al. 1982). Of the above listed genera, only *Arthrobotrys robustus*, in a commercial formulation called Royale 300, has been used to control the nematode *Ditylenchus myceliophagous* on commercial mushrooms (Cayrol et al. 1978), and another *Arthrobotrys* sp. (Royale 350) has been used to control light infestations of *Meloidogyne* spp. on tomatoes in the field (Cayrol and Frankowski 1979).

Members of the nematode-destroying group are more diverse. The group includes 140 species (Sayre 1986) classified as Chytri-diomycetes, Oomycetes, Hyphomycetes, Basidiomycetes, Mycelia sterilia, and Yeast. These fungi parasitize both the eggs and adult stages of nematodes. The impact of many of the members of this group on nematode populations in soil is minor, but a few have shown more drastic effects. The fungi *Nematophthora gynophila, Catenaria auxillaris*, and *Verticillium chlamydosporium* and also a lagenidiaceous fungus parasitized both the eggs and adults of the cereal cyst nematode in nematode-suppressive soils in England (Kerry et al. 1980). Kerry (1980) indicated that the preferential attack of these fungi on adult females greatly reduced the nematode population by destruction of its reproductive capacity. An ultrastructural study of the parasitism of *Meloidogyne arenaria* eggs by *V. chlamydosporium* showed that the parasite colonized the eggs by direct hyphal penetration (Morgan-Jones et al. 1983). Hyphae proliferated readily within the egg and reemerged through the shell. Chitin and lipid layers in the shell and larval cuticle became disorganized. Isolates of *V. chlamydosporium* grown on ground oat grain and introduced into soil reduced the numbers of *Heterodera avenae*, the cereal cyst nematode, by 26 to 80% (Kerry et al. 1984). *V. chlamydosporium* could be isolated from the soil 6 months after treatment, and some isolates colonized the roots of wheat without causing lesions or affecting dry weight of the plant. Results of two experiments on a potato farm infested with the cyst nematode *Globodera rostochiensis* showed that *Paecilomyces lilacinus* added as a tuber-dip and soil-mix combination significantly controlled the cyst nematode until harvest (Davide and Zorilla 1983). Control was generally comparable to that obtained with the nematocides Ethoprop and Carbofuran, but not with Phenamiphos. Potato yield with the combined treatment was comparable to that obtained with Carbofuran and Ethoprop treatments, and significantly better than that of the control.

Ecology of Fungal Biocontrol Agents

One of the salient characteristics of biological control agents that distinguishes them from chemical control agents is the fact that they are living entities. They therefore have nutritional requirements, and they are subject to virtually all of the environmental parameters that govern the existence and activities of plant pathogens. In recent years fungal biocontrol agents that have shown efficacy in disease control under prescribed conditions have been widely reported. The vast majority of these have either failed completely or have failed to perform up to expectations when taken to the field or orchard, where conditions are much more variable and cannot be controlled. Virtually all of these failures can be ascribed to our lack of knowledge concerning the environmental factors involved and their effect on the activities and longevity of the biocontrol agent and/or pathogen.

A number of reports in the literature note the existence of soils that are suppressive to the activities of one plant pathogen or another (Cook and Baker 1983), and this suppressiveness has been ascribed to the presence in that soil of one or more antagonists of the pathogen. If soils are suppressive to pathogens because of certain members of the indigenous microflora, it follows that soils may also be suppressive to the activities of biocontrol agents due to the presence of microorganisms that are antagonistic to them.

Physical and chemical factors are among the primary environmental factors affecting biocontrol agents. In a study of factors affecting the parasitic activity of *Gliocladium virens* on *Sclerotinia sclerotiorum*, Philips (1985) observed that the antagonist was active over a broad range of soil moisture levels and over the entire agricultural soil pH range. The main limiting factor in its use as a biocontrol agent was its temperature requirements. Active parasitism of sclerotia took place over the range of 15 to 35°C, but parasitism was greatly reduced at 15°C and little parasitism occurred below that temperature. The author concluded that the use of *G. virens* as a biocontrol agent was restricted unless strains with lower temperature requirements could be selected.

Fravel and Marois (1986) carried out an extensive investigation of the edaphic parameters associated with the establishment of *Talaromyces flavus* in soil. They monitored the populations of *T. flavus* for 13 weeks in 25 freshly collected field soils, and found that 5 of 23 physical, chemical, and biological parameters measured were related to survival and proliferation of the fungus. These parameters were cation exchange capacity; potassium, sodium, and zinc concentrations; and total soil-bacteria population sizes. Potassium and zinc concentrations were positively correlated with *T. flavus* survival, while sodium concentration was inversely related. The relationships of cation exchange capacity and soil-bacteria population size to *T. flavus* survival were not elucidated.

The mycoparasitic activities of *T. flavus* and *Trichoderma* spp. are also affected by soil temperature. Boosalis (1956) noted that only 8% of the hyphae of *Rhizoctonia solani* were parasitized by either of these two fungi at 18°C, whereas at 28°C over 18% of the host's hyphae were attacked. High soil temperature apparently favors *T. flavus*. Katan (1985) observed that the normally dormant ascospores of this fungus were activated to germinate by a heat treatment of 53°C for 15 minutes, and they survived 70°C for 1 hour. Heat treatment of soil infested with *T. flavus* increased recovery of the fungus from 5.9 to 103%.

Perhaps one of the most important factors, if not the most important, governing the activities of a fungal biocontrol agent in soil is the qualitative and quantitative makeup of the soil microflora. In a number of instances, antagonists of pathogens have been introduced into pasteurized soil where they functioned quite well as biocontrol agents. Attempts to introduce these same antagonists into natural soil have met with failure or only qualified success (Cook and Baker 1983). Since most of the environmental parameters remain the same, the difference in biocontrol efficacy can probably be ascribed to the competing microflora. Chao et al. (1986) studied the ability of various fungi and bacteria to move from inoculated seeds to developing roots. When roots were grown in sterile soil, *Trichoderma harzianum* colonized the upper half of the roots, while *Enterobacter cloacae* colonized the entire rhizosphere. In nonsterile soil, none of several fungi and bacteria tested could be detected more than 3 cm from the planted seed. In autoclaved soil to which fungi had been added, *E. cloacae* colonized the roots well, while *T. harzianum* colonization was inhibited. In autoclaved soil to which bacteria had been added, *E. cloacae* was inhibited and *T. harzianum* grew well. Apparently biocontrol agents are more inhibited by members of their own class than they are by those from others.

This phenomenon does not extend to intraspecific competition, however. Marois and Locke (1985) observed that the addition of a *Trichoderma viride* mutant to steamed plant growth medium already infested with another isolate of *T. viride* increased the total population of the biocontrol agent in soil. This was also correlated with a reduction in the population of the pathogen *Fusarium oxysporum* f. sp. *chrysanthemi* and in the incidence of *Fusarium* wilt of chrysanthenum.

Interspecific competition is another matter. Vajna (1985) reported mutual parasitism between isolates of *Trichoderma hamatum* (Tha-2) and *Trichoderma pseudokoningii* (Tp-1). He observed that the hyphae of Tp-1 coiled around and penetrated the hyphae of Tha-2. Tha-2, on the other hand, did very little coiling. Its hyphae terminated in a great number of thin, penetrating hyphae that penetrated the cells of Tp-1. Penetration resulted in disintegration of the host-cell cytoplasm. The author hypothesized that this kind of mutual parasitism also occurs in nature, and if so, introduction of a given *Trichoderma* biocontrol isolate into soil might result in interference from indigenous species.

In some cases the source of interference for a fungal biocontrol agent may be fungi that are taxonomically very different from it. Vesely and Hejdanek (1984) reported that the biocontrol agent *Pythium oligandrum*, which parasitizes or inhibits a number of pathogens, was itself inhibited in vitro by *Mucor heterosporum*, *Rhizopus arrhizus*, and a *Mortierella* species. *Drechslera* sp. and *Mucor piriformis* were even more effective, and the biocontrol agent disappeared from mixed cultures. This inhibitory activity, however, could not be confirmed in vivo, where nutritional differences and other biotic and abiotic factors may interfere.

In some soils, the failure of a fungal biocontrol agent to prevent or hinder the progress of a plant disease may be due to a combination of factors. Hubbard et al. (1983) found that an isolate of *Trichoderma hamatum* that had prevented seed rot of peas incited by *Pythium* spp. in Colorado, failed to protect seeds in New York soils with low iron availability. A *Pseudomonas* sp. was isolated from lysed *T. hamatum* germlings in this soil and was shown to be responsible for the failure of the biocontrol agent to prevent disease. Addition of the *Pseudomonas* sp. to seed treated with *T. hamatum* and planted in steamed soil low in iron caused the fungus to be ineffective in disease control. Addition of iron (8 micrograms per gram of soil) to this same system resulted in effective control of the disease. Apparently, a combination of low iron availability and the presence of the bacterium renders the biocontrol agent ineffective. The bacterium may have inhibited the fungus in low-iron soil by co-opting all available iron through siderophore production, or low-iron conditions may have stimulated antibiotic production by the bacterium. A follow-up experiment in low-iron New York soils by Hadar et al. (1984) demonstrated that *Trichoderma* spp. isolated from that soil were not affected by seed-colonizing pseudomonads. One strain of *Trichoderma* produced a siderophore in its own right, and the other was less sensitive to low iron levels. It would be interesting to see if these conditions would have the same effect on the fungal biocontrol agent *Gliocladium virens*. This fungus has been reported to produce large quantities of siderophore under low-iron conditions (Jalal et al. 1986).

The success or failure of a fungal biocontrol agent in controlling disease may often be influenced by differences in the species or even variety of the host. Bourbos and Skoudridakis (1987) found that the *Verticillium*-wilt-resistant tomato variety GC 204 fostered the presence of *Aspergillus alutaceous, Paecilomyces lilacinus, Penicillium herquei, Penicillium nigricans,* and *Trichoderma viride* in its rhizosphere—all fungi that have been implicated in the biocontrol of many plant diseases. The wilt-sensitive variety "Early pack" did not favor the presence of these fungi. The authors speculated that the early establishment of *Penicillium chrysogenum* and *Penicillium funiculosum* in the rhizosphere of wilt-susceptible "Early pack" might facilitate the colonization of this zone by the other biocontrol fungi and thus render this variety resistant.

In order to function effectively as disease antagonists, fungal biocontrol agents must be able to maintain themselves in the soil or plant environment. Most are good saphrophytes and can subsist at a low level, but they may require a supplementary food base to pro-liferate and function efficiently as biocontrol agents. This is proba-bly more true of those that depend exclusively or partially on the production of antibiotics to effect antagonism than of those that are pure mycoparasites. Lewis and Papavizas (1984) found that popu-lation densities of *Trichoderma viride* and *Tricoderma harzianum* increased by factors of 10^4 and 10^3, respectively, during 3 weeks in-cubation in natural soil when the antagonists were added as wheat-bran mycelial preparations. When conidia were added to soil, with and without bran, population increases did not occur. A wide vari-ety of *Trichoderma* isolates as well as *Talaromyces flavus, Gliocladium virens, Gliocladium roseum, Gliocladium catenulatum*, and *Aspergillus ochraceous* gave similar results when added to soil in the same fashion. Beagle-Ristaino and Papavizas (1985) discov-ered that although *Trichoderma* spp. and *G. virens* conidia were sensitive to soil fungistasis, chlamydospores produced from a fer-mentation system readily germinated in soil. Both fungi grew ver-tically down through the soil, and the number of propagules in-creased 100-fold when either fungus was added to soil in fermentor biomass consisting mostly of chlamydospores and containing traces of food. Populations of both fungi increased in soil planted with cot-ton; however, the rhizosphere and nonrhizosphere populations were not significantly different, indicating that the food base was prob-ably most responsible for the increase.

One form of substrate that may contribute to the population level and longevity of a mycoparasitic biocontrol agent in soil is the host pathogen itself. Van Den Boogert and Jager (1983) observed that the addition of live mycelial fragments of *Rhizoctonia solani* to potato-field soil resulted in the accumulation of the hyperparasites *Gliocladium roseum* and *Verticillium biguttatum*, with the latter pre-dominating. Disease ratings on sprouts from infected potato seedpieces, planted in untreated or *R. solani*–activated soil, showed that activation dramatically reduced the number and extent of sprout infections. A similar technique was used by Zogg and Joeggi (1974) to induce suppression of *Gaeumannomyces graminis*, the incitant of take-all disease in wheat, although a specific antagonist was not found.

Another nutrition-related phenomenon in fungal biocontrol agents is the effect of substrate on secondary metabolite production. This af-fects not only antibiotic production, but also the production of phytotoxic compounds by the fungus. When Howell and Stipanovic (1984) attempted to culture the fungal biocontrol agent *Gliocladium virens* on a polished-rice substrate, in lieu of a lighter peat moss car-rier, the fungus produced large quantities of a potent phytotoxin that destroyed the radicles of cotton seedlings treated with the rice-fungus mixture. The phytotoxin was isolated and identified as viridiol, a steroid compound. Viridiol is a reduced form of the potent antifungal compound viridin, which is also produced by strains of this fungus. Viridin exhibits strong antifungal activity, but weak phytotoxicity, whereas viridiol has virtually no antifungal or antibacterial activity, but is strongly phytotoxic. The authors found that viridiol had a wide phytotoxic spectrum, including many weed species. When the air-dried *G. virens*–rice culture was worked into the soil surface above planted cotton seed, it successfully controlled pigweed and morning glory without harm to emerging cotton seedlings. Viridiol is also produced by *G. virens* in wheat-bran culture, a common carrier for *Trichoderma* and *Gliocladium* biocontrol strains (Howell, unpublished).

In a follow-up study on viridiol production by *G. virens*, Jones and Hancock (1987) determined that viridin was the precursor of viridiol. They also found that reduction of viridin to viridiol was independent of culture pH, carbon source, or nitrogen source and quantity. A simple production system consisting of peat moss amended with dextrose and calcium nitrate supported the production of 86 micrograms of viridiol per gram of peat, and the authors suggested that viridiol might have value as an herbicide.

The soil or phylloplane environment may influence both the production of secondary metabolites by fungal biocontrol agents and the fate of these compounds once they are elaborated. Boudreau and Andrews (1987) observed that the apple scab antagonist *Chaetomium globosum* failed to control the pathogen *Venturia inaequalis* on apple seedlings when it was applied several days in advance of the pathogen. The authors showed that control was effected by the diffusion of antibiotics from ascospores onto the phylloplane, and that antibiotics could be readily degraded by high pH and air drying. They suggested that control loss was due to abiotic degradation of the antibiotics on the phylloplane.

Apparently not all the secondary metabolites produced by fungal biocontrol agents are harmful either to other microbes or plants. Windham et al. (1986) discovered that the addition of *Trichoderma harzianum* or *Trichoderma koningii* to autoclaved soil increased the rate of emergence of tomato and tobacco seedlings. Eight weeks after planting, root and shoot dry weights of tomato and tobacco were increased 213 to 275% and 259 to 318%, respectively. The population densities of soil microflora were similar in soils infested with *Trichoderma* spp. and noninfested controls. Radish plants grown under gnotobiotic conditions with *T. harzianum* were larger than plants grown under similar conditions without the fungus. Seed germination increased when seeds were separated from *Trichoderma* spp. by a cellophane membrane. The authors concluded that the *Trichoderma* spp. produced a growth-regulating factor that increased the rate of seed germination and dry weight of shoots and roots.

In a similar experiment, Chang et al. (1986) found that steamed or raw soil infested with *T. harzianum* hastened the flowering of periwinkle, increased the number of blooms per plant of chrysanthemum, and increased the heights and weights of other plants. Responses occurred consistently at population densities of 10^5 or more colony-forming units per gram of soil.

The results of the experiments described above indicate that some fungal biocontrol agents are capable of directly influencing the growth of plants without going through the intermediate step of modifying the activities of other microorganisms.

Production and Application of Fungal Biocontrol Agents

The use of fungi as biocontrol agents will require the development of mass production techniques to obtain the huge quantities of inoculum required. Experience gained by the pharmaceutical industry in the large-scale production of antibiotics and other useful metabolites will be beneficial, but these techniques have been geared toward the production of fungal metabolic products rather than the fungi themselves, and they may require minor or even extensive modification.

A number of delivery systems for fungal biocontrol agents have been developed in recent years. Chet et al. (1979) pioneered the use of *Trichoderma*-colonized wheat bran as a medium for the biocontrol of soilborne plant pathogens, while blackstrap molasses-enriched clay granules were used by Backman and Rodriguez-Kabana (1975) as a carrier for *Trichoderma harzianum*. Howell (1982) reported the use of a *Gliocladium virens*–colonized peat moss and Czapek's broth mixture to control damping-off in cotton seedlings. He also showed that the fungus produced large numbers of chlamydospores in this medium, and that these propagules retained viability for a lengthy period when air-dried in the carrier. A mixture of peat, soil, leaf compost, and sand amended with Czapek's broth was used by Moody and Gindrat (1977) as a growth medium for *Gliocladium roseum* inoculum. They also formed ground *G. roseum*–colonized barley kernels, conifer bark, and barley flour into pellets for use as an inoculum preparation. Both forms of inoculum reduced the incidence of cucumber black root rot. Papavizas et al. (1984) developed a technique for producing large batches of biomass of *G. virens*, *Trichoderma* spp., and *Talaromyces flavus* in liquid-fermentation vessels. After 15 days incubation, 75% of the spores in the *Trichoderma* and *Gliocladium* cultures were mature chlamydospores. When air-dried, ground with pyrophylite carrier, and added to soil at 10^3 colony-forming units per gram, the chlamydospores of both fungi proliferated greatly, increasing from 2×10^6 to 6×10^6 propagules per gram. *Talaromyces flavus* produced no spores, but the biomass had 44% survival at room temperature after 5 months. A technique for delivering fungal biocontrol agents to vegetable seeds through fluid drilling was developed by Conway et al. (1982), and the efficacy of the technique was confirmed by Mihuta-Grimm and Rowe (1986), who used *Trichoderma* isolates in drill fluid to control *Rhizoctonia* damping-off of radish. Fluid drilling is the extrusion of seed suspended in a gel into the furrow.

One of the more novel delivery systems for fungal biocontrol agents developed in recent years has been the encapsulation of fungal propagules. Fravel et al. (1985*b*) amended suspensions of 1% sodium alginate and 10% Pyrax with the ascospores or conidia of *Talaromyces flavus*; conidia of *Gliocladium virens, Penicillium oxalicum*, or *Trichoderma viride*; or cells of *Pseudomonas cepacia*. The suspension was dispensed dropwise from Pasteur pipettes into a solution of $0.1M$ calcium gluconate, which caused the formation of solid aggregates. The aggregates were dried overnight and stored at room temperature. Initial populations ranged from 10^5 to 10^8 colony-forming units per milliliter and these declined 10-fold to 100-fold after 4 weeks.

An important consideration in the development of fungal antagonists for use in disease control is the compatibility of these fungi with other methods of disease control. A fungal biocontrol agent alone may sometimes be inadequate for disease control. However, in combination with host-plant resistance, physical or chemical soil fumigation, or treatment of seed or soil with fungicides, a fungal biocontrol agent may be very effective. Henis et al. (1978) and Hadar et al. (1979), using radish and eggplant, respectively, combined treatments of *Trichoderma harzianum* with the fungicide pentachloronitrobenzene (PCNB) to control disease incited by *Rhizoctonia solani*. They found that the combined treatments increased disease control over either component alone. Lee and Wu (1986) combined applications of the fungicide dichloronitroaniline (DCNA) with *Trichoderma viride* or *Gliocladium virens* to control sclerotinia disease of sunflower. They observed that the combined treatments gave better control than either the fungicide or the fungi separately. Biological control of *Rhizoctonia solani* on potato tubers by *Vertcillium biguttatum* was aided by treatment with the fungicide Pencycuron at 20% of the recommended dosage (Jager and Velvis 1986). Treatment resulted in a substantial reduction in the number of sclerotia on the tubers.

The results of combined treatments are not always so straightforward. Fravel et al. (1985a) tested the compatibility of *Talaromyces flavus* with five fungicides recommended for use on potato seedpieces, in vitro or in the field. *T. flavus* was recovered most frequently from seedpieces treated with the fungus alone or combined with thiabendazole. It was least tolerant of captan. Tolerance of fungicides in vitro was not consistent with field compatibility. Apparently the field-compatibility phenomenon is more complex than simple fungicide tolerance. It may involve the effect of both abiotic and biotic factors in the soil.

Combinations of physical, chemical, and biological control techniques have been used effectively in the control of soilborne plant diseases in recent years. Solarization or fumigation are excellent short-term methods of disease control, but they are essentially nondiscriminatory; these techniques usually remove beneficials along with the pathogens, and they allow reintroduction of pathogens without natural competition. The introduction of biocontrol agents into soil subsequent to these treatments helps to retard this process, and aids in the destruction of any remaining pathogens in the soil.

Elad et al. (1980) found that combining solar heating or methyl bromide fumigation with the introduction of *Trichoderma harzianum* into soil improved efficiency of both treatments in control of potato diseases caused by *Rhizoctonia solani* and *Verticillium dahliae*. A combination of heat treatment with *T. harzianum* introduction resulted in 90 to 100% control of *Sclerotium rolfsii* on beans. Reduction of *S. rolfsii* disease by *T. harzianum* added to untreated soils was only 50%.

Solarization may enhance the populations and activities of fungal biocontrol agents that are already present in the soil. Katan (1985) found that the ascospores of *Talaromyces flavus* would survive heat treatment in an aqueous suspension for 1 hour at 70°C, and that heat treatment of soil for 30 minutes at 56°C increased the recovery of *T. flavus* from between 3.3 and 5.9% to between 87 and 103%. Such results and the work of others (Kodama et al. 1980, Tjamos and Paplomatas 1986) may help to explain the observation that soil solarization sometimes induces suppressiveness to reintroduction of the pathogen (Greenberger et al. 1987).

The resistance to fumigants that is found in certain strains of fungal biocontrol agents may be used to advantage. Strashnow et al. (1985) found that isolate TH-203 of *Trichoderma harzianum* was tolerant of up to 20,000 ppm (vol/vol) methyl bromide (MB), whereas the plant pathogen *Rhizoctonia solani* was susceptible to a dose of less than 9,000 ppm (vol/vol). Exposure to sublethal concentrations of MB had no effect on the in vitro antagonistic ability of *T. harzianum*, and soil fumigation with the equivalent of a commercial dose of 500 kg/ha did not reduce the population of *Trichoderma* in soil. In fact, it allowed rapid *Trichoderma* colonization to develop in the soil. Under greenhouse conditions, a combination of *T. harzianum* and a reduced dose of MB (equivalent to 200 kg/ha) completely controlled *R. solani* on bean seedlings. Under field conditions, the same combination gave a significant synergistic effect on damping-off of carrot seedlings caused by *R. solani*. The fungus-fumigant combination had a similar effect on growth, yield, and disease control to that of the recommended dose of MB, and *T. harzianum* was able to prevent reinfestation of the fumigated soil by *R. solani*.

Biotechnology and Fungal Biocontrol Agents

Fungal biocontrol agents or biocontrol agents in general are supremely well equipped to maintain themselves and to perform the role that nature has intended. However, high levels of disease control in genetically uniform crop plants exposed to massive quantities of pathogen inoculum is not exactly what nature had in mind. Characters that may be of immense benefit to an organism in the natural state may be of little use or even detrimental when employed in agriculture. The tailoring of fungal biocontrol agents to meet the requirements of current agricultural practices will most likely entail genetic alteration of the microbes. Although lagging somewhat behind the advances made in the genetic manipulation of bacteria, the process in fungi has already begun.

Several researchers have used mutants of fungal biocontrol agents, either to answer questions about mechanisms in the biocontrol process (Howell and Stipanovic 1983, Howell 1987) or to enhance the efficacy of a biocontrol agent by making it resistant to commonly used fungicides (Papavizas et al. 1982, Katan et al. 1984, Fravel et al. 1985a). The difficulty in making mutants of fungi is that mutagenesis is a relatively nonspecific process. All too often many silent mutations are made in addition to the one desired. A mutant strain may perform well when assayed for a given trait in vitro, such as antibiotic resistance or antibiotic production, but it may be unable to sustain itself in nature because other undetected mutations make it less competitive.

Some of the techniques currently employed for genetic alteration of fungi may help scientists to avoid the problems associated with mutagenesis. One such technique is "protoplast fusion." Stasz et al. (1988) have recently employed this method to combine the genomes of auxotrophic strains T12 and T95 of *Trichoderma harzianum*. Protoplasts were obtained from mycelia and immature conidia by digestion of the cell walls with NovoZym 234. Fusion of lysine-requiring and histidine-requiring auxotrophs of T95 resulted in about 10% prototrophs. These were apparently balanced heterokaryons, since conidia from these colonies were nearly all auxotrophs and required lysine or histidine in approximately equal numbers. However, fusion between T12-his and T95-lys gave a prototrophic frequency of less than 0.01%, and gave rise to unbalanced heterokaryons in which T12 predominated. Various nonparental progeny were subsequently recovered, and many remained unchanged when propagated from single conidia, indicating that they were homokaryotic. Others yielded a variety of parental and nonparental single-spore isolates, indicating complex heterokaryosis. Nonparental progeny may have resulted from karogamy and genetic recombination, from cytoplasmically inherited characters, or from nuclear-cytoplasmic interactions.

A similar technique has been used (C. M. Kenerley and M. D. Thomas, personal communication) to fuse the protoplasts of auxotrophic strains of *Gliocladium roseum*. Nonparental fusion progeny were recovered at a frequency of 1×10^{-3}. Stable recombinants were isolated after several successive single-conidial isolations on selective media. Prototrophic recombinants also contained a color marker derived from one of the auxotrophic parents. No reversion was observed in the parental markers.

Transformation has recently been accomplished in two *Gliocladium* spp. (M. D. Thomas and C. M. Kenerley, personal communication). Protoplasts of *G. roseum* and *G. virens* were formed from 18-hour-old germinating conidia by digestion with a mixture of lyticase, chitinase, and cellulase. The protoplasts of *G. roseum* were transformed with genomic DNA from *G. virens* by adding the DNA to an osmoticum containing heat-shocked protoplasts, followed by slowly adding PEG-MOPS (3-[N-Morpholino] propanesulfonic acid) buffer solution, incubating the suspension for 20 minutes at room temperature, and then plating the protoplasts on selective medium. The transformation of *G. virens* protoplasts to hygromycin resistance was accomplished by electroporation of the protoplasts in an osmoticum containing pH1S plasmid DNA, followed by addition of high-molecular-weight PEG. Transformants were then detected on a hygromycin-containing selective medium. In most cases, the DNA conferring hygromycin resistance from the plasmid was integrated into the high-molecular-weight DNA of the protoplast. This indicates that the resistance trait will be much more stable in the transformed genome, and that it will be subject to Mendelian inheritance.

If the above described techniques can be developed to allow the genetic transfer of desirable traits associated with biocontrol from one fungal strain or species to another, a giant step will have been taken in the development of more effective biocontrol agents.

Future Prospects for Fungal Biocontrol Agents

Research on biocontrol has provided us with a wealth of information—and, along with it, given us a healthy respect for how little we actually know about the process. This is particularly true with respect to soilborne diseases. After a biocontrol agent is placed in soil, we know virtually nothing about what processes take place. If successful strategies for the biological control of plant diseases are to be developed, we must obtain answers to the following questions:

1. What mechanisms are responsible for the observed effects of biocontrol agents on plant pathogens?
2. What are the effects of the biotic and abiotic environment on the viability and metabolism of the biocontrol agent, and on the activities of the mechanisms involved in the biocontrol process?
3. What genes are responsible for the production of biocontrol mechanisms, and can these genes be isolated and/or transferred to other genomes?
4. Will genetic manipulation of biocontrol agents result in the formation of superior biocontrol strains?
5. How can biocontrol agents be mass produced most efficiently and economically?
6. What are the best carrier formulations for storage, application, and activity of biocontrol agents?
7. What are the optimum timings, frequencies of application, and treatment concentrations for biocontrol agents?
8. How readily can biocontrol agents be integrated with other means of disease control and with crop production practices?

Once these questions have been answered, agriculture should be able to make practical and economic use of fungal and other biocontrol agents to control plant diseases. This may be achieved by the biocontrol agents alone, or in combination with other means of disease control.

The following U.S. patents have been awarded for the use of fungi or related procedures in the biocontrol of plant diseases:

1. Ayers, W. A., and Adams, P. B. 1981. Biological control system. U.S. Patent 4,246,258. Jan. 20, 1981.
2. Papavizas, G. C. 1984. Strain of *Trichoderma viride* to control Fusarium wilt. U.S. Patent 4,489,161. Dec. 18, 1984.
3. Baker, R., and Lifshitz, R. 1986. Isolates of *Pythium* species which are antagonistic to *Pythium ultimum*. U.S. Patent 4,534,965. May 4, 1986.
4. Lewis, J. A., Papavizas, G. C., and Connick, W. J., Jr. 1987. Prep-

aration of pellets containing fungi and nutrient for control of soilborne plant pathogens. U.S. Patent 4,668,512. May 26, 1987.

5. Ricard, J. J. L. 1987. Method of using commensals. U.S. Patent 4,678,669. July 7, 1987.

6. Chet, I., Sivan, A., and Elad, Y. 1987. Novel isolate of *Trichoderma*, fungicidal compositions containing said isolate and use thereof. U.S. Patent 4,713,342. Dec. 15, 1987.

7. Marois, J. J., Fravel, D. R., Connick, W. J., Jr., Walker, H. L., and Quimby, P. C., Jr. 1988. Preparation of pellets containing fungi for control of soilborne diseases. U.S. Patent 4,724,147. Feb. 9, 1988.

To date, none of the above patented materials are used in commercial production. However, they serve as a basis for further research and development, and in the author's opinion, it is only a matter of time before the changing needs of agriculture make it not only feasible, but necessary, for these or similar products to be used in the field.

References

Adams, P. B., and Ayers, W. A. 1980. Factors affecting parasitic activity of *Sporidesmium sclerotivorum* on sclerotia of *Sclerotinia minor* in soil. *Phytopathology* 70:366–368.

Adams, P. B., and Ayers, W. A. 1981. *Sporidesmium sclerotivorum*: distribution and function in natural biological control of sclerotial fungi. *Phytopathology* 71:90–93.

Ahmed, A. H. M., and Tribe, H. T. 1977. Biological control of white rot of onion (*Sclerotium cepivorum*) by *Coniothyrium minitans*. *Plant Pathol.* 26:75–78.

Aluko, M. O., and Hering, T. F. 1970. The mechanisms associated with the antagonistic relationship between *Corticium solani* and *Gliocladium virens*. *Trans. Br. Mycol. Soc.* 55:173–179.

Ayers, W. A., and Adams, P. B. 1979. Mycoparasitism of sclerotia of *Sclerotinia* and *Sclerotium* species by *Sporidesmium sclerotivorum*. *Can. J. Microbiol.* 25:17–23.

Ayers, W. A., and Adams, P. B. 1981. Mycoparasitism of sclerotial fungi by *Teratosperma oligocladum*. *Can. J. Microbiol.* 27:886–892.

Ayers, W. A., and Adams, P. B. 1985. Interaction of *Laterispora brevirama* and the mycoparasites *Sporidesmium sclerotivorum* and *Teratosperma oligocladum*. *Can. J. Microbiol.* 31:786–792.

Backman, P. A., and Rodriguez-Kabana, R. 1975. A system for the growth and delivery of biological control agents to the soil. *Phytopathology* 65:819–821.

Barak, R., Elad, Y., Mirelman, D., and Chet, I. 1985. Lectins: a possible basis for specific recognition in the interaction of *Trichoderma* and *Sclerotium rolfsii*. *Phytopathology* 75:458–462.

Barnett, H. L., and Binder, F. L. 1973. The fungal host-parasite relationship. *Annu. Rev. Phytopathol.* 11:273–292.

Beagle-Ristaino, J. E., and Papavizas, G. C. 1985. Survival and proliferation of propagules of *Trichoderma* spp. and *Gliocladium virens* in soil and in plant rhizospheres. *Phytopathology* 75:729–732.

Boosalis, M. G. 1956. Effect of soil temperature and green manure amendment of unsterilized soil on parasitism of *Rhizoctonia solani* by *Penicillium vermiculatum* and *Trichoderma* sp. *Phytopathology* 46:473–478.

Boudreau, M. A., and Andrews, J. H. 1987. Factors influencing antagonism of

Chaetomium globosum to *Venturia inaequalis*: a case study in failed biocontrol. *Phytopathology* 77:1470–1475.

Bourbos, V. A., and Skoudridakis, M. T. 1987. The action of some fungal antagonists in the rhizosphere of resistant and susceptible tomato plants in the greenhouse. *J. Phytopathol.* 120:193–198.

Brewer, D., Duncan, J. M., Jerram, W. A., Leach, C. K., Safe, S., Taylor, A., Vining, L. C., Archibald, R. M., Stevenson, R. G., Mirocha, C. J., and Christensen, C. M. 1972. Ovine ill-thrift in Nova Scotia. V. The production and toxicology of chetomin, a metabolite of *Chaetomium* spp. *Can. J. Microbiol.* 18:1129–1137.

Brian, P. W., and Hemming, H. G. 1945. Gliotoxin, a fungistatic metabolic product of *Trichoderma viride*. *Ann. Appl. Biol.* 32:214–220.

Brian, P. W., and McGowan, J. G. 1945. Viridin: a highly fungistatic substance produced by *Trichoderma viride*. *Nature* 156:144.

Bruce, A., and King, B. 1983. Biological control of wood decay by *Lentinus lepideus* (Fr.) produced by *Scytalidium* and *Trichoderma* residues. *Mater.-Org. Berlin W. Ger.: Duncker and Humblot* 18:171–181.

Burdsall, H. H., Hoch, H. C., and Boosalis, M. G. 1980. *Laetisaria arvalis* sp. nov., a possible biological control agent for *Rhizoctonia solani* and *Pythium* spp. *Mycologia* 72:728–736.

Cardoso, J. E., and Echandi, E. 1987. Nature of protection of bean seedlings from Rhizoctonia root rot by a binucleate *Rhizoctonia*-like fungus. *Phytopathology* 77:1548–1551.

Cayrol, J. C., and Frankowski, J. P. 1979. Une Methode de lutte biologique contre les nematodes a galles des racines appartenante au genre *Meloidogyne*. *Rev. Hortic.* 193:15–23.

Caylor, J. C., Frankowski, J. P., Laniece, A., d'Hordemare, and Talon, J. P. 1978. Contre les nematodes en champignonniere. Mise au point d'une methode de lutte biologique a l'aide d'un hyphomycete predater *Arthrobotrys robusta* souche antipolis (Royale 300). *Rev. Hortic.* 184:23–30.

Chang, Y. C., Chang, Y. C., Baker, R., Kleifeld, O., and Chet, I. 1986. Increased growth of plants on the presence of the biological control agent *Trichoderma harzianum*. *Plant Dis.* 70:145–148.

Chao, W. L., Nelson, E. B., Harman, G. E., and Hoch, H. C. 1986. Colonization of the rhizosphere by biological control agents applied to seeds. *Phytopathology* 76:60–65.

Chet, I., Hadar, Y., Elad, Y., Katan, J., and Henis, Y. 1979. Biological control of soilborne plant pathogens by *Trichoderma harzianum*. In *Soil-Borne Plant Pathogens*, B. Schippers and W. Gams (eds.). Academic, New York, pp. 585–592.

Chet, I., Harman, G. E., and Baker, R. 1981. *Trichoderma hamatum*, its hyphal interactions with *Rhizoctonia solani* and *Pythium* spp. *Microb. Ecol.* 7:29–38.

Claydon, N., Allan, M., Hanson, J. R., and Avent, A. G. 1987. Antifungal alkyl pyrones of *Trichoderma harzianum*. *Trans. Br. Mycol. Soc.* 88:503–513.

Cohen, Y., and Kuc, J. 1980. Evaluation of systemic resistance to blue mold induced in tobacco leaves by prior stem inoculation by *Peronospora hyoscyami* f. sp. tabacina. *Phytopathology* 71:783–789.

Conway, K. E., Fisher, C. G., and Motes, J. E. 1982. A new technique for delivery of biological agents with germinated vegetable seeds. *Phytopathology* 72:987 (abstr.).

Cook, R. J., and Baker, J. F. 1983. *The Nature and Practice of Biological Control of Plant Pathogens*. American Phytopathological Society, St. Paul, Minnesota, 593 pp.

Cullen, D., Berbee, F. M., and Andrews, J. H. 1984. *Chaetomium globosum* antagonizes the apple scab pathogen, *Venturia inaequalis*, under field conditions. *Can. J. Bot.* 62:1819–1823.

Cullen, D., and Andrews, J. H. 1984. Evidence for the role of antibiosis in the antagonism of *Chaetomium globosum* to the apple scab pathogen, *Venturia inaequalis*. *Can. J. Bot.* 62:1819–1823.

Davide, R. G. 1983. Evaluation of a fungus, *Paecilomyces lilacinus* (Thom.) Samson, for the biological control of the potato cyst nematode *Globodera rostochiensis* Woll. as compared with some nematicides. *Phil. Agr.* 66:397–404.

Deacon, J. W. 1976. Studies on *Pythium oligandrum*, an aggressive parasite of other fungi. *Trans. Br. Mycol. Soc.* 66:383–391.

Dennis, C., and Webster, J. 1971a. Antagonistic properties of species-groups of *Trichoderma*. I. Production of non-volatile antibiotics. *Trans. Br. Mycol. Soc.* 57:25–39.

Dennis, C., and Webster, J. 1971b. Antagonistic properties of species-groups of *Trichoderma*. II. Production of volatile antibiotics. *Trans. Br. Mycol. Soc.* 57:41–48.

Domsch, K. H., Gams, W., and Anderson, T. H. 1980. *Compendium of Soil Fungi*, vol. I. Academic, New York.

Dos Santos, A. F., and Dhingra, O. D. 1982. Pathogenicity of *Trichoderma* spp. on the sclerotia of *Sclerotinia sclerotiorum*. *Can. J. Bot.* 60:472–475.

Dutta, B. K. 1981. Studies on some fungi isolated from the rhizosphere of tomato plants and the consequent prospect for the control of Verticillium wilt. *Plant and Soil* 63:209–216.

Dwived, R. 1984. Biocontrol of fusarial wilt by *Trichoderma harzianum* Rifai. *Indian J. Agric. Sci.* 54:513–514.

Elad, Y., Barak, R., and Chet, I. 1983. Possible role of lectins in mycoparasitism. *J. Bacteriol.* 154:1431–1435.

Elad, Y., Chet, I., Boyle, P., and Henis, Y. 1982. Parasitism of *Trichoderma* spp. on *Rhizoctonia solani* and *Sclerotium rolfsii*—scanning electron microscopy and fluorescence microscopy. *Phytopathology* 73:85–88.

Elad, Y., Katan, J., and Chet, I. 1980. Physical, biological, and chemical control integrated for soilborne diseases in potatoes. *Phytopathology* 70:418–422.

Elad, Y., Lifshitz, R., and Baker, R. 1985. Enzymatic activity of the mycoparasite *Pythium nunn* during interaction with host and non-host fungi. *Physiol. Plant Pathol.* 27:131–148.

Elad, Y., Zvieli, Y., and Chet, I. 1986. Biological control of *Macrophomina phaseolina* (Tassi) Goid by *Trichoderma harzianum*. *Crop Protec.* 5:288–292.

Elliston, J., Kuc, J., and Williams, E. 1976. Protection of bean against anthracnose by *Colletotrichum* species nonpathogenic on bean. *Phytopathol. Z.* 86:117–126.

Foley, M. F., and Deacon, J. W. 1986. Susceptibility of *Pythium* spp. and other fungi to antagonism by the mycoparasite *Pythium oligandrum*. *Soil Biol. Biochem.* 18:91–95.

Fravel, D. R., Marois, J. J., Dunn, M. T., and Papavizas, G. C. 1985a. Compatibility of *Talaromyces flavus* with potato seedpiece fungicides. *Soil Biochem.* 17:163–166.

Fravel, D. R., Marois, J. J., Lumsden, R. D., and Connick, W. J. 1985b. Encapsulation of potential biocontrol agents in an alginate-clay matrix. *Phytopathology* 75:774–777.

Fravel, D. R., and Marois, J. J. 1986. Edaphic parameters associated with establishment of the biocontrol agent *Talaromyces flavus*. *Phytopathology* 76:643–646.

Freeman, S., Sztejnberg, A., and Chet, I. 1986. Evaluation of *Trichoderma* as a biocontrol agent for *Rosellinia necatrix*. *Plant and Soil* 94:163–170.

Fuska, J., Fuskova, A., and Nemec, P. 1979a. Vermistatin, an antibiotic with cytotoxic effects, produced from *Penicillium vermiculatum*. *Biologia* (Bratislava) 34:735–739.

Fuska, J., Nemec, P., and Fuskova, A. 1979b. Vermicillin, a new metabolite from *Penicillium vermiculatum* inhibiting tumor cells in vitro. *J. Antibiot.* 32:667–669.

Fuska, J., Nemec, P., and Kuhr, I. 1972. Vermiculine, a new antiprotozoal antibiotic from *Penicillium vermiculatum*. *J. Antibiot.* 25:208–211.

Gilman, J. C. 1966. *A Manual of Soil Fungi*, 2nd edition. The Iowa State University Press, Ames, Iowa.

Greenberger, A., Yogev, A., and Katan, J. 1987. Induced suppressiveness in solarized soils. *Phytopathology* 77:1663–1667.

Grosclaude, C., Ricard, J., and Dobos, B. 1973. Inoculation of *Trichoderma viride* spores via pruning shears for biological control of *Stereum purpureum* on plum tree wounds. *Plant Dis. Rep.* 57:25–28.

Hadar, Y., Chet, I., and Henis, Y. 1979. Biological control of *Rhizoctonia solani* damping-off with wheat bran culture of *Trichoderma harzianum*. *Phytopathology* 69:64–68.

Hadar, Y., Harman, G. E., and Taylor, A. G. 1984. Evaluation of *Trichoderma koningii* and *T. harzianum* from New York soils for biological control of seed rot caused by *Pythium* spp. *Phytopathology* 74:106–110.

Hashioka, Y., and Fukita, T. 1969. Ultrastructural observations on mycoparasitism of *Trichoderma, Gliocladium* and *Acremonium* to phytopathogenic fungi. *Rep. Tottori Mycol. Inst.* (Japan) 7:8–18.

Henis, Y., Ghaffar, A., and Baker, R. 1978. Integrated control of *Rhizoctonia solani* of radish: effect of successive plantings, PCNB, and *Trichoderma harzianum* on pathogen and disease. *Phytopathology* 68:900–907.

Hepperly, P. R. 1986. Hansfordia sp.: a parasitic pathogen of Dematiaceous plant pathogenic fungi in Puerto Rico. *J. Agric. Univ. P.R.*, University of Puerto Rico Agricultural Experiment Station 70:113–119.

Heye, C. C., and Andrews, J. H. 1983. Antagonism of *Athelia bombacina* and *Chaetomium globosum* to the apple scab pathogen Venturia inaequalis. *Phytopathology* 73:650–654.

Hoch, H. C., and Abawi, G. S. 1979. Biological control of Pythium root rot of table beet with *Corticium* sp. *Phytopathology* 69:417–419.

Howell, C. R. 1982. Effect of *Gliocladium virens* on *Pythium ultimum, Rhizoctonia solani* and damping-off of cotton seedlings. *Phytopathology* 72:496–498.

Howell, C. R. 1987. Relevance of mycoparasitism in the biological control of *Rhizoctonia solani* by *Gliocladium virens*. *Phytopathology* 77:992–994.

Howell, C. R., and Stipanovic, R. D. 1983. Gliovirin, a new antibiotic from *Gliocladium virens*, and its role in the biological control of *Pythium ultimum*. *Can. J. Microbiol.* 29:321–324.

Howell, C. R., and Stipanovic, R. D. 1984. Phytotoxicity to crop plants and herbicidal effects on weeds of viridiol produced by *Gliocladium virens*. *Phytopathology* 74:1346–1349.

Huang, H. C. 1978. *Gliocladium catenulatum*: hyperparasite of *Sclerotinia sclerotiorum* and *Fusarium* species. *Can. J. Bot.* 56:2243–2246.

Huang, H. C. 1980. Control of sclerotinia wilt of sunflower by hyperparasites. *Can. J. Plant Pathol.* 2:26–32.

Huang, H. C., and Hoes, J. A. 1976. Penetration and infection of *Sclerotinia sclerotiorum* by *Coniothyrium minitans*. *Can. J. Bot.* 54:406–410.

Hubbard, J. P., Harman, G. E., and Hadar, Y. 1983. Effect of soilborne *Pseudomonas* spp. on the biological control agent *Trichoderma hamatum*, on pea seeds. *Phytopathology* 73:655–659.

Hughes, S. J. 1951. Studies of microfungi. IX. *Mycol. Pap.* 43:13–25.

Jager, G., and Velvis, H. 1983. Suppression of *Rhizoctonia solani* in potato fields. I. Occurrence. *Neth. J. Plant Pathol.* 89:21–29.

Jager, G., and Velvis, H. 1984. Biological control of *Rhizoctonia solani* on potatoes by antagonists. 2. Sprout protection against soil-borne *R. solani* through seed inoculation with *Verticillium biguttatum*. *Neth. J. Plant Pathol.* 90:29–33.

Jager, G., and Velvis, H. 1986. Biological control of *Rhizoctonia solani* on potatoes by antagonists. 5. The effectiveness of three isolates of *Verticillium biguttatum* as inoculum for seed tubers and of a soil treatment with a low dosage of pencycuron. *Neth. J. Plant Pathol.* 92:231–238.

Jalal, M. A. F., Love, S. K., and van der Helm, D. 1986. Siderophore mediated iron(III) uptake in *Gliocladium virens*. 1. Properties of cis-fusarinine, trans-fusarinine, dimerum acid, and their ferric complexes. *J. Inorg. Biochem.* 28:417–430.

Jones, R. W., and Hancock, J. G. 1987. Conversion of viridin to viridiol by viridin-producing fungi. *Can. J. Microbiol.* 33:963–966.

Karhuvaara, L. 1960. On the parasites of the sclerotia of some fungi. *Acta Agric. Scand.* 10:127–134.

Katan, T. 1985. Heat activation of dormant ascospores of *Talaromyces flavus*. *Trans. Br. Mycol. Soc.* 4:748–750.

Katan, T., Dunn, M. T., and Papavizas, G. C. 1984. Genetics of fungicide resistance in *Talaromyces flavus*. *Can. J. Microbiol.* 30:1079–1087.

Kerry, B. R. 1980. Biocontrol: fungal parasites of female cyst nematodes. *J. Nematol.* 12:253–259.

Kerry, B. R., Crump, D. H., and Mullen, L. A. 1980. Parasitic fungi, soil moisture and the multiplication of the cereal cyst nematode, *Heterodera avenae*. *Nematologica* 26:57–68.

Kerry, B. R., Simon, A., and Rovira, A. D. 1984. Observations on the introduction of *Verticillium chlamydosporium* and other parasitic fungi into soil for control of the cereal cyst-nematode *Heterodera avenae*. *Ann. Appl. Biol.* 105:509–516.

Kim, K. K., Fravel, D. R., and Papavizas, G. C. 1988. Identification of a metabolite produced by *Talaromyces flavus* as glucose oxidase and its role in the biocontrol of *Verticillium dahliae*. *Phytopathology* 78:488–492.

Kodama, T., Kukui, T., and Matsumoto, Y. 1980. Solar heating sterilization in closed vinyl house against soil-borne diseases. III. Influence of treatment on the population level of soil microflora and on the behavior of strawberry yellows pathogen, *Fusarium oxysporum* f. sp. fragariae. *Bull. Nara Pref. Agric. Exp. Stn.* 11:21–25.

Kommedahl, T., and Chang-Mew, I. 1975. Biocontrol of corn root infection in the field by seed treatment with antogonists. *Phytopathology* 65:296–300.

Kuc, J., and Preisig, C. 1984. Fungal regulation of disease resistance mechanisms in plants. *Mycologia* 76:767–784.

Kuhlman, E. G., Carmichael, J. W., and Miller, T. 1976. *Scytalidium uredinicola*, a new mycoparasite of *Cronartium fusiforme* on *Pinus*. *Mycologia* 68:1188–1194.

Kuhlman, E. G., Matthews, F. R., and Tillerson, H. P. 1978. Efficacy of *Darluca filum* for biological control of *Cronartium fusiforme* and *C. strobilinum*. *Phytopathology* 68:507–511.

Lee, Y. A., and Wu, W. S. 1986. Chemical and biological controls of sunflower sclerotinia disease. *Plant Protec. Bull.* 28:101–109.

Lewis, J. A., and Papavizas, G. C. 1980. Integrated control of Rhizoctonia fruit rot of cucumber. *Phytopathology* 70:85–89.

Lewis, J. A., and Papavizas, G. C. 1984. A new approach to stimulate population proliferation of *Trichoderma* species and other potential biocontrol fungi introduced into natural soil. *Phytopathology*: 74:1240–1244.

Lifshitz, R., Dupler, M., Elad, Y., and Baker, R. 1984a. Hyphal interactions between a mycoparasite, *Pythium nunn*, and several soil fungi. *Can. J. Microbiol.* 30:1482–1487.

Lifshitz, R., Sneh, B., and Baker, R. 1984b. Soil suppressiveness to *Pythium ultimum* induced by antagonistic *Pythium* species. *Phytopathology* 74:1054–1061.

Lim, T. K., and Chan, L. G. 1986. Parasitism of *Phytophthora palmivora* by *Gliocladium roseum*. *J. Plant Dis. and Protec.* 93:509–514.

Liu, S. H., and Vaughan, E. K. 1965. Control of *Pythium* infection in table beet seedlings by antagonistic microorganisms. *Phytopathology* 55:986–989.

Lutchmeah, R. S., and Cooke, R. C. 1984. Aspects of antagonism by the mycoparasite *Pythium oligandrum*. *Trans. Br. Mycol. Soc.* 83:696–700.

Lutchmeah, R. S., and Cooke, R. C. 1985. Pelleting of seed with the antagonist *Pythium oligandrum* for biological control of damping-off. *Plant Pathol.* 34:528–531.

Makkonen, R., and Pohjakallio, O. 1960. On the parasites attacking the sclerotia of some fungi pathogenic to higher plants and on the resistance of these sclerotia to their parasites. *Acta Agric. Scand.* 10:105–126.

Marois, J. J., Johnston, S. A., Dunn, M. T., and Papavizas, G. C. 1982. Biological control of *Verticillium* wilt of eggplant in the field. *Plant Dis.* 66:1166–1168.

Marois, J. J., and Locke, J. C. 1985. Population dynamics of *Trichoderma viride* in steamed plant growth medium. *Phytopathology* 75:115–118.

Martin, F. N., and Hancock, J. G. 1984. The use of *Pythium oligandrum* for the biological control of *P. ultimum*. *Phytopathology* 74:835 (Abstr.).

Martin, F. N., and Hancock, J. G. 1987. The use of *Pythium oligandrum* for biological control of preemergence damping-off caused by *P. ultimum*. *Phytopathology* 77:1013–1020.

Martyniuk, S., and Myskow, W. 1984. Control of the take-all fungus by *Phialophora* sp. (lobed hyphopodia) in microplot experiments with wheat. *Zbl. Mikrobiol.* 139:575–579.

McLaren, D. L., Huang, H. C., and Rimmer, S. R. 1985. *Talaromyces flavus*, an effective mycoparasite of *Sclerotinia sclerotiorum*. *Phytopathology* 75:1328 (abstr.).

McLaren, D. L., Huang, H. C., and Rimmer, S. R. 1986. Hyperparasitism of *Sclerotinia sclerotiorum* by *Talaromyces flavus*. *Can. J. Plant Pathol.* 8:43–48.

Mihuta-Grimm, L., and Rowe, R. C. 1986. *Trichoderma* spp. as biocontrol agents of Rhizoctonia damping-off of radish in organic soil and comparison of four delivery systems. *Phytopathology* 76:306–312.

Mizuno, K., Yagi, A., Takada, M., Matsuura, K., Yamaguchi, K., and Asano, K. 1974. A new antibiotic, Talaron. *J. Antibiot.* 27:560–563.

Moody, A. R., and Gindrat, D. 1977. Biological control of cucumber black rot by *Gliocladium roseum*. *Phytopathology* 67:1159–1162.

Moore, W. L. 1981. Recent advances in the biological control of bacterial plant diseases. In *Biological Control in Crop Production*, G. C. Papavizas (ed.). Allanheld, Osmun, Totowa, New Jersey, pp. 375–390.

Morgan-Jones, G., White, J. F., Rodriguez-Kabana, R. 1983. Phytonematode pathology: ultrastructural studies. I. Parasitism of *Meloidogyne arenaria* eggs by *Verticillium chlamydosporium*. *Nematropica* 13:245–260.

Moubasher, A. H., Elnage, M. A., and Megala, S. E. 1970. Fungi isolated from sclerotia of *Sclerotium cepivorum* and from soil, and their effects upon the pathogen. *Plant and Soil* 33:305–312.

Nordbring-Hertz, B., and Mattiasson, B. 1979. Action of a nematode-trapping fungus shows lectin-mediated host-microorganism interaction. *Nature* 281:477–479.

Nordbring-Hertz, B., Friman, B., and Mattiasson, B. 1982. A recognition mechanism in the adhesion of nematodes to nematode-trapping fungi. In *Lectins, Biology, Biochemistry and Clinical Biochemistry*, vol. II, T. C. Bog-Hansen (ed.). Berlin, W. de Gruyter, pp. 83–90.

Odunfa, S. A. 1982. *Penicillium* species antagonistic to cowpea root disease fungi. *Trop. Grain-Legume Bull*. Ibadan Nigeria: International Grain Legume Information Center 26:10–15.

Odvody, G. N., Boosalis, M. G., and Kerr, E. D. 1980. Biological control of *Rhizoctonia solani* with a soil inhabiting basidiomycete. *Phytopathology* 70:655–658.

Odvody, G. M., Boosalis, M. G., Lewis, J. A., and Papavizas, G. C. 1977. Biological control of *Rhizoctonia solani*. *Proc. Am. Phytopathol. Soc.* 4:158 (Abstr.).

Okuda, T., Fujiwara, A., and Fujiwara, M. 1982. Correlation between species of *Trichoderma* and production patterns of isonitrile antibiotics. *Agric. Biol. Chem.* 46:1811–1822.

Overeem, J. C., and Mackor, A. 1973. Scytalidic acid, a novel compound from *Scytalidium* species. *Recueil* 92:349–359.

Pachenari, A., and Dix, N. J. 1980. Production of toxins and wall degrading enzymes by *Gliocladium roseum*. *Trans. Br. Mycol. Soc.* 74:561–566.

Papavizas, G. C., Dunn, M. T., Lewis, J. A., and Beagle-Ristaino, J. 1984. Liquid fermentation technology for experimental production of biocontrol fungi. *Phytopathology* 74:1171–1175.

Papavizas, G. C., Lewis, J. A., and Abd-El Moity. 1982. Evaluation of new biotypes of *Trichoderma harzianum* for tolerance to benomyl and enhanced biocontrol capabilities. *Phytopathology* 72:126–132.

Peresse, M., and Le Picard, D. 1980. *Hansfordia pulvinata*, mycoparasite destructeur du *Cladosporium fulva*. *Mycopathologia* 71:22–30.

Philips, A. J. L. 1986. Factors affecting the parasitic activity of *Gliocladium virens* on sclerotia of *Sclerotinia sclerotiorum* and a note on its host range. *J. Phytopathol.* 116:212–220.

Sayre, R. M. 1986. Pathogens for biological control of nematodes. *Crop Protec.* 5:268–276.

Sivan, A., Elad, Y., and Chet, I. 1984. Biological control effects of a new isolate of *Trichoderma harzianum* on *Pythium aphanidermatum*. *Phytopathology* 74:498–501.

Sivan, A., and Chet, I. 1986. Biological control of *Fusarium* spp. in cotton, wheat and muskmelon by *Trichoderma harzianum*. *J. Phytopathol.* 116:39–47.

Stasz, T. E., Harman, G. E., and Weeden, N. F. 1988. Protoplast preparation and fusion in two biocontrol strains of *Trichoderma* harzianum. *Mycologia* 80:141–150.

Stillwell, M. A., Wall, R. E., and Strunz, G. M. 1973. Production, isolation and antifungal activity of scytalidin, a metabolite of *Scytalidium* species. *Can. J. Microbiol.* 19:597–602.

Strashnow, Y., Elad, Y., Sivan, A., and Chet, I. 1985. Integrated control of *Rhizoctonia solani* by methyl bromide and *Trichoderma harzianum*. *Plant Pathol.* 34:146–151.

Sundheim, L. 1982. Control of cucumber powdery mildew by the hyperparasite *Ampelomyces quisqualis* and fungicides. *Plant Pathol.* 31:209–214.

Sundheim, L., and Amundsen, T. 1982. Fungicide tolerance in the hyperparasite *Ampelomyces quisqualis* and integrated control of cucumber powdery mildew. *Acta Agric. Scand.* 32:349–355.

Swendsrud, D. P., and Calpouzos, L. 1972. Effect of inoculation sequence and humidity on infection of *Puccinia recondita* by the mycoparasite *Darluca filum*. *Phytopathology* 62:931–932.

Sztejnberg, A. 1979. Biological control of powdery mildew by *Ampelomyces quisqualis*. *Phytopathology* 69:1047 (Abstr.).

Taber, R. A., Pettit, R. E., McGee, R. E., and Smith, D. H. 1981. Potential for biological control of Cercosporidium leafspot of peanuts by *Hansfordia*. *Phytopathology* 71:260 (Abstr.).

Tirilly, Y., Kloosterman, J., Spima, G., and Kettenes-Van Den Bosch. 1983. A fungitoxic sesquiterpene from *Hansfordia pulvinata*. *Phytochemistry* 22:2082–2083.

Trutman, P., Keane, P. J., and Merriman, P. R. 1982. Biological control of *Sclerotinia sclerotiorum* on aerial parts of plants by the hyperparasite *Coniothyrium minitans*. *Trans. Br. Mycol. Soc.* 78:521–529.

Tsuneda, A., Hiratsuka, Y., and Maruyama, P. J. 1980. Hyperparasitism of *Scytalidium uredinicola* on western gall rust, *Endocronartium harknessii*. *Can. J. Bot.* 58:1154–1159.

Tu, J. C. 1980. *Gliocladium virens*, a destructive mycoparasite of *Sclerotinia sclerotiorum*. *Phytopathology* 70:670–674.

Tu, J. C., and Vaartaja, O. 1981. The effect of the hyperparasite (*Gliocladium virens*) on *Rhizoctonia solani* and on Rhizoctonia root rot of white beans. *Can. J. Bot.* 59:22–27.

Turner, G. J., and Tribe, H. T. 1976. On *Coniothyrium minitans* and its parasitism of *Sclerotinia* species. *Trans. Br. Mycol. Soc.* 66:97–105.

Tuzun, S., and Kuc, J. 1985. Movement of a factor in tobacco infected with *Peronospora tabacina* Adam which systemically protects against blue mold. *Physiol. Plant Pathol.* 26:321–330.

Tuzun, S., and Kuc, J. 1987. Persistence of induced systemic resistance to blue mold in tobacco plants derived via tissue culture. *Phytopathology* 77:1032–1035.

Tveit, M., and Wood, R. K. S. 1955. The control of *Fusarium* blight in oat seedlings with antagonistic species of *Chaetomium*. *Ann. Appl. Biol.* 43:538–552.

Uecker, F. A., Ayers, W. A., and Adams, P. B. 1978. A new hyphomycete on sclerotia of *Sclerotinia sclerotiorum*. *Mycotaxon* 7:275–282.

Uecker, F. A., Ayers, W. A., and Adams, P. B. 1980. *Teratosperma oligocladum*, a new hyphomyceteous mycoparasite on sclerotia of *Sclerotinia sclerotiorum*, *S. trifoliorum*, and *S. minor*. *Mycotaxon* 10:421–427.

Vajna, L. 1985. Mutual parasitism between *Trichoderma hamatum* and *Trichoderma pseudokoningii*. *Phytopathol. Z.* 113:300–303.

Van Den Boogert, P. H. J. F., and Jager, J. 1983. Accumulation of hyperparasites of *Rhizoctonia solani* by addition of live mycelium of *R. solani* to soil. *Neth. J. Plant Pathol.* 89:223–228.

Vannacci, G., and Harman, G. E. 1987. Biocontrol of seed-borne *Alternaria raphani* and *A. brassicicola*. *Can. J. Microbiol.* 38:850–856.

Velvis, H., and Jager, G. 1983. Biological control of *Rhizoctonia solani* on potatoes by antagonists. 1. Preliminary experiments with *Verticillium biguttatum*, a sclerotium-inhibiting fungus. *Neth. J. Plant Pathol.* 89:113–123.

Vesely, D., and Hejdanek, S. 1984. Microbial relations of *Pythium oligandrum* and problems in the use of this organism for the biological control of damping-off in sugarbeet. *Zentralbl. Bakteriol. Mikrobiol. Hyg.* 139:257–265.

Webster, J., and Lomas, N. 1964. Does *Trichoderma viride* produce gliotoxin and viridin? *Trans. Br. Mycol. Soc.* 47:535–540.

Weindling, R., and Emerson, O. H. 1936. The isolation of a toxic substance from the culture filtrate of *Trichoderma*. *Phytopathology* 26:1068–1070.

Wells, H. D., Bell, D. K., and Jaworski, C. A. 1972. Efficacy of *Trichoderma harzianum* as a biological control for *Sclerotium rolfsii*. *Phytopathology* 62:442–447.

Windels, C. E. 1981. Growth of *Penicillium oxalicum* as a biological seed treatment on pea seed in soil. *Phytopathology* 71:929–933.

Windham, M. T., Elad, Y., and Baker, R. 1986. A mechanism for increased plant growth induced by *Trichoderma* spp. *Phytopathology* 76:518–521.

Wynn, A. R., and Epton, H. A. S. 1979. Parasitism of oospores of *Phytophthora erythroseptica* in soil. *Trans. Br. Mycol. Soc.* 73:255–259.

Zogg, H., and Joeggi, W. 1974. Studies on the biological soil disinfection. 7. Contribution to the take-all decline (*Gaeumannomyces graminis*) initiated by means of laboratory trials and some of its possible mechanisms. *Phytopathol. Z.* 81:160–169.

Biological Control of Plant Root Diseases by Bacteria

W. D. Gould

CANMET
Department of Energy, Mines and Resources
555 Booth Street
Ottawa, Ontario K1A 0G1
Canada

Introduction

Due to more stringent environmental regulations and the weaknesses of chemical control, the biological control of root diseases has become more attractive (Deacon 1983, Lynch 1987). Table 1 lists a number of microorganisms that have been considered as biological disease-control agents for various root diseases. The diversity of both the biocontrol agents used and the crops they protect is a result of the greatly increased activity that has occurred in this field during the past ten years. Although most of this work is still at the experimental

TABLE 1 Examples of Microorganisms That Have Been Used to Control Root Diseases in Various Crops

Biocontrol organism	Crop	Causal organism or disease	References
Pseudomonas fluorescens	Wheat	Take-all	Weller 1983
Pseudomonas fluorescens	Cotton	*Pythium ultimum*	Loper 1988
Pseudomonas fluorescens	Cotton	Damping-off	Howell and Stipanovic 1979, 1980
Pseudomonas putida	Flax, radish, cucumber	*Fusarium* wilt	Scher and Baker 1982
Pseudomonas putida	Beans	*Fusarium solani*	Anderson and Guerra 1985
Fluorescent pseudomonads	Potatoes	*Erwinia carotovora*	Xu and Gross 1986a,b
Fluorescent pseudomonads	Potatoes	Nonspecific pathogens	Kloepper et al. 1980a Burr et al. 1978 Kloepper and Schroth 1981a
Fluorescent pseudomonads	Potatoes	*Erwinia carotovora*	Kloepper 1983
Fluorescent pseudomonads	Wheat	Take-all	Weller and Cook 1983
Fluorescent pseudomonads	Tobacco	*Thielaviopsis basicola*	Stutz et al. 1986
Fluorescent pseudomonads	Radish	Nonspecific pathogens	Kloepper and Schroth 1981b
Cytophaga sp.	Conifer seedlings	Damping-off	Hocking and Cook 1972
Trichoderma hamatum	Radish and pea seedlings	*Pythium* sp. and *Rhizoctonia solani*	Harman et al. 1980
Trichoderma harzianum	Radish seedlings	*Alternaria* spp.	Vannacci and Harman 1987
Penicillium oxalicum	Peas	Root rot	Kommedahl and Windels 1978
Nonpathogenic isolate of *Fusarium oxysporum*	Cucumber	*Fusarium* wilt	Paulitz et al. 1987
Trichoderma viride	Tomatoes	*Verticillium* wilt	Dutta 1981
Bacillus subtilis	Corn	*Fusarium roseum*	Chang and Kommedahl 1968
Chaetomium globosum	Sugar beets	Damping-off	Kommedahl and Mew 1975, Walther and Gindrat 1988
A mixture of fungi, *Aspergillus* sp.	Tomatoes	Crown rot	Marois et al. 1981
Bacillus sp., *Penicillium* sp., *Pseudomonas* sp., *Alcaligenes* sp.	Cherry seedlings	Crown gall *Agrobacterium tumefaciens*	Cooksey and Moore 1980
Bacillus sp., *Pseudomonas* sp.	Carnations	*Fusarium oxysporum*	Yuen et al. 1985

stage, there have been some commercial successes (Deacon 1983, Schroth and Hancock 1981) and there is considerable potential for future exploitation.

The following advantages of biological disease-control organisms have been noted: (1) these organisms are considered safer than many of the chemicals now in use; (2) they do not accumulate in the food chain; (3) self-replication circumvents repeated application; (4) target organisms seldom develop resistance as happens when chemical agents are used; (5) where less effective than a chemical control agent, the two can sometimes be combined; and (6) properly developed biocontrol agents are not considered harmful to the ecology. The major disadvantages include variability of field performance, and the necessity for precautions to ensure survival and delivery of the product. Also, the effectiveness of a given biocontrol agent may be restricted to a specific location, due to the effects of soil and climate.

There exists some controversy regarding the definition of "biological disease control" (Deacon 1983). Most definitions assume a particular mechanism of action, but recent work indicates that biological disease-control mechanisms are very complex. Bacterial inocula have frequently been added to antagonize root pathogens, but the positive effects observed with many bacterial inoculants could be enhanced by other mechanisms, such as the bacterial production of plant-growth-promoting compounds (Brown 1974). For the purposes of this review, "biological disease control" will be defined as the direct or indirect use of other microorganisms both to reduce the effects of plant pathogens and to improve plant survival and growth. Thus, this article will discuss most aspects of plant-growth-promoting bacterial inoculants, with particular emphasis on the control of root diseases. The possible mechanisms of biological control, methods for selection and use of biological control agents, potential for genetic engineering, and future applications will also be addressed.

A natural model for biological disease control is the decline of take-all disease (Deacon 1983, Weller 1983). Wheat or barley can be infected with the take-all organism, *Gaeumannomyces graminis*, during the early years of monoculture. After 3 or 4 years the disease reaches a peak, whereupon it proceeds to decline. The addition of suppressive soil to a soil conducive to the disease renders the conducive soil suppressive (Scher and Baker 1980, Schroth and Hancock 1982, Simon et al. 1987). Steam treatment of the suppressive soil eliminated the suppressive activity, suggesting a biological factor (Scher and Baker 1980). Sivasithamparam et al. (1979) found fluorescent pseudomonads antagonistic to *G. graminis* to be more numerous in the rhizosphere of wheat plants growing in infected soil than in the bulk soil. Fluorescent pseudomonads are generally considered to be responsible for take-all decline (Charigkapakorn and Sivasithamparam 1987, Cook and Rovira 1976), although several *Bacillus* isolates have been used for the biological control of take-all (Campbell and Clor 1985).

The biological control of root diseases can be accomplished by several means: (1) direct inoculation of bacteria on seeds or stem cuttings (Henis and Chet 1975, Schroth and Hancock 1981); (2) incorporation into the soil (Batra 1981, Brown 1974); (3) indirect control by the use of cultural changes to enhance naturally occurring disease-control organisms already in the soil (Batra 1981, Deacon 1973, Smiley 1978*a*,*b*, 1979); and (4) integrated control, such as the use of a seed inoculum plus chemicals or any other combination of control measures (Batra 1982).

An effective biological disease-control organism should do some of the following: (1) rapidly colonize the root zone (Lifshitz et al. 1986, Loper et al. 1984, 1985, Weller 1983); (2) produce antibiotics to antagonize pathogenic microorganisms (Brathwaite and Cunningham 1982, Howell and Stipanovic 1979, 1980); (3) produce high-affinity iron-chelating compounds, called "siderophores," that make iron less available to the pathogens (Scher et al. 1984*a*, Schroth and Hancock 1981, 1982); (4) compete for substrates that are essential for growth of the pathogens (Brown 1974, Papavizas and Lumsden 1980); (5) compete with the pathogens for infection sites (Asher 1978, Deacon 1976, Gutteridge and Slope 1978, Wong and Siviour 1979); and (6) produce plant-growth-promoting compounds, such as gibberellin-like substances or indolyl-3-acetic acid (Brown 1974). Other characteristics that may be advantageous are induction of resistance to plant pathogens, inhibition or displacement of nonpathogenic inhibitory rhizosphere bacteria (Elliott and Lynch 1984), and ability to increase the availability of soil nutrients to plants (Brown 1974).

Mechanisms of Biological Disease Control

Antibiosis

One of the major mechanisms postulated for the biological control of plant root diseases is the production of antimicrobial compounds by the disease-control agent (Deacon 1983, Howell and Stipanovic 1979, 1980, 1983, 1984, Stipanovic and Howell 1982). Antibiosis may be mediated by toxic secondary metabolites (Howell and Stipanovic 1979, 1980) or specific toxins such as bacteriocins (Kerr and Htay 1974). Antimicrobial compounds may act on plant-pathogenic fungi by inducing fungistasis, inhibition of germination, lysis of fungal mycelia, or by exerting fungicidal effects. The factors mitigating antibiotic effectiveness in soil are inactivation by binding with soil colloids (Howell and

Stipanovic 1980), biological degradation (Williams and Vickers 1986), chemical degradation, development of resistance by the target organism (Alconero 1980, Moore and Warren 1979), and inadequate production of the antibiotic by the biocontrol organism.

The fluorescent pseudomonads have frequently been considered for the biological control of plant root diseases. They produce a large number of secondary metabolites, many of which possess antimicrobial activity (Ingram and Blackwood 1970, Leisinger and Margraff 1979). The major classes of these compounds are illustrated in Figure 1. A number of other pseudomonads have been shown to produce potent antibiotics, such as tropolone (Figure 2), a wide-spectrum antibiotic (Lindberg 1981, Lindberg et al. 1980). Tropolone was found to be either bacteriostatic or bactericidal for a wide range of bacterial species (Trust 1975). Tropolone is also a potent fungicide, inducing rapid lysis of established fungal colonies (Lindberg 1981). A *Pseudomonas cepacia* isolate was found to produce two acetylenic antibiotics, cepacins A and B (Parker et al. 1984); another pseudomonad was capable of oxidizing glycine to hydrogen cyanide (Wissing 1974).

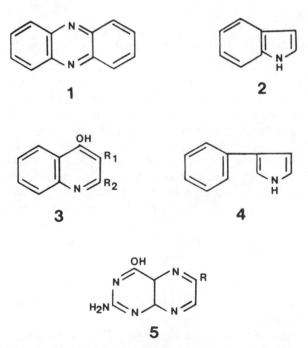

Figure 1 Some of the more important classes of secondary metabolites produced by fluorescent pseudomonads. (1) Phenazines. (2) Indoles. (3) Pyo compounds. (4) Phenylpyrroles. (5) Pterines.

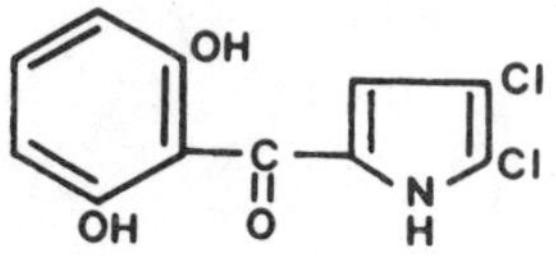

Figure 2 Representative antibiotics that are produced by pseudomonads. (1) Pyoleutorin. (2) Pyrrolnitrin. (3) Tropolone. (4) Pyocyanin. (5) Cepacin A.

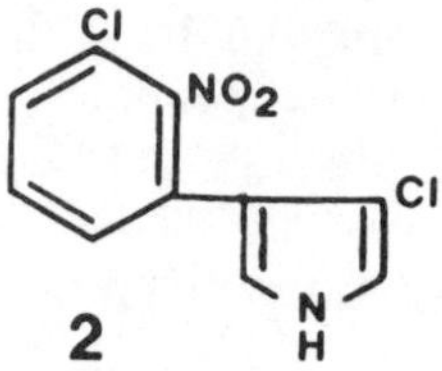

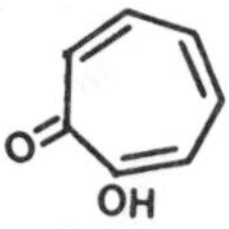

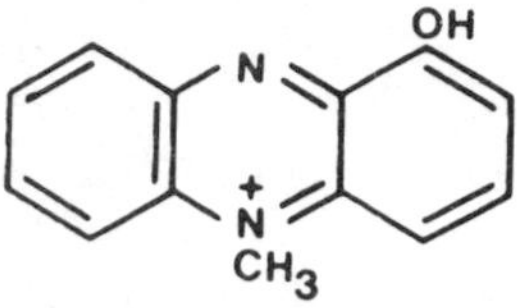

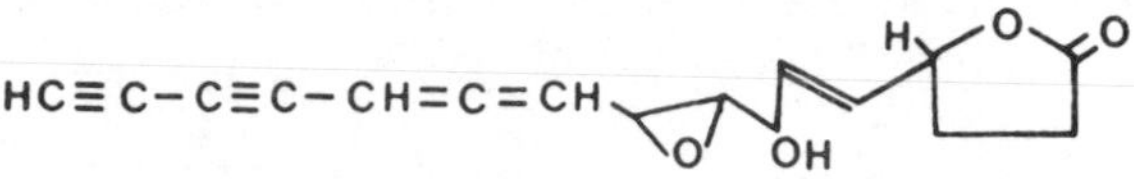

The chemical and nutritional status of a soil can also influence the production of secondary metabolites by disease-control bacteria. In particular, the production of secondary metabolites can be affected by pH, inorganic nutrients, culture age, and the concentrations of the precursors of these compounds (Chang et al. 1981, Elander et al. 1968). The production of three antibiotics by one *Pseudomonas fluorescens* (HV37a) isolate was found to be regulated by glucose (James and Gutterson 1986). Both antibiotic synthesis and production of secondary metabolites are inhibited by orthophosphate (Martin 1977). It has been observed in batch culture that secondary metabolites are produced only after the inorganic nutrients have been exhausted, during the latter half of the stationary growth phase.

Although antibiosis has been shown to play a role in the biological control of *Pythium ultimum* by *Gliocladium virens* (Howell and Stipanovic 1983), experimental evidence for the direct involvement of antibiotics in disease control is difficult to obtain. The production of antibiotics by a soil microorganism would be expected to confer on it a competitive advantage. In spite of numerous attempts to demonstrate the production of antibiotics in soil, there is no conclusive evidence for their presence in natural soil (Williams and Vickers 1986). Moreover, the fact that a biocontrol agent produces an antibiotic in culture does not indicate that the agent will perform likewise in situ. However, indirect evidence for the role of antibiosis in biological disease control was obtained by Howell and Stipanovic (1979, 1980). They isolated from the rhizosphere of cotton seedlings a strain of *Pseudomonas fluorescens* (pf-5) that was antagonistic to both *Rhizoctonia solani* and *Pythium ultimum*. The isolate produced two phenylpyrrole antibiotics: pyoleutorin, which was active against *P. ultimum*, and pyrrolnitrin, which was active against *R. solani*. When inoculated onto cotton seeds, the bacterial isolate protected cotton seedlings against damping-off caused by both fungal pathogens. Each antibiotic, when applied as a seed treatment, protected the cotton seedlings from the fungal pathogen sensitive to it. Howell and Stipanovic (1983) also obtained direct evidence for the role of antibiosis in biological disease control. *Gliocladium virens*, a fungus that protects cotton seedlings against damping-off, produces an antibiotic, gliovirin, which is active against *P. ultimum* (Stipanovic and Howell 1982). A gliovirin-deficient mutant of *G. virens* was not inhibitory to *P. ultimum* in vitro, and was also ineffective in preventing damping-off of cotton seedlings (Howell and Stipanovic 1983). The parent strain and a gliovirin-enhanced mutant were equally effective in protecting cotton seedlings against *P. ultimum*.

Streptomyces hygroscopus var. *geldanus*, which produces the antibiotic geldanamycin, controlled rhizoctonia root rot of pea in sterilized soil (Rothrock and Gottleib 1984). Geldanamycin was inhibitory to *R. solani* both in vitro and in vivo. Sterilized soil inoculated with *S. hygroscopicus* contained 88 micrograms of geldanamycin per gram of soil after 7 days of incubation (Rothrock and Gottleib 1984). However, Rothrock and Gottlieb did not measure antibiotic accumulation in unsterilized soil, and antibiotic stability and persistence are much lower in natural soils (Williams 1982).

Pseudomonas fluorescens 2-79 (NRRL-B-15132), which has been used to control take-all disease of wheat (Weller and Cook 1983), produces a phenazine antibiotic with potent antifungal activity (Gurusiddaiah et al. 1986). The antibiotic has subsequently been identified as phenazine-1-carboxylic acid (Brisbane et al. 1987). *Tn5* mutants of *P. fluorescens* 2-79 unable to produce phenazine-1-carboxylic acid were not inhibitory to *Gaeumannomyces graminis* in vitro and were also less effective at protecting wheat from take-all disease (Thomashow and Weller 1988). Since the mutants defective in phenazine synthesis were as effective as the parent strain in their root-colonizing ability, the antibiotic must play a major role in the suppression of *G. graminis*. Production of phenazine-1-carboxylic acid would probably not assist the *P. fluorescens* isolate in colonizing the root, because that antibiotic has been shown to have only moderate activity against a number of bacterial isolates (Gurusiddaiah et al. 1986).

Another important class of antibiotics produced by bacteria are the bacteriocins. Bacteria not closely related to a particular biocontrol organism are immune from the effects of the bacteriocins of that organism. One of the first commercial applications of biological control for root diseases has been the use of *Agrobacterium radiobacter* K84 to control the causal organism of crown gall, *Agrobacterium tumefaciens* (New and Kerr 1972). Tate et al. (1979) found that *A. radiobacter* produces the nucleotide antibiotic "agrocin 84" (Figure 3). The antibiotic specifically inhibits *A. tumefaciens* (Tate et al. 1979), due to which specificity it is classified as a bacteriocin (Moore and Warren 1979). Its synthesis is encoded by a 47.7-kb plasmid, pAgK84 (Ellis et al. 1979). Resistant strains of *A. tumefaciens* are considered to constitute the main threat to the success of biological control of crown gall using *A. radiobacter* (Alconero 1980, Moore and Warren 1979). One of the

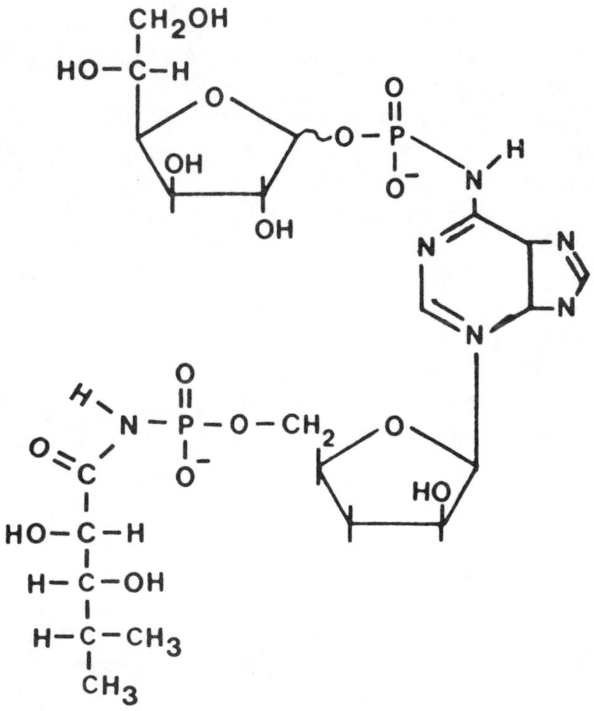

Figure 3   Structure of agrocin 84, the bacteriocin produced by *Agrobacterium radiobacter*.

mechanisms by which resistant strains arise is the transfer of the plasmid that encodes for agrocin 84 production from *A. radiobacter* to *A. tumefaciens* (Farrand et al. 1985). In order to overcome this deficiency, Shim et al. (1987) constructed two transfer-deficient *Tn5* insertion mutants of *A. radiobacter*. Except in locations where naturally occurring resistant strains of *A. tumefaciens* are found, the transfer-deficient mutant should be a more efficient biocontrol agent than the parent strain.

Shim et al. (1987) also constructed two mutants of *A. radiobacter* both of which overproduced agrocin 84. The first, of K84 genetic background, was an effective biocontrol agent and readily colonized the roots of the host plant. The second, derived from an ineffective parent strain, was not only an ineffective biocontrol organism, but a poor root colonizer as well. Agrocin 84 probably does not confer an advantage in root-colonizing ability to the organism possessing it, because most of the other rhizosphere bacteria are not sensitive to the antibiotic.

Siderophore production and competition for iron

Although iron is abundant in soils, it is often unavailable to both plants and microorganisms because its solubility is very low at neutral and alkaline pH values (Akers 1983, Cline et al. 1982, 1983, Powell et al. 1980). Many microorganisms have evolved high-affinity iron uptake systems, which are induced by low iron concentrations and repressed by high iron levels (Marugg et al. 1985, 1987, Meyer and Abdallah 1978, Neilands 1981). Microorganisms excrete compounds (siderophores) that have a very high affinity specifically for ferric iron ($K \approx 10^{25}$ to 10^{40}) (Cody and Gross 1987, Meyer and Abdallah 1978, 1980, Ong et al. 1979). These microorganisms possess permeases specific for the ferric siderophore complexes (de Weger et al. 1986, Hohnadel and Meyer 1988, Magazin et al. 1986, Meyer and Hornsperger 1978). Lodge et al. (1982) investigated the ferria-grobactin reductase system of *A. tumefaciens*, and postulated that intracellular reduction of the ferric siderophore occurs prior to utilization of the iron. Siderophores have a much lower affinity for ferrous than ferric iron. They retain ferric iron until its reduction; the resulting ferrous iron is then released. Gaines et al. (1981) extracted a ferrisiderophore reductase activity associated with an aromatic biosynthetic enzyme complex from *Bacillus subtilis*. The reductase served a second function: that of activating chorismate synthase, a key enzyme in aromatic biosynthesis. Gaines et al. (1981) postulated that the reductase played a role in the regulation of the synthesis of the siderophore (2,3-dihydroxybenzoic acid), which is synthesized via chorismate. In the presence of high iron levels, the reductase acts pri-

marily as a ferrisiderophore reductase; less of its activity can therefore be involved in activating chorismate synthase. In the presence of
low iron levels, more activity is directed towards activating
chorismate synthase, to produce more chorismate for the synthesis of
additional siderophore. *Fusarium roseum* and a *Penicillium* sp. were
both found to produce an enzyme that hydrolyzed ornithine ester
bonds (Emery 1976). Emery (1976) suggested that these ornithy-
lesterases, in conjunction with ferrisiderophore reductases, function
in the removal of iron from ferric siderophores. Ferrous iron is
readily reoxidized to ferric iron in the cell. The hydrolysis of the
ornithine ester bonds inactivates the siderophore, preventing it
from recomplexing the iron.

Numerous bacterial (Leong and Neilands 1982, Misaghi et al. 1982)
and fungal (Frederick et al. 1981, Szaniszlo et al. 1981) species have
been shown to produce siderophores. Two major types of siderophores
are produced by microorganisms: hydroxamate and catechol compounds (Figure 4). Hydroxamate siderophores usually contain N^δ-
hydroxyornithine as the ligand involved in the chelation of iron. However, Cody and Gross (1987) found that β-hydroxyaspartic acid
replaced N^δ-hydroxyornithine in pyoverdin pss extracted from a
Pseudomonas syringae isolate. Typical hydroxamate siderophores are
schizokinin (Mullis et al. 1971), nocardamine (Meyer and Abdallah
1980), N,N^1,N^{11}-triacetylfusigen (Moore and Emery 1976), and
aerobactin (Gibson and Magrath 1969). The other major group consists of siderophores that contain catechol functional groups as the
iron-binding moieties, and have higher affinity constants than do the
hydroxamate siderophores (Ong et al. 1979, Powell et al. 1982).
Agrobactin, a typical catechol siderophore (Figure 4), is produced by
A. tumefaciens under iron-limiting conditions (Ong et al. 1979).
Pseudobactin, which is produced by a plant-growth-promoting fluorescent pseudomonad (strain B10) is a mixed-function siderophore
(Teintze et al. 1981). Pseudobactin consists of a linear hexapeptide in
which the N^δ-OH nitrogen of the C-terminal ornithine is cyclized with
the C-terminal carboxyl group, and the N^e-amino group of the lysine
is linked via an amide bond to a fluorescent quinoline derivative. The
iron-chelating groups consist of a hydroxamate group derived from N^δ-
hydroxyornithine, an α-hydroxy acid derived from β-hydroxyaspartic
acid, and the o-dihydroxy aromatic group of the quinoline moiety
(Teintze et al. 1981).

One of the major mechanisms by which biological disease-control organisms exert their effects is the production of siderophores to deny
iron to plant pathogens. Hydroxamate siderophores have been extracted from a wide variety of soils (Akers 1983, Powell et al. 1980,
1982). It has been suggested that they make iron more available to

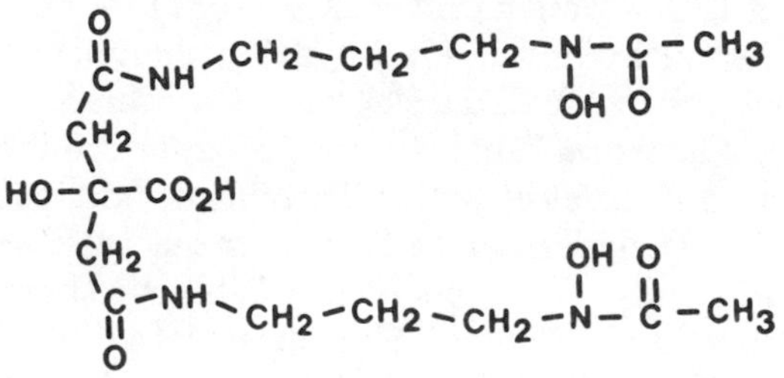

1

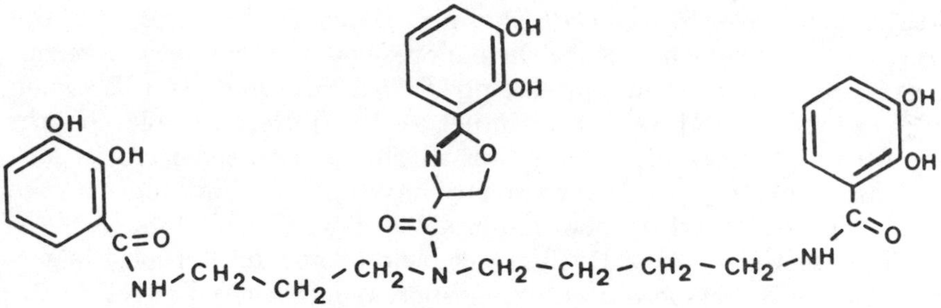

2

Figure 4   Several typical siderophores produced by microorganisms: (1) A hydroxamate siderophore, schizokinen. (2) A typical catechol siderophore, agrobactin.

plants by maintaining chelated iron in solution (Erich et al. 1987, Powell et al. 1980, 1982). Powell et al. (1982), using ^{55}Fe, demonstrated the uptake and translocation of ferrichrome and a number of ferrated fungal hydroxamate siderophores by monocot seedlings. Misaghi et al. (1982) extracted the water-soluble pigments from numerous fluorescent pseudomonads, and found the pigments to be inhibitory to *Geotrichum candidum*. They also isolated, from *Pseudomonas fluorescens*, a fluorescent, iron-binding pigment that inhibited *Pythium aphanidermatum*. However, the inhibition could be reversed by the addition of iron. Kloepper et al. (1980*b*) added cultures of *Pseudomonas* B10 and of its siderophore, pseudobactin, to soils conducive to take-all disease of wheat. These investigators observed that either amendment rendered the soils suppressive, but that the soils returned to their original conducive state with the addition of iron. Lowering the pH of suppressive soils eliminated the suppressive effect, probably due to the increased availability of iron at lower pH values (Scher and Baker 1980). Scher and Baker (1982) found that the addition of EDDHA (ethylenediaminedi-o-hydroxyphenylacetic acid), FeEDDHA (the ferrated form of EDDHA), or a *Pseudomonas putida* strain isolated from a *Fusarium*-suppressive soil, to a conducive soil renders it suppressive to *Fusarium* wilt pathogens of flax, cucumber, and radish. Because FeEDDHA is already complexed with iron, it is unable to chelate additional iron, and thus unable to suppress the pathogenic fungi. Scher and Baker (1982) postulated that in the rhizosphere, ferrated EDDHA first releases iron to the plant and then complexes additional iron, making the iron unavailable to the pathogen. The work of Elad and Baker (1985a) supports this hypothesis. EDDHA inhibited the germination of *Fusarium oxysporum* chlamydospores in bulk soil, but FeEDDHA did not. However, both chelators inhibited the germination of *F. oxysporum* chlamydospores in the cucumber rhizosphere (Elad and Baker 1985a). The optimum iron (Fe^{3+}) concentrations for the suppression of *F. oxysporum* germination by a siderophore-producing *P. putida* isolate are between $10^{-22}M$ and $10^{-27}M$ Fe^{3+} (Simeoni et al. 1987). Because siderophores are 10 to 50 times more abundant in rhizosphere soil than in surrounding soil, they would presumably have a marked influence on the composition of the rhizosphere microflora (Powell et al. 1982).

Siderophores produced by fluorescent pseudomonads inhibit the germination of *F. oxysporum* chlamydospores, but have no effect on the germination of *Fusarium solani* (Elad and Baker 1985b). The chlamydospores of *F. solani* are much larger than those of *F. oxysporum* and therefore less likely to be susceptible to iron starvation, as they would require less exogenous iron to germinate.

A number of studies have shown that siderophore-minus mutants lose the ability to inhibit plant pathogens (Elad and Baker 1985b, Loper 1988, Simeoni et al. 1987). A *P. fluorescens* strain isolated from a cotton rhizosphere soil protected cotton from preemergence damping-off caused by *Pythium ultimum* (Loper 1988). Fourteen nonfluorescent *Tn5* insertion mutants of the *P. fluorescens* isolate had no effect on either the colonization of the cotton roots by *P. ultimum* or the preemergence damping-off of the cotton seedlings. All these mutants were found to be equivalent to the parent strain in their ability to colonize cotton roots (Loper 1988). Since the lack of siderophore production by the mutants is the only major difference between the mutants and the parent strain, that study provides strong evidence for the role of siderophores in biological control of root diseases.

Buyer and Leong (1986) found that fluorescent pseudomonads that enhanced the growth of bean plants were able to inhibit many of the fluorescent pseudomonads that were detrimental to bean plants. They also noted that those of the detrimental strains that were inhibited by the beneficial strains could not use iron chelated by the siderophores produced by the latter; those that were not inhibited could.

Recently, Gill and Warren (1988) isolated two genetically distinct iron-binding compounds from a plant-growth-promoting fluorescent pseudomonad. One of the compounds was a classical siderophore; the other had a high affinity for iron, but was not involved in iron nutrition. The latter compound was antagonistic to several phytopathogenic fungi at low iron concentrations. It appears to function in a fashion analogous to that of the antimicrobial agents released by other microorganisms.

Other mechanisms

There have been proposed a number of alternate explanations for the increased plant growth and resistance to disease observed in plants inoculated with biocontrol bacteria. The production of potent extracellular lytic enzymes, which are capable of destroying the integrity of fungal cell walls, has been suggested as a possible disease-control mechanism (Mitchell and Alexander 1961, 1963, Mitchell and Hurwitz 1965). Mitchell and Alexander (1961) found that the ability of bacteria to lyse the hyphal walls of a *Fusarium* sp. was correlated with chitinase activity, and that *Pythium debaryanum*, known to have chitin-deficient walls, was resistant to the chitinolytic bacteria (Mitchell and Alexander 1963). Ordentlich et al. (1988) found that the hyphae of *Sclerotium rolfsii* can be degraded by either an isolate of *Serratia marcescens* or a cell-free *Serratia marcescens* extract containing chitinolytic activity. The addition of chitin to soil has been shown to increase the numbers of bacteria and actinomycetes in the soil (Peterson et al. 1965). The addition of chitin and an inoculum of myxobacteria increased only the bacterial numbers. The amount of chitin required to alter the composition of soil microflora would probably render direct application to the soil impractical. Relatively small quantities of chitin, combined with a chitinolytic inoculum, may encourage the establishment of a self-sustaining chitinolytic population. Hocking and Cook (1972) isolated several *Cytophaga* strains that both colonized the roots of four species of conifer and protected the conifer seedlings from fungal damping-off pathogens. These isolates produce very active proteolytic and chitinolytic enzymes. However, a related *Cytophaga* isolate also produces a potent phenazine-type antibiotic (Peterson et al. 1966), and thus antibiosis may also play a role in the observed disease control.

It has been suggested that plant growth can be inhibited by both parasitic and nonparasitic microbes not recognized as plant pathogens (Elliott and Lynch 1984, Suslow and Schroth 1982a). Suslow and

Schroth (1982a) isolated from the rhizosphere of sugar beets a number of bacteria that inhibited the growth of sugar beet seedlings and enhanced the colonization of sugar beet roots by pathogenic fungi. When sugar beets were coinoculated with the deleterious rhizobacteria and plant-growth-promoting rhizobacteria, the former were excluded from the rhizosphere (Suslow and Schroth 1982a). These workers suggested that one of the mechanisms of biological disease control is the exclusion of minor pathogens from the rhizosphere by the disease-control bacteria. Broadbent et al. (1977) obtained yield increases after inoculating sorghum and cabbage with a *Bacillus subtilis* strain A13. They suggested that the growth response was a result of the exclusion of deleterious rhizobacteria by the exogenous strain. It is likely that the deleterious rhizosphere bacteria inhibit plant growth by producing toxic metabolites. Schippers et al. (1987) suggested that one of the major groups of deleterious bacteria are the HCN-producing pseudomonads. Because iron is necessary for HCN production by these microorganisms, they also suggested that plant-growth-promoting bacteria exert their beneficial effect by competing for iron with the deleterious bacteria.

Nonpathogenic strains of pathogenic fungi have been shown to prevent infection by the related pathogen (Ichielevich-Auster et al. 1985, Paulitz et al. 1987). Paulitz et al. (1987) found that nonpathogenic strains of *Fusarium oxysporum* prevented the infection of cucumber plants with *F. oxysporum* f.sp. *cucumerinum*. It was suggested that the nonpathogenic isolates compete with the pathogens for similar ecological niches in soil. However, Ichielevich-Auster et al. (1985) suggested that an induced-resistance mechanism is responsible for the suppression of damping-off of cotton, radish, and wheat seedlings by nonpathogenic isolates of the damping-off fungi.

Resistance to infection by a particular pathogen can be conferred upon a plant by treating it with a nonpathogenic microorganism related to the pathogen (Gessler and Kuc 1982, Ichielevich-Auster 1985) or a culture filtrate of the pathogen (Tjamos 1979). This phenomenon is called "induced resistance." There is, to date, no evidence indicating that resistance to fungal pathogens can be induced by bacterial inocula.

The inhibition of fungal germination and growth by volatile compounds in the soil "atmosphere" has occasionally been observed (Lockwood 1977). These compounds probably result from the metabolic activities of soil bacteria. Ethylene, which inhibits the germination of fungal spores, has frequently been detected in soil atmospheres (Lockwood 1977). However, after measuring the concentrations of various soil gases, Pavlica et al. (1978) concluded that ammonia is the only gas present in sufficient concentrations to inhibit soil fungi. The production of volatile inhibitors by disease-control bacteria is therefore probably not a significant mechanism in the control of root pathogens.

A large number of soil bacteria produce plant-growth-promoting compounds, such as indole acetic acid (Brown 1972, 1974, Tien et al. 1979), gibberellin, and cytokinin-like substances (Tien et al. 1979). As the production of these compounds has been demonstrated in culture media only, the importance of this mechanism for the promotion of plant growth has yet to be determined (Gaskins et al. 1984).

Another potential biocontrol mechanism is competition among the disease-control bacteria and pathogenic fungi for nutrients. Competition for available iron has been discussed in the section on siderophores. Elad and Chet (1987) found a correlation between the ability of various bacterial strains to compete for nutrients with germinating oospores of *Pythium aphanidermatum*, and their efficiency as biocontrol agents against damping-off caused by *P. aphanidermatum*. Maurer and Baker (1964) found that the addition of chitin residues to soil inhibits the development of bean root rot. Benson and Baker (1970) suggested that the low C/N ratio of chitin would cause additional carbon to be immobilized, in which case carbon would become the limiting nutrient, and competition for simple carbon substrates would be responsible for the disease suppression observed when chitin is added to the soil. However, the addition of chitin would also stimulate chitinolytic bacteria, which have been shown to have disease-control properties. It is very likely that more than one mechanism is involved in pathogen suppression when bacterial inoculants are used for control of root diseases. Pathogenic fungi that are able to grow saprophytically in soil can colonize nutrient-poor zones in soil by hyphal extension, relying on the nutrient reserves of the organic matter that was colonized initially (Stack and Miller 1985). Biological disease-control bacteria may also contribute to plant growth by rendering more soil nutrients available to the plant (Barea et al. 1976, Brown 1974, Lifshitz et al. 1986, 1987). Any activity that contributes to the general health of a plant (e.g., production of growth hormones, rendering nutrients available to the plant) will also decrease the susceptibility of that plant to disease. A number of studies have examined for potential as bacterial inoculants, root-colonizing bacteria that solubilize phosphate (Brown 1974, Lifshitz et al. 1987) and nonsymbiotically fix nitrogen (Brown 1974, Lifshitz et al. 1986). The contribution made to biological control of root diseases by these indirect mechanisms is still uncertain.

Legumes have shown improved nodulation after having been co-inoculated with biological disease-control bacteria and symbiotic nitrogen-fixing bacteria (Grimes and Mount 1984, Polonenko et al.

1987). Polonenko et al. (1987) examined co-inoculation of soybeans with 18 root-colonizing bacterial strains, of which 17 were inhibitory to *Bradyrhizobium japonicum* in vitro. In a soil-perlite mixture, however, these bacteria had either no effect or, in the case of six strains, increased the total nodule mass on the plant roots. The results of this study are typical of the difficulty encountered in extrapolating in vitro results to soil. Because the inclusion of root-colonizing bacteria with legume inoculants is a recent innovation, the future possibilities of this practice have yet to be determined.

Role of Root Colonization in Biological Disease Control

The ability of an introduced disease-control organism to colonize the rhizosphere of the host plant is a major requirement for the biological control of root disease (Lifshitz et al. 1986, Loper et al. 1984, 1985, Weller 1983). It does not follow, however, that every good root colonizer is an effective disease-control organism (Randhawa and Schaad 1985). In order to colonize a plant root, a microorganism must (1) move to the appropriate location on the root, (2) adhere to the root, (3) utilize the available substrates, (4) compete with the indigenous microflora in the soil, and (5) multiply on the root.

A disease-control bacterium added as a seed inoculum must establish itself in the spermosphere of the germinating seed prior to becoming established on the emerging radicle (Kloepper et al. 1985). Motility and chemotaxis also help bacteria to become established on the elongating root (Scher et al. 1985, Schmidt 1979). Scher et al. (1985) found that a number of fluorescent pseudomonads exhibited chemotaxis toward soybean root exudates. They observed chemotaxis toward the amino acids in the exudates, but not toward the sugars in the exudates. The dispersal of bacteria along plant roots can also be caused by the flow of water through the soil (Chao et al. 1986, Parke et al. 1986). Howie et al. (1987) studied, at various levels of soil matric potential, the colonization of wheat roots suppressive to take-all. Active colonization along the developing root was observed at moisture levels at which there occurred no downward movement of water. Nonmotile mutants colonized the wheat roots as readily as the motile parent strain (Howie et al. 1987). As downward transport in water flow and motility were not significant mechanisms of dispersal for the bacteria, Howie et al. (1987) suggested the bacteria were dispersed along the root by passive transport with root extension through the soil. Exudates from fungal propagules may also act as attractants for bacteria in soil (Lim and Lockwood 1988). Microbial activity in the vicinity of fungal propagules may increase nutrient stress, thereby reducing propagule viability and virulence (Lim and Lockwood 1988). Anderson et al. (1988) suggested that binding of the biocontrol organism to root-surface glycoproteins is a critical step in root colonization. They found that a root-colonizing isolate of *P. putida* could be agglutinated by a root-surface glycoprotein, whereas agglutination-minus mutants of the same isolate were less effective root colonizers than the parent strain (Anderson et al. 1988, Tari and Anderson 1988). James et al. (1985) found there to be no correlation between the hydrophobicity of various bacterial strains and their ability to bind to sterile radish roots. Divalent cations enhanced the binding of a *P. fluorescens* isolate to radish roots, suggesting that ionic interactions are more important in binding than hydrophobic interactions (James et al. 1985). Although short-term binding does not necessarily result in long-term colonization of the root, it is clear that effective root-colonizing bacteria possess mechanisms for initial binding. Antibiosis has been shown to affect the initial colonization of potato tubers by plant-pathogenic *Erwinia* spp. (Axelrood et al. 1988). *Erwinia carotovora* subsp. *betavasculorum* strains, which produced a bacteriocidal antibiotic in vitro, were able to displace competing, sensitive *Erwinia carotovora* subsp. *carotovora* strains in the infection court of potato (Axelrood et al. 1988).

The quantity and nature of the root exudates determines the composition of the rhizosphere microflora (Bowen and Rovira 1969, 1976, Brown 1976). The quality and composition of root exudates, and sloughed-off root material, is governed by environmental conditions, plant genotype, and stage of growth (Rovira 1965, Rovira and McDougall 1967). It has been demonstrated that the numbers and types of rhizosphere microorganisms are influenced by genetic variations in the host (Atkinson et al. 1975, Bird 1982, Bowen and Rovira 1976, Elkan 1962). Single-chromosome substitutions in the host plant can induce in the rhizosphere microflora of wheat alterations that increase the numbers of nitrogen-fixing bacilli (Neal and Larson 1976) and bacteria antagonistic to a root pathogen (Neal et al. 1970, 1973). Root exudates can also influence a plant's susceptibility to disease. Cotton seedlings subjected to cold stress exhibited increased root education and enhanced susceptibility to both *Verticillium* wilt and *Rhizoctonia* root rot (Shao and Christiansen 1982).

Various selective media have been used to isolate and enumerate various genera of root-colonizing bacteria. For example, the fluorescent pseudomonads can be isolated using detergent-based (Kado and Heskett 1970) and antibiotic-based selective media (Sands and Rovira 1970, 1971, Simon et al. 1973). Desoxycholate combined with various defined carbon sources (Katoh and Itoh 1983), and selective media combining detergents and antibiotics (Burr and Katz 1982, Gould et al. 1985), have both been used to isolate fluorescent pseudomonads. A number of studies have sought to evaluate the effects of soil type and other variables on the colonization of plant roots by specific disease-control isolates marked with antibiotic resistance (Dupler and Baker 1984, Frederickson and Elliott 1985, Loper et al. 1984, Weller 1983, Yuen and Schroth 1986, Yuen et al. 1986). Antibiotic resistance in marked strains has the advantage of allowing their enumeration and recovery in the presence of other bacteria. Antibiotic-resistant variants have been successfully used in the study of *Rhizobium* ecology, and also as biological tracers for the study of water movement (Hagedorn 1986).

The rhizosphere microflora is dominated by Gram-negative bacteria (Rovira and McDougall 1967, Rouatt and Katznelson 1961, Scher et al. 1984b), in particular by fluorescent pseudomonads (Scher et al. 1984b, Vancura 1980). Due to their ability to colonize the plant root, the fluorescent pseudomonads have frequently been considered as biological control agents against various root diseases (Anderson and Guerra 1985, Howell and Stipanovic 1979, 1980, Weller 1984, Weller and Cook 1983). The ability of an introduced organism to colonize a root is less dependent on plant type than on the ability of the organism to survive in a given soil and compete with the indigenous microflora (Dupler and Baker 1984, Nesmith and Jenkins 1985, Weller 1983). The following soil factors have been shown to influence root colonization: (1) soil texture and soil-water matric potential (Dupler and Baker 1984, Nesmith and Jenkins 1985), (2) soil nutrient levels and soil pH (Turner et al. 1985, West et al. 1985, Yuen et al. 1985), (3) soil calcium levels (Kao and Ko 1986a,b), and (4) composition of the indigenous soil bacterial populations (Bashan 1986, Chao et al. 1986). After 4 months of examining the survival of a disease-control bacterial strain on carnation roots in several soil types, Yuen et al. (1986) found that some soils supported significant rhizosphere populations of the strain, and others did not. In view of the major role that soil properties exert on the colonization of roots by specific bacterial strains, it may be necessary to develop different bacterial inocula for various soil types or geographical regions.

Development of Biological Disease-Control Agents

Although many examples of the successful use of disease-control organisms for root diseases have been published in the scientific literature, there has been little commercial development of these organisms. Of the commercially successful, two excellent examples are the biological control of crown gall on fruit trees and the control of *Heterobasidium annosum* infection of pine trees (Deacon 1983). Currently, a *Pseudomonas*-based biofungicide called "Dagger G" is being marketed for the prevention of the damping-off of cotton seedlings.

By choosing the appropriate system for biological disease control, one can improve the chances of success. Biological control is most likely to control a disease if the plant to be protected is susceptible only for a short time, and if the infection court of the plant can be readily colonized by an introduced strain. Among the most likely candidates for biological control are the damping-off diseases. Seedlings are generally susceptible to damping-off during the first 4 weeks of growth, and the infection is usually restricted to the vicinity of the infection court.

The screening of several thousand isolates in a selection program demands an efficient and rapid selection technique. Both in vitro and *in planta* assays have been developed. In vitro screening techniques usually include assays for siderophore (Kloepper et al. 1980c, Vandenburgh et al. 1983) and antibiotic (Kloepper and Schroth 1981c) production. Methods have been developed to measure spermosphere (Kloepper et al. 1985) and rhizosphere (Juhnke et al. 1987) colonization. Juhnke et al. (1987) suggested that the most effective biocontrol agents would be obtained by isolating effective root colonizers, and genetically introducing either antibiotic or siderophore production into the bacterial isolates. As the characteristics governing root-colonizing ability are complex and poorly understood, it would be premature to attempt to genetically engineer good root colonizers. Xu and Gross (1986a) screened 293 fluorescent pseudomonads for in vitro antibiotic and siderophore production, after which they assayed them for *in planta* disease control in the greenhouse. They used King's B medium (King et al. 1954) for the siderophore assay, as it is low in iron, and potato-dextrose agar for the determination of antibiotic production, as it is high in iron and low in phosphate. Media with high iron and low phosphate concentrations have been shown to encourage antibiotic production (Gross 1985). Xu and Gross (1986a) found that because in vitro antagonism does not always correlate well with *in planta* assays, a greenhouse screen is necessary for the selection of control agents. Kennerley and Stack (1987) used three techniques to assess fungal antagonists of the sclerotium-forming fungus *Phymatotrichum ommivorum*. Each technique yielded somewhat different results. In order to overcome the inconsistencies of bioassays, Randhawa and Schaad (1985) developed seedling bioassay chambers, which allowed them to determine the levels of both root colonization and antagonism between the biocontrol organism and the pathogen. In general, the use of in vitro bioassays on plates to obtain biocontrol organisms is less effective than simple *in planta* assays in the greenhouse (Burr et al. 1978, Xu and Gross 1986*a,b*).

Xu and Gross (1986*b*) examined a number of fluorescent pseudomonads marked with antibiotic resistance for two sets of activities. The first set was root colonization and inhibition of potato soft rot caused by *Erwinia caratovora*; the second, the in vitro production of antibiotics and siderophores in plate assays. They found no correlation between the two.

Suslow and Schroth (1982*b*) evaluated selected strains (strains SH5 and B4) of fluorescent pseudomonads as plant-growth-promoting inoculants for sugar beets. They found each strain was most effective in a particular geographical region. One strain (SH5) increased yields in California soils, but was unsuccessful in Idaho soils. The opposite was found to be true for strain B4 (Suslow and Schroth 1982*b*).

A biological disease-control product is likely to be successful if (1) it contains enough viable microorganisms to provide sufficient inoculum, (2) it has an adequate shelf-life, (3) it is convenient to handle and apply in the field (i.e., it is compatible with standard agricultural equipment), and (4) it is economically attractive. Bacterial or fungal inoculants can be prepared most easily by liquid fermentation (Papavizas et al. 1984). The inoculant can be applied as a liquid slurry directly to the seed (Harman et al. 1980, Kloepper and Schroth 1981*a*), or added to a solid carrier and applied in a powdered or granular form (Backman and Rodriguez-Kabana 1975). Solid carriers generally perform better than liquid slurries in the field, and have a longer shelf life. Windels (1981) applied conidia of *Penicillium oxalicum* directly to pea seeds. The fungal conidia germinated and coated the pea seeds, but did not colonize the roots. It should be noted that some techniques that are effective in the field may not be commercially feasible. The most commonly used technique for preparing a biocontrol inoculum comprises (1) the addition of a bacterial suspension to a powdered or granular carrier, (2) the addition of nutrients to allow proliferation of the biocontrol agent, and (3) the addition of an adhesive to bind the inoculant to the seed.

A wide variety of solid carriers have been used, such as diatomaceous earth (Backman and Rodriguez-Kabana 1975), talc (Kloepper and Schroth 1981*a*), clay (Campbell and Ephgrave 1983), calcium alginate (Fravel et al. 1985), lignite (Jones et al. 1984), and peat (Suslow and Schroth 1982*b*). Campbell and Ephgrave (1983) studied the effect of clay on the interactions between *Gaeumannomyces graminis* and two biocontrol bacteria, using a single bacterial strain for each experiment. They found that the presence of the clay enhanced the growth of the pathogen, and suggested that the clay binds some of the bacterial toxins, thus protecting the fungus. Calcium alginate–clay mixtures were found to be satisfactory carriers for a number of fungal biocontrol agents, although one *Pseudomonas cepacia* isolate did not survive the gelling procedure (Campbell and Ephgrave 1983). Peat has been a satisfactory carrier of *Rhizobium* inoculants for many years (Davidson and Reuszer 1978, Roughley and Vincent 1967).

The addition of excess nutrients to a biocontrol inoculant can potentially benefit the pathogen (Backman 1978). In an attempt to control *Phytopthora cinnamomi*, Kelley (1976) prepared a biocontrol inoculant of *Trichoderma harzianum* in clay granules containing additional nutrients. He found that the biocontrol agent increased, rather than decreased, the activity of the pathogen, particularly at high moisture levels. Kelley (1976) suggested that nutrients may have been leached out of the clay granules and used by the pathogen.

Xanthan gums (Kloepper and Schroth 1981*a*) and methylcellulose (Harman et al. 1980, Suslow and Schroth 1982*b*) have frequently been used as adhesives. Xanthan gums enhance both the adhesion of the inoculant to the seed and the survival of the bacteria (Kloepper and Schroth 1981*a*).

Henis et al. (1978) studied the integrated control of damping-off caused by *Rhizoctonia solani*, using a combination of *Trichoderma harzianum* and pentachloronitrobenzene (PCNB). They found the combination of the fungicide and the biocontrol agent the most effective means of suppression. However, fungicides may suppress certain pathogens while allowing others to proliferate (Gibson et al. 1961), and may suppress microorganisms that are inhibitory to the pathogen (Gibson et al. 1961, Henis et al. 1978).

Conclusions and Future Possibilities

Because bacterial inoculants have less impact on the environment than chemical control agents do, the future of the biological control of root diseases is very promising. Although considerable research into this area has been carried out, commercial development has only recently been initiated.

It has been shown that antibiosis and the production of siderophores by the biocontrol organism are two of the major mechanisms in the biological control of root diseases. There exists indirect evidence that lytic enzymes, the production of growth hormones, the release of soil nutrients to the plant, and the reduction of nonspecific pathogens by biocontrol organisms all have a role in biological disease control. Disease-control organisms have been used successfully in acidic soils, which presumably contain enough iron to protect the pathogens from the effects of siderophore production. Other mechanisms must therefore be involved in these soils.

Soil properties have been shown to have a major effect on the ability of biocontrol agents to colonize roots. The nature of the plant and the environmental conditions affcct the quantity and composition of the root exudates, which in turn determine the population of the rhizosphere microflora. The inconsistent performance of disease-control agents in different geographical regions is probably due to the effect of soil type on root colonization.

There is a critical need for the development of a production and delivery system for biological disease-control agents. The most satisfactory procedure for the production of a bacterial inoculant involves liquid fermentation to produce a cell suspension, followed by the addition of the resulting liquid to neutralized peat. The peat-based inoculant can then be applied either as a seed coating, or in furrow during planting.

Genetic manipulation of biocontrol bacteria has the potential both to improve their effectiveness and to increase the range of crops and soils in which they can be applied. Genes controlling siderophore (Magazin et al. 1986, Marugg et al. 1985, 1988) and antibiotic (Gutterson et al. 1986) synthesis in several organisms have been identified. *Pseudomonas fluorescens* strains that overproduce phenylpyrrole antibiotics have been developed by the selection of analogue-resistant mutants (Elander et al. 1968). Shim et al. (1987) created transfer-deficient *Tn5* insertion mutants of *Agrobacterium radiobacter* in order to obtain strains with a lower probability of transferring resistance to the pathogen. However, the improvement of other antibiotic-producing strains by molecular biology techniques may be more difficult since many of the antibiotics are secondary metabolites, and their biosynthetic pathways are complex. Because genes for root-colonizing ability have not yet been identified for any of the biocontrol bacteria, one is restricted to isolating efficient colonizers and cloning other genes of interest into these isolates (Juhnke 1987).

Further research into the following areas of the biological control of plant root diseases is required: (1) elucidation of biological control mechanisms in a variety of differing soil types; (2) identification of environmental parameters that affect root colonization; (3) development of improved strain-selection procedures and carrier systems; (4) identification of key genes coding for siderophore production, antibiotic production, and root colonization; and (5) genetic modification of naturally occurring strains to produce superior biocontrol agents. If undertaken, this research will serve to increase the scope of the biological control of root diseases.

Acknowledgment

The editorial assistance of Ms. D. Limoges is gratefully acknowledged.

References

Akers, H. A. 1983. Multiple hydroxamic acid microbial iron chelators (siderophores) in soils, *Soil Sci.* 135:156–159.

Alconero, R. 1980. Crown gall of peaches from Maryland, South Carolina, and Tennessee and problems with biological control, *Plant Dis.* 64:835–838.

Anderson, A. J., and Guerra, D. 1985. Responses of bean to root colonization with *Pseudomonas putida* in a hydroponic system, *Phytopathology* 75:992–995.

Anderson, A. J., Habibzadegah-Tari, P., and Tepper, C. S. 1988. Molecular studies on the role of a root surface agglutinin in adherence and colonization by *Pseudomonas putida*, *Appl. Environ. Microbiol.* 54:375–380.

Asher, M. J. C. 1978. Interactions between isolates of *Gaeumannomyces graminis*, var. *tritici. Trans. Br. Mycol. Soc.* 71:367–373.

Atkinson, T. G., Neal, J. L., Jr., and Larson, R. I. 1975. Genetic control of the rhizosphere microflora of wheat. In *Biology and Control of Soil-Borne Plant Pathogens* D. W. Bruehl (ed.). American Phytopathological Society, St. Paul, Minnesota, pp. 116–122.

Axelrood, P. E., Rella, M., and Schroth, M. N. 1988. Role of antibiosis in competition of *Erwinia* strains in potato infection courts, *Appl. Environ. Microbiol.* 54:1222–1229.

Backman, P. A. 1978. Fungicide formulation: relationship to biological activity, *Annu. Rev. Phytopathol.* 16:211–237.

Backman, P. A., and Rodriguez-Kabana, R. 1975. A system for the growth and delivery of biological control agents to the soil, *Phytopathology* 65:819–821.

Baker, R., Hanchey, P., and Dottarar, S. D. 1978. Protection of carnation against fusarium stem rot by fungi, *Phytopathology* 68:1495–1501.

Barea, J. M., Navarro, E., and Montoya, E. 1976. Production of plant growth regulators by rhizosphere phosphate-solubilizing bacteria, *J. Appl. Bacteriol.* 40:129–134.

Bashan, Y. 1986. Enhancement of wheat root colonization and plant development by *Azospirillum brasilense* Cd. following temporary depression of rhizosphere microflora, *Appl. Environ. Microbiol.* 51:1067–1070.

Batra, S. W. T. 1982. Biological control in agroecosystems, *Science* 215:134–139.

Benson, D. M., and Baker, R. 1970. Rhizosphere competition in model soil systems, *Phytopathology* 60:1058–1061.

Bird, L. S. 1982. The MAR (Multi-Adversity Resistance) system for genetic improvement of cotton, *Plant Dis.* 66:172–176.

Bowen, G. D., and Rovira, A. D. 1969. The influence of microorganisms on growth and metabolism of plant roots. In *Root Growth*, Proceedings of the 15th Easter School in Agriculture Science, W. J. Whittington (ed.). Butterworths, London, pp. 170–201.

Bowen, G. D., and Rovira, A. D. 1976. Microbial colonization of plant roots, *Annu. Rev. Phytopathol.* 14:121–144.

Brathwaite, C. W. D., and Cunningham, H. G. A. 1982. Inhibition of *Sclerotium rolfsii* by *Pseudomonas aeruginosa*, *Can. J. Bot.* 60:237–239.

Brisbane, P. G., Janik, L. J., Tate, M. E., and Warren, R. F. O. 1987. Revised structure for the phenazine antibiotic from *Pseudomonas fluorescens* 2-79 (NRRL B-15132), *Antimicrob. Agents Chemother.* 31:1967–1971.

Broadbent, P., Baker, K. F., Franks, N., and Holland, J. 1977. Effect of *Bacillus* spp. on increased growth of seedlings in steamed and in non-treated soil, *Phytopathology* 67:1027–1034.

Brown, M. E. 1972. Plant growth substances produced by micro-organisms of soil and rhizosphere, *J. Appl. Bacteriol.* 35:443–451.

Brown, M. E. 1974. Seed and root bacterization, *Annu. Rev. Phytopathol.* 12:181–197.

Brown, M. E. 1976. Rhizosphere microorganisms—opportunists, bandits or benefactors. In *Soil Microbiology*, Walker (ed.). Wiley, New York and Toronto, p. 262.

Burr, T. J., and Katz, B. 1982. Evaluation of a selective medium for detecting *Pseudomonas syringae* pv *papulans* and *P. syringae* pv *syringae* in apple orchards, *Phytopathology* 72:564–567.

Burr, T. J., Schroth, M. N., and Suslow, T. 1978. Increased potato yields by treatment of seedpieces with specific strains of *Pseudomonas fluorescens* and *P. putida, Phytopathology* 68:1377–1383.

Buyer, J. S., and Leong, J. 1986. Iron transport-mediated antagonism between plant growth-promoting and plant-deleterious *Pseudomonas* strains, *J. Biol. Chem.* 261:791–794.

Campbell, R., and Clor, A. 1985. Soil moisture affects the interaction between *Gaeumannomyces graminis* var. *tritici* and antagonistic bacteria, *Soil Biol. Biochem.* 17:441–446.

Campbell, R., and Ephgrave, J. M. 1983. Effect of bentonite clay on the growth of *Gaeumannomyces graminis* var. *tritici* and on its interactions with antagonistic bacteria, *J. Gen. Microbiol.* 129:771–777.

Chang, C. J., Floss, H. D., Hook, D. J., Mabe, J. A., Manni, P. E., Martin, L. K., Schroder, K., and Shieh, T. L. 1981. The biosynthesis of the antibiotic pyrrolnitrin by *Pseudomonas aureofaciens, J. Antibiot.* 34:555–566.

Chang, I.-P., and Kommedahl, T. 1968. Biological control of seedling blight of corn by coating kernels with antagonistic microorganisms, *Phytopathology* 58:1395–1401.

Chao, W. L., Nelson, E. B., Harman, G. E., and Hoch, H. C. 1986. Colonization of the rhizosphere by biological control agents applied to seeds, *Phytopathology* 76:60–65.

Charigkapakorn, N., and Sivasithamparam, K. 1987. Changes in the composition and population of fluorescent pseudomonads on wheat roots inoculated with successive generations of root-piece inoculum of the take-all fungus, *Phytopathology* 77:1002–1007.

Cline, G. R., Powell, P. E., Szaniszlo, P. J., and Reid, C. P. P. 1982. Comparison of the abilities of hydroxamic, synthetic, and other natural organic acids to chelate iron and other ions in nutrient solution, *Soil Sci. Soc. Am. J.* 46:1158–1164.

Cline, G. R., Powell, P. E., Szaniszlo, P. J., and Reid, C. P. P. 1983. Comparison of the abilities of hydroxamic and other natural organic acids to chelate iron and other ions in soil, *Soil Sci.* 136:145–157.

Cody, Y. S., and Gross, D. G. 1987. Characterization of pyoverdin pss, the fluorescent siderophore produced by *Pseudomonas syringae* pv *syringae, Appl. Environ. Microbiol.* 53:928–934.

Cook, R. J., and Rovira, A. D. 1976. The role of bacteria in the biological control of *Gaeumannomyces graminis* by suppressive soils, *Soil Biol. Biochem.* 8:269–273.

Cooksey, D. A., and Moore, L. W. 1980. Biological control of crown gall with fungal and bacterial antagonists, *Phytopathology* 70:506–509.

Davidson, F., and Reuszer, H. W. 1978. Persistence of *Rhizobium japonicum* on the soybean seed coat under controlled temperature and humidity, *Appl. Environ. Microbiol.* 35:94–96.

de Weger, L. A., van Boxtel, R., van der Burg, B., Gruters, R. A., Geels, F. P., Schippers, B., and Lugtenberg, B. 1986. Siderophores and outer membrane proteins of antagonistic, plant growth-stimulating, root-colonizing *Pseudomonas* spp., *J. Bacteriol.* 165:585–594.

Deacon, J. W. 1973. Control of the take-all fungus by grass leys in intensive cereal cropping, *Plant Pathol.* 22:88–94.

Deacon, J. W. 1976. Biological control of the take-all fungus, *Gaeumannomyces graminis*, by *Phialophora radicola* and similar fungi, *Soil Biol. Biochem.* 8:275–283.

Deacon, J. W. 1983. *Microbial Control of Plant Pests and Diseases*, American Society for Microbiology, Washington, D.C.

Dupler, M., and Baker, R. 1984. Survival of *Pseudomonas putida*, a biological control agent, in soil, *Phytopathology* 74:195–200.

Dutta, B. K. 1981. Studies on some fungi isolated from the rhizosphere of tomato plants and the consequent prospect for the control of verticillium wilt, *Plant and Soil* 63:209–216.

Elad, Y., and Baker, R. 1985a. Influence of trace amounts of cations and siderophore-producing pseudomonads on chlamydospore germination of *Fusarium oxysporum*, *Phytopathology* 75:1047–1052.

Elad, Y., and Baker, R. 1985b. The role of competition for iron and carbon in suppression of chlamydospore germination of *Fusarium* spp. by *Pseudomonas* spp., *Phytopathology* 75:1053–1059.

Elad, Y., and Chet, I. 1987. Possible role of competition for nutrients in biocontrol of Pythium damping-off by bacteria, *Phytopathology* 77:190–195.

Elander, R. P., Mabe, J. A., Hamill, R. L., and Gorman, M. 1968. Biosynthesis of pyrrolnitrins by analogue-resistant mutants of *Pseudomonas fluorescens*, *Appl. Microbiol.* 16:156–165.

Elkan, G. H. 1962. Comparison of rhizosphere microorganisms of genetically related nodulating and non-nodulating soybean lines, *Can. J. Microbiol.* 8:79–87.

Elliott, L. F., and Lynch, J. M. 1984. Pseudomonads as a factor in the growth of winter wheat (*triticum aestivum* L.), *Soil Biol. Biochem.* 16:69–71.

Ellis, J. G., Kerr, A., van Montagu, M., and Schell, J. 1979. Agrobacterium: genetic studies on agrocin 84 production and the biological control of crown gall, *Physiol. Plant Pathol.* 15:215–223.

Emery, T. 1976. Fungal ornithine esterases: relationship to iron transport, *Biochemistry* 15:2723–2728.

Erich, M. S., Duxbury, J. M., Bouldin, D. R., and Cary, E. 1987. The influence of organic complexing agents on iron mobility in a simulated rhizosphere, *Soil Sci. Soc. Am. J.* 51:1207–1214.

Farrand, S. K., Slota, J. E., Shim, J.-S., and Kerr, A. 1985. TN5 insertions in the agrocin 84 plasmid: the conjugal nature of pAgK84 and the locations of determinants for transfer and agrocin 84 production, *Plasmid* 13:106–117.

Fravel, D. R., Marois, J. J., Lumsden, R. D., and Connick, W. J., Jr. 1985. Encapsulation of potential biocontrol agents in an alginate-clay matrix, *Phytopathology* 75:774–777.

Frederick, C. B., Szaniszlo, P. J., Vickrey, P. E., Bentley, M. D., and Shive, W. 1981. Production and isolation of siderophores from the soil fungus *Epicoccum purpurascens*, *Biochemistry* 20:2432–2436.

Frederickson, J. K., and Elliot, L. F. 1985. Colonization of winter wheat roots by inhibitory rhizobacteria, *Soil Sci. Soc. Am. J.* 49:1172–1177.

Gaines, C. G., Lodge, J. S., Arceneaux, J. E. L., and Byers, B. R. 1981. Ferrisiderophore reductase activity associated with an aromatic biosynthetic enzyme complex in *Bacillus subtilis*, *J. Bacteriol.* 148:527–533.

Gaskins, M. H., Albrecht, S. L., and Hubbell, D. H. 1985. Rhizosphere bacteria and their use to increase plant productivity: a review, *Agric. Ecosystems Environ.* 12:99–116.

Gessler, C., and Kuc, J. 1982. Induction of resistance to Fusarium wilt in cucumber by root and foliar pathogens, *Phytopathology* 72:1439–1441.

Gibson, F., and Magrath, D. I. 1969. The isolation and characterization of a hydroxamic acid (aerobactin) formed by Aerobacter aerogenes 62-1, *Biochim. Biophys. Acta* 192:175–184.

Gibson, I. A. S., Ledger, M., and Boehm, E. 1961. An anomalous effect of pentachloronitrobenzene on the incidence of damping-off caused by a *Phythium* sp., *Phytopathology* 51:531–533.

Gill, P. R., and Warren, G. J. 1988. An iron-antagonized fungistatic agent that is not required for iron assimilation from a fluorescent rhizosphere pseudomonad, *J. Bacteriol.* 170:163–170.

Gould, W. D., Hagedorn, C., Bardinelli, T. R., and Zablotowicz, R. M. 1985. New selective media for enumeration and recovery of fluorescent pseudomonads from various habitats, *Appl. Environ. Microbiol.* 49:28–32.

Grimes, H. D., and Mount, M. S. 1984. Influence of *Pseudomonas putida* on nodulation of *Phaseolus vulgaris*, *Soil Biol. Biochem.* 16:27–30.

Gross, D. C. 1985. Regulation of syringomycin synthesis in *Pseudomonas syringae* pv. *syringae* and defined conditions for its production, *J. Appl. Bacteriol.* 58:167–174.

Gurusiddaiah, S., Weller, D. M., Sarkar, A., and Cook, R. J. 1986. Characterization of an antibiotic produced by a strain of *Pseudomonas fluorescens* inhibitory to

Gaeumannomyces graminis var. *tritici* and *Pythium* spp., *Antimicrob. Agents Chemother.* 29:488–495.

Gutteridge, R. J., and Slope, D. B. 1978. Effect of inoculating soils with *Phialophora radicola* var. graminicola on take-all disease of wheat, *Plant Pathol.* 27:131–135.

Gutterson, N. I., Layton, T. J., Ziegle, J. S., and Warren, G. J. 1986. Molecular cloning of genetic determinants for inhibition of fungal growth by a fluorescent pseudomonad, *J. Bacteriol.* 165:696–703.

Hagedorn, C. 1986. Role of genetic variants in autecological research. In *Microbial Autecology: A Method for Environmental Studies*, R. L. Tate III (ed.). Wiley, New York, pp. 61–73.

Harman, G. E., Chet, I., and Baker, R. 1980. *Trichoderma hamatum* effects on seed and seedling disease induced in radish and pea by *Pythium* spp. or *Rhizoctonia solani*, *Phytopathology* 70:1167–1172.

Henis, Y., and Chet, I. 1975. Microbiological control of plant pathogens, *Adv. Appl. Microbiol.* 19:85–111.

Henis, Y., Ghaffar, A., and Baker, R. 1978. Integrated control of *Rhizoctonia solani* damping-off of radish: Effect of successive plantings, PCNB, and *Trichoderma harzianum* on pathogen and disease, *Phytopathology* 68:900–907.

Hocking, D., and Cook, F. D. 1972. Myxobacteria exert partial control of damping-off and root disease in container-grown tree seedlings, *Can. J. Microbiol.* 18:1557–1560.

Hohnadel, D., and Meyer, J.-M. 1988. Specificity of pyoverdine-mediated iron uptake among fluorescent *Pseudomonas* strains, *J. Bacteriol.* 170:4865–4873.

Howell, C. R., and Stipanovic, R. D. 1979. Control of *Rhizoctonia solani* on cotton seedlings with *Pseudomonas fluorescens* and with an antibiotic produced by this bacterium, *Phytopathology* 69:480–482.

Howell, C. R., and Stipanovic, R. D. 1980. Suppression of *Pythium ultimum*-induced damping-off of cotton seedlings by *Pseudomonas fluorescens* and its antibiotic, pyoleutorin, *Phytopathology* 70:712–715.

Howell, C. R., and Stipanovic, R. D. 1983. Gliovirin, a new antibiotic from *Gliocladium virens* and its role in the biological control of *Pythium ultimum*, *Can. J. Microbiol.* 29:321–324.

Howell, C. R., and Stipanovic, R. D. 1984. Phytotoxicity to crop plants and herbicidal effects on weeds of viridiol produced by *Gliocladium virens*, *Phytopathology* 74:1346–1349.

Howie, W. J., Cook, R. J., and Weller, D. M. 1987. Effects of soil matric potential and cell motility on wheat root colonization by fluorescent pseudomonads suppressive to take-all, *Phytopathology* 77:286–292.

Ichielevich-Auster, M., Sneh, B., Koltin, Y., and Barash, I. 1985. Suppression of damping-off caused by *Rhizoctonia* species by a nonpathogenic isolate of *R. solani*, *Phytopathology* 75:1080–1084.

Ingram, J. M., and Blackwood, A. C. 1970. Microbial production of phenazines, *Adv. Appl. Microbiol.* 13:267–282.

James, D. W., Jr., and Gutterson, N. I. 1986. Multiple antibiotics produced by *Pseudomonas fluorescens* HV37a and their differential regulation by glucose, *Appl. Environ. Microbiol.* 52:1183–1189.

James, D. W., Jr., Suslow, T. V., and Steinback, K. E. 1985. Relationship between rapid, firm adhesion and long-term colonization of roots by bacteria, *Appl. Environ. Microbiol.* 50:392–397.

Jones, R. W., Pettit, R. E., and Taber, R. A. 1984. Lignite and stillage: carrier and substrate for application of fungal biocontrol agents to the soil, *Phytopathology* 74:1167–1170.

Juhnke, M. E., Mathre, D. E., and Sands, D. C. 1987. Identification and characterization of rhizosphere-competent bacteria of wheat, *Appl. Environ. Microbiol.* 53:2793–2799.

Kado, C. I., and Heskett, M. G. 1970. Selective media for isolation of *Agrobacterium*, *Corynebacterium*, *Erwinia*, *Pseudomonas*, and *Xanthomonas*, *Phytopathology* 60:969–976.

Kao, C. W., and Ko, W. H. 1986a. Suppression of *Pythium splendens* in a Hawaiian soil by calcium and microorganisms, *Phytopathology* 76:215–220.

Kao, C. W., and Ko, W. H. 1986b. The role of calcium and microorganisms in suppression of cucumber damping-off caused by *Pythium splendens* in a Hawaiian soil, *Phytopathology* 76:221–225.

Katoh, K., and Itoh, K. 1983. New selective media for *Pseudomonas* strains producing fluorescent pigment, *Soil Sci. Plant Nutr.* 29:525–532.

Kelley, W. D. 1976. Evaluation of *Trichoderma harzianum*—impregnated clay granules as a biocontrol for damping-off of pine seedlings caused by *Phytophthora cinnamomi*, *Phytopathology* 66:1023–1027.

Kenerley, C. M., and Stack, J. P. 1987. Influence of assessment methods on selection of fungal antagonists of the sclerotium-forming fungus *Phymatotrichum omnivorum*, *Can. J. Microbiol.* 33:632–635.

Kerr, A., and Htay, K. 1974. Biological control of crown gall through bacteriocin production, *Physiol. Plant Pathol.* 4:37–44.

King, E. O., Ward, M. K., and Raney, D. E. 1954. Two simple media for the demonstration of pyocyanin and fluorescin, *J. Lab. Clin. Med.* 44:301–307.

Kloepper, J. W. 1983. Effect of seed piece inoculation with plant growth-promoting rhizobacteria on populations of *Erwinia carotovora* on potato roots and in daughter tubers, *Phytopathology* 73:217–219.

Kloepper, J. W., and Schroth, M. N. 1981a. Development of a powder formulation of rhizobacteria for inoculation of potato seed pieces, *Phytopathol.* 71:590–592.

Kloepper, J. W., and Schroth, M. N. 1981b. Plant growth-promoting rhizobacteria and plant growth under gnotobiotic conditions, *Phytopathology* 71:642–644.

Kloepper, J. W., and Schroth, M. N. 1981c. Relationship of in vitro antibiosis of plant growth-promoting rhizobacteria to plant growth and the displacement of root microflora, *Phytopathology* 71:1020–1024.

Kloepper, J. W., Schroth, M. N., and Miller, T. D. 1980a. Effects of rhizosphere colonization by plant growth-promoting rhizobacteria on potato plant development and yield, *Phytopathology* 70:1078–1082.

Kloepper, J. W., Leong, J., Teintze, M., and Schroth, M. N. 1980b. *Pseudomonas* siderophores: a mechanism explaining disease-suppressive soils, *Curr. Microbiol.* 4:317–320.

Kloepper, J. W., Leong, J., Teintze, M., and Schroth, M. N. 1980c. Enhanced plant growth by siderophores produced by plant growth-promoting rhizobacteria, *Nature* 286:885–886.

Kloepper, J. W., Scher, F. M., Laliberté, M., and Zaleska, I. 1985. Measuring the spermosphere colonizing capacity (spermosphere competence) of bacterial inoculants, *Can. J. Microbiol.* 31:926–929.

Kommedahl, T., and Mew, I.-P. C. 1975. Biocontrol of corn root infection in the field by seed treatment with antagonists, *Phytopathology* 65:296–300.

Kommedahl, T., and Windels, C. E. 1978. Evaluation of biological seed treatment for controlling root diseases of pea, *Phytopathology* 68:1087–1095.

Leisinger, T., and Margraff, R. 1979. Secondary metabolites of the fluorescent pseudomonads, *Microbiol. Rev.* 43:422–442.

Leong, S. A., and Neilands, J. B. 1982. Siderophore production by phytopathogenic microbial species, *Arch. Biochem. Biophys.* 218:351–359.

Lifshitz, R., Kloepper, J. W., Scher, F. M., Tipping, E. M., and Laliberté, M. 1986. Nitrogen-fixing pseudomonads isolated from roots of plants grown in the Canadian high arctic, *Appl. Environ. Microbiol.* 51:251–255.

Lifshitz, R., Kloepper, J. W., Kozlowski, M., Simonson, C., Carlson, J., Tipping, E. M., and Zaleska, I. 1987. Growth promotion of canola (rapeseed) seedlings by a strain of *Pseudomonas putida* under gnotobiotic conditions, *Can. J. Microbiol.* 33:390–395.

Lim, W. C., and Lockwood, J. L. 1988. Chemotaxis of some phytopathogenic bacteria to fungal propagules in vitro and in soil, *Can. J. Microbiol.* 34:196–199.

Lindberg, G. D. 1981. An antibiotic lethal to fungi, *Plant Dis.* 65:680–683.

Lindberg, G. D., Larkin, J. M., and Whaley, H. A. 1980. Production of tropolone by a *Pseudomonas*, *J. Natural Products* 43:592–594.

Lockwood, J. L. 1977. Fungistasis in soils, *Biol. Rev.* 52:1–43.

Lodge, J. S., Gaines, C. D., Arceneaux, J. E. L., and Byers, B. R. 1982. Ferrisiderophore reductase activity in *Agrobacterium tumefaciens*, *J. Bacteriol.* 149:771–774.

Loper, J. E. 1988. Role of fluorescent siderophore production in biological control of *Pythium ultimum* by a *Pseudomonas fluorescens* strain, *Phytopathology* 78:166–172.

Loper, J. E., Suslow, T. V., and Schroth, M. N. 1984. Log-normal distribution of bacterial populations in the rhizosphere, *Phytopathology* 74:1454–1460.

Loper, J. E., Haack, C., and Schroth, M. N. 1985. Population dynamics of soil pseudomonads in the rhizosphere of potato (*Solanum tuberosum* L.), *Appl. Environ. Microbiol.* 49:416–422.

Lynch, J. M. 1987. Soil biology: accomplishments and potential, *Soil Sci. Soc. Am. J.* 51:1409–1412.

Magazin, M. D., Moores, J. C., and Leong, J. 1986. Cloning of the gene coding for the outer membrane receptor protein for ferric pseudobactin, a siderophore from a plant growth-promoting *Pseudomonas* strain, *J. Biol. Chem.* 261:795–799.

Marois, J. J., Mitchell, D. J., and Sonoda, R. M. 1981. Biological control of Fusarium crown rot of tornato under field conditions, *Phytopathology* 71:1257–1260.

Martin, J. F. 1977. Control of antibiotic synthesis by phosphate, *Adv. Biochem. Eng.* 6:105–127.

Marugg, J. D., van Spanje, M., Hoekstra, W. P. M., Schippers, B., and Weisbeck, P. J. 1985. Isolation and analysis of genes involved in siderophore biosynthesis in plant growth-stimulating *Pseudomonas putida* WCS358, *J. Bacteriol.* 164:563–570.

Marugg, J. D., Nielander, H. B., Horreboets, A. J., van Megen, I., van Genderen, I., and Weisbeek, P. J. 1988. Genetic organization and transcriptional analysis of a major gene cluster involved in siderophore biosynthesis in *Pseudomonas putida* WCS358, *J. Bacteriol.* 170:1812–1819.

Maurer, C. L., and Baker, R. 1964. Ecology of plant pathogens in soil. I. Influence of chitin and lignin amendments on development of bean root rot, *Phytopathology* 54:1425–1426.

Meyer, J.-M., and Abdallah, M. A. 1978. The fluorescent pigment of *Pseudomonas fluorescens*: biosynthesis, purification and physicochemical properties, *J. Gen. Microbiol.* 107:319–328.

Meyer, J.-M., and Abdallah, M. A. 1980. The siderochromes of non-fluorescent pseudomonads: production of nocardamine by *Pseudomonas stutzeri*, *J. Gen. Microbiol.* 118:125–129.

Meyer, J.-M., and Hornsperger, J. M. 1978. Role of pyoverdine pf, the iron-binding fluorescent pigment of *Pseudomonas fluorescens*, in iron transport, *J. Gen. Microbiol.* 107:329–331.

Misaghi, I. J., Stowell, L. J., Grogan, R. G., and Spearman, L. C. 1982. Fungistatic activity of water-soluble fluorescent pigments of fluorescent pseudomonads, *Phytopathology* 72:33–36.

Mitchell, R., and Alexander, M. 1961. The mycolytic phenomenon and biological control of *Fusarium* in soil, *Nature* 190:109–110.

Mitchell, R., and Alexander, M. 1963. Lysis of soil fungi by bacteria, *Can. J. Microbiol.* 9:169–177.

Mitchell, R., and Hurwitz, E. 1965. Suppression of *Pythium debaryanium* by lytic rhizosphere bacteria, *Phytopathology* 55:156–158.

Moore, R. E., and Emery, T. 1976. N$^\alpha$-acetylfusarinines isolation, characterization and properties, *Biochemistry* 15:2719–2723.

Moore, L. W., and Warren, G. 1979. *Agrobacterium radiobacter* strain 84 and biological control of crown gall, *Annu. Rev. Phytopathol.* 17:163–179.

Mullis, K. B., Pollack, J. R., Neilands, J. B. 1971. Structure of schizokinen, an iron-transport compound from *Bacillus megaterium*, *Biochemistry* 10:4894–4898.

Neal, J. L., and Larson, R. I. 1976. Acetylene reduction by bacteria isolated from the rhizosphere of wheat, *Soil Biol. Biochem.* 8:151–155.

Neal, J. L., Jr., Atkinson, T. G., and Larson, R. I. 1970. Changes in the rhizosphere microflora of spring wheat induced by disomic substitution of a chromosome, *Can. J. Microbiol.* 16:153–158.

Neal, J. L., Jr., Larson, R. I., and Atkinson, T. G. 1973. Changes in rhizosphere populations of selected physiological groups of bacteria related to substitution of specific pairs of chromosomes in spring wheat, *Plant and Soil* 39:209–212.

Neilands, J. B. 1981. Microbial iron compounds, *Annu. Rev. Biochem.* 50:715–731.

Nesmith, W. C., and Jenkins, S. F., Jr. 1985. Influence of antagonists and controlled matric potential on the survival of *Pseudomonas solonacearum* in four North Carolina soils, *Phytopathology* 75:1182–1187.

New, P. B., and Kerr, A. 1972. Biological control of crown gall: field measurements and glasshouse experiments, *J. Appl. Bacteriol.* 35:279–287.

Ong, S. A., Peterson, T., and Neilands, J. M. 1979. Agrobactin, a siderophore from *Agrobacterium tumefaciens*, *J. Biol. Chem.* 254:1860–1865.

Ordentlich, A., Elad, Y., and Chet, I. 1988. The role of chitinase of *Serratia marcesens* in biocontrol of *Sclerotium rolfsii*, *Phytopathology* 78:84–88.

Papavizas, G. C., and Lumsden, R. D. 1980. Biological control of soilborne fungal propagules, *Annu. Rev. Phytopathol.* 18:389–413.

Papavizas, G. C., Dunn, M. T., Leurs, J. A., and Beagle-Ristiano, J. 1984. Liquid fermentation technology for experimental production of biocontrol fungi, *Phytopathology* 74:1171–1175.

Parke, J. L., Moen, R., Rovira, A. D., and Bowen, G. D. 1986. Soil water flow affects the rhizosphere distribution of a seed-borne biological control agent, *Pseudomonas fluorescens*, *Soil Biol. Biochem.* 18:583–588.

Parker, W. L., Rathnum, M. L., Seiner, V., Tregjo, W. H., Principe, P. A., and Sykes, R. B. 1984. Cepacin A and cepacin B two new antibiotics produced by *Pseudomonas cepacia*, *J. Antibiot.* 37:431–440.

Paulitz, T. C., Park, C. S., and Baker, R. 1987. Biological control of Fusarium wilt of cucumber with nonpathogenic isolates of *Fusarium oxysporum*, *Can. J. Microbiol.* 33:349–353.

Pavlica, D. A., Hora, T. I., Bradshaw, J. J., Skogerboe, R. K., and Baker, R. 1978. Volatiles from soil influencing activities of soil fungi, *Phytopathology* 68:758–765.

Peterson, E. A., Katznelson, H., and Cook, F. D. 1965. The influence of chitin and myxobacters on numbers of actinomycetes in soil, *Can. J. Microbiol.* 11:595–596.

Peterson, E. A., Gillespie, D. C., and Cook, F. D. 1966. A wide spectrum antibiotic produced by a species of *Sorangium*, *Can. J. Microbiol.* 12:221–230.

Polonenko, D. R., Scher, F. M., Kloepper, J. W., Singleton, C. A., Laliberté, M., and Zaleska, I. 1987. Effects of root colonizing bacteria on modulation of soybean roots by *Bradyrhizobium japonicum*, *Can. J. Microbiol.* 33:498–503.

Powell, P. E., Cline, G. R., Reid, C. P. P., and Szaniszlo, P. J. 1980. Occurrence of hydroxamate siderophore iron chelators in soil, *Nature* 287:833–834.

Powell, P. E., Szaniszlo, P. J., Cline, G. R., and Reid, C. P. P. 1982. Hydroxamate siderophores in the iron nutrition of plants, *J. Plant Nutr.* 5:653–673.

Randhawa, P. S., and Schaad, N. W. 1985. A seedline bioassay chamber for determining bacterial colonization and antagonism on plant roots, *Phytopathology* 75:254–259.

Rothrock, C. S., and Gottlieb, D. 1984. Role of antibiosis in antagonism of *Streptomyces hygroscopicus* var. *geldanus* to *Rhizoctonia solani* in soil, *Can. J. Microbiol.* 30:1440–1447.

Rouatt, J. W., and Katznelson, H. 1961. A study of the bacteria on the root surface and in the rhizosphere soil of crop plants, *J. Appl. Bacteriol.* 24:164–171.

Roughley, R. J., and Vincent, J. M. 1967. Growth and survival of *Rhizobium* spp. in peat culture, *J. Appl. Bacteriol.* 30:362–376.

Rovira, A. D. 1965. Interactions between plant roots and soil microorganisms, *Annu. Rev. Microbiol.* 19:241–266.

Rovira, A. D., and McDougall, B. 1967. Microbiological and biochemical aspects of the rhizosphere. In *Soil Biochemistry*, A. D. McLaren and D. H. Peterson (eds.), vol. 1. Marcel Dekker, New York, pp. 417–463.

Sands, D. C., and Rovira, A. D. 1970. Isolation of fluorescent pseudomonads with a selective medium, *Appl. Microbiol.* 20:513–514.

Sands, D. C., and Rovira, A. D. 1971. *Pseudomonas fluorescens* biotype G, the dominant fluorescent pseudomonad in south Australian soils and wheat rhizospheres, *J. Appl. Bacteriol.* 34:261–275.

Scher, F. M., and Baker, R. 1980. Mechanism of biological control in a Fusarium-suppressive soil, *Phytopathology* 70:412–417.

Scher, F. M., and Baker, R. 1982. Effect of *Pseudomonas putida* and a synthetic iron chelator on induction of soil suppressiveness to Fusarium wilt pathogens, *Phytopathology* 72:1567–1573.

Scher, F. M., Dupler, M., and Baker, R. 1984a. Effect of synthetic iron chelates on population densities of *Fusarium oxysporum* and the biological control agent *Pseudomonas putida* in soil, *Can. J. Microbiol.* 30:1271–1275.

Scher, F. M., Ziegle, J. S., and Kloepper, J. W. 1984b. A method for assessing the root-colonizing capacity of bacteria on maize, *Can. J. Microbiol.* 30:151–157.

Scher, F. M., Kloepper, J. W., and Singleton, C. A. 1985. Chemotaxis of fluorescent *Pseudomonas* spp. to soybean seed exudates in vitro and in soil, *Can. J. Microbiol.* 31:570–574.

Schippers, B., Bakker, A. W., and Bakker, P. A. H. M. 1987. Interactions of deleterious and beneficial rhizosphere microorganisms and the effect of cropping practices, *Annu. Rev. Phytopathol.* 25:339–358.

Schmidt, E. L. 1979. Initiation of plant root-microbe interactions, *Annu. Rev. Microbiol.* 33:355–376.

Schroth, M. N., and Hancock, J. G. 1981. Selected topics in biological control, *Annu. Rev. Microbiol.* 35:453–476.

Schroth, M. N., and Hancock, J. G. 1982. Disease-suppressive soil and root-colonizing bacteria, *Science* 216:1376–1381.

Shao, F. M., and Christiansen, M. N. 1982. Cotton seedling radicle exudates in relation to susceptibility to *Verticillium* wilt and *Rhizoctonia* root rot, *Phytopathol. Z.* 105:351–359.

Shim, J.-S., Farrand, S. K., and Kerr, A. 1987. Biological control of crown gall: construction and testing of new biocontrol agents, *Phytopathology* 77:463–466.

Simeoni, L. A., Lindsay, W. L., and Baker, R. 1987. Critical iron level associated with biological control of Fusarium wilt, *Phytopathology* 77:1057–1061.

Simon, A., Rovira, A. D., and Sands, D. C. 1973. An improved selective medium for isolating fluorescent pseudomonads, *J. Appl. Bacteriol.* 36:141–145.

Simon, A., Sivasithamparam, K., and MacNish, G. C. 1987. Biological suppression of the saprophytic growth of *Gaeumannomyces graminis* var. *tritici* in soil, *Can. J. Microbiol.* 33:515–519.

Sivasithamparam, K., Parker, C. A., and Edwards, C. S. 1979. Bacterial antagonists to the take-all fungus and fluorescent pseudomonads in the rhizosphere of wheat, *Soil Biol. Biochem.* 11:161–165.

Smiley, R. W. 1978a. Antagonists of *Gaeumannomyces graminis* from the rhizoplane of wheat in soils fertilized with ammonium- or nitrate-nitrogen, *Soil Biol. Biochem.* 10:169–174.

Smiley, R. W. 1978b. Colonization of wheat roots by *Gaeumannomyces graminis* inhibited by specific soils, microorganisms and ammonium-nitrogen, *Soil Biol. Biochem.* 10:175–179.

Smiley, R. W. 1979. Wheat-rhizoplane pseudomonads as antagonists of *Gaeumannomyces graminis*, *Soil Biol. Biochem.* 11:371–376.

Stack, J. P., and Miller, R. L. 1985. Competitive colonization of organic matter in soil by *Phytophthora megasperma* f. sp. *medicaginis*, *Phytopathology* 75:1020–1025.

Stipanovic, R. D., and Howell, C. R. 1982. The structure of gliovirin, a new antibiotic from *Gliocladium virens*, *J. Antibiot.* 35:1326–1330.

Stutz, E. W., Défago, G., and Kern, H. 1986. Naturally occurring fluorescent pseudomonads involved in suppression of black root rot of tobacco, *Phytopathology* 76:181–185.

Suslow, T. V., and Schroth, M. N. 1982a. Role of deleterious rhizobacteria as minor pathogens in reducing crop growth, *Phytopathology* 72:111–115.

Suslow, T. V., and Schroth, M. N. 1982b. *Rhizobacteria* of sugar beets: effects of seed application and root colonization on yield, *Phytopathology* 72:199–206.

Szaniszlo, P. J., Powell, P. E., Reid, C. P. P., and Cline, G. R. 1981. Production of hydroxamate siderophore iron chelators by ectomycorrhizal fungi, *Mycologia* 72:1158–1174.

Tari, P. H., and Anderson, A. J. 1988. Fusarium wilt suppression and agglutinability of *Pseudomonas putida*, *Appl. Environ. Microbiol.* 54:2037–2041.

Tate, M. E., Murphy, P. J., Roberts, W. P., and Kerr, A. 1979. Adenine N^6-substituent of agrocin 84 determines its bacteriocin-like specificity, *Nature* 280:697–699.

Teintze, M., Hossain, M. B., Barnes, C. L., Leong, J., and van der Helm, D. 1981. Structure of ferric pseudobactin, a siderophore from a plant growth promoting *Pseudomonas, Biochemistry* 20:6446–6457.

Thomashow, L. S., and Weller, D. M. 1988. Role of a phenazine antibiotic from *Pseudomonas fluorescens* in biological control of *Gaeumannomyces graminis* var. *tritici, J. Bacteriol.* 170:3499–3508.

Tien, T. M., Gaskins, M. H., and Hubbell, D. H. 1979. Plant growth substances produced by *Ozospirillum brasilense* and their effect on the growth of pearl millet (*Pennisetum americanum* L.), *Appl. Environ. Microbiol.* 37:1016–1024.

Tjamos, E. C. 1979. Induction of resistance to Verticillium wilt in cucumber (*Cucumis sativas*), *Physiol. Plant Pathol.* 15:223–227.

Trust, T. J. 1975. Antibacterial activity of tropolone, *Antimicrob. Agents Chemother.* 7:500–506.

Turner, S. M., Newman, E. I., and Campbell, R. 1985. Microbial population of ryegrass root surfaces: influence of nitrogen and phosphorus supply, *Soil Biol. Biochem.* 17:711–715.

Vancura, V. 1980. Fluorescent pseudomonads in the rhizosphere of plants and their relation to root exudates, *Folia Microbiol.* 25:168–173.

Vandenbergh, P. A., Gonzalez, C. F., Wright, A. M., and Kunka, B. S. 1983. Iron-chelating compounds produced by soil pseudomonads: correlation with fungal growth inhibition, *Appl. Environ. Microbiol.* 46:128–132.

Vannacci, G., and Harman, G. E. 1987. Biocontrol of seed-borne *Alternaria raphani* and *A. brassicola, Can. J. Microbiol.* 33:850–856.

Walther, D., and Gindrat, D. 1988. Biological control of damping-off of sugar-beet and cotton with *Chaetomium globosum* or a fluorescent *Pseudomonas* sp., *Can. J. Microbiol.* 34:631–637.

Weller, D. M. 1983. Colonization of wheat roots by a fluorescent pseudomonad suppressive to take-all, *Phytopathology* 73:1548–1553.

Weller, D. M. 1984. Distribution of a take-all suppressive strain of *Pseudomonas fluorescens* on seminal roots of winter wheat, *Appl. Environ. Microbiol.* 48:897–899.

Weller, D. M., and Cook, R. J. 1983. Suppression of take-all of wheat by seed treatments with fluorescent pseudomonads, *Phytopathology* 73:463–469.

West, A. W., Burges, H. D., Dixon, T. J., and Wyborn, C. H. 1985. Survival of *Bacillus thuringiensis* and *Bacillus cereus* spore inocula in soil: effects of pH, moisture, nutrient availability and indigenous microorganisms, *Soil Biol. Biochem.* 17:657–665.

Williams, S. T. 1982. Are antibiotics produced in soil? *Pedobiologia* 23:427–435.

Williams, S. T., and Vickers, J. C. 1986. The ecology of antibiotic production, *Microb. Ecol.* 12:43–52.

Windels, C. E. 1981. Growth of *Penicillium oxalicum* as a biological seed treatment on pea seed in soil, *Phytopathology* 71:929–933.

Wissing, F. 1974. Cyanide formation from oxidation of glycine by a *Pseudomonas* species, *J. Bacteriol.* 117:1289–1294.

Wong, P. T. W., and Siviour, T. R. 1979. Control of ophiobolus patch in *Agrostis* turf using avirulent fungi and take-all suppressive soils in pot experiments, *Ann. Appl. Biol.* 92:191–197.

Xu, G.-W., and Gross, D. C. 1986a. Selection of fluorescent pseudomonads antagonistic to *Erwinia carotovora* and suppressive of potato seed piece decay, *Phytopathology* 76:414–422.

Xu, G.-W., and Gross, D. C. 1986b. Field evaluations of the interactions among fluorescent pseudomonads, *Erwinia carotovora*, and potato yields, *Phytopathology* 76:423–430.

Yuen, G. Y., and Schroth, M. N. 1986. Interactions of *Pseudomonas fluorescens* strain E6 with ornamental plants and its effect on the composition of root-colonizing microflora, *Phytopathology* 76:176–180.

Yuen, G. Y., Schroth, M. N., and McCain, A. H. 1985. Reduction of Fusarium wilt of carnation with suppressive soils and antagonistic bacteria, *Plant Dis.* 69:1071–1075.

EPA Regulations Governing Release of Genetically Engineered Microorganisms

Elizabeth A. Milewski

U.S. Environmental Protection Agency
401 M Street S.W.
Washington, D.C. 20460

Introduction

For thousands of years humans have been employing microorganisms to produce useful substances. The making of beer, wine, bread, and cheese are examples of the utilization of useful microorganisms.

Throughout the millenia during which microorganisms have been used, human beings have developed techniques to manipulate the genetic materials of useful organisms. Recently, they have developed powerful new techniques in molecular biology, which permit them to manipulate the genetic materials of organisms in ways that heretofore were not possible. Cognizant of the equation that powerful new technologies may be associated with dramatic consequences, the public has called for some type of oversight to ensure the safe application of this new technology.

This chapter will explore how the Federal government of the United States is responding to this challenge. Particular emphasis will be placed on the program developed by the U.S. Environmental Protection Agency (EPA).

U.S. Government Approach to Regulation of Biotechnology

The EPA is one of several U.S. agencies currently regulating biotechnology. In order to better understand the EPA's activities, some description of the framework of regulation established by the U.S. government, and the EPA's role within that framework, is needed.

In 1984, recognizing its responsibility to address issues raised by the use of biotechnology, the U.S. government formed an interagency working group under the White House Cabinet Council on Natural Resources and the Environment. The working group sought to achieve a balance between regulation adequate to ensure human health and environmental safety and the maintenance of sufficient regulatory flexibility to avoid impeding the growth of the biotechnology industry.

The group examined existing laws and concluded that, for the most part, these laws would adequately address regulatory needs for biotechnology. A matrix describing applicable laws and responsible agencies was published in the *Federal Register* on November 14, 1985 (10). Table 1 shows some of the laws the United States stated it could use to regulate biotechnology products.

Subsequently, the Coordinated Framework for Regulation of Biotechnology was published on June 26, 1986 (11). The Coordinated Framework includes statements of regulatory policy from the agencies principally responsible for such regulation, describes how the framework provides jurisdiction over both research and development (R&D) and commercial products, and spells out the basic federal philosophy for regulating biotechnology.

The basic philosophy underlying the Coordinated Framework is that biotechnology encompasses a large and varied collection of techniques and activities, and that the potential products resulting from

TABLE 1   Laws under Which the United States Can Regulate Products of Biotechnology

Agency	Law
EPA	Federal Insecticide, Fungicide, and Rodenticide Act (7 USC 136–136y)
	Toxic Substances Control Act (5 USC 2601–2929)
	Food, Drug, and Cosmetic Act (21 USC 301–392)
USDA	Virus Serum Toxin Act of 1913 (21 USC 151–158)
	Federal Plant Pest Act of May 23, 1957 (7 USC 150aa–150jj)
	Plant Quarantine Act of August 20, 1912 (7 USC 1551–1564, 1566, 1567)
	Organic Act of September 21, 1944 (7 USC 147a)
	Noxious Weed Act of 1974 (7 USC 2801 et seq.)
	Federal Seed Act (7 USC 551 et seq.)
	Plant Variety Act (7 USC 2321 et seq.)
	Federal Meat Inspection Act (21 USC 451 et seq.)
	Poultry Products Inspection Act (921 USC 451 et seq.)
FDA	Food, Drug, and Cosmetic Act (21 USC 301–392)
	Public Health Service Act (42 USC 3262, 353)
OSHA	Occupational Safety and Health Act (29 USC 651 et seq.)

SOURCE: Ref. 10.

this technology will cover a wide spectrum of uses. The Coordinated Framework provides that biotechnology products will be regulated in the United States as are products of other technologies; that is, by the various regulatory agencies on the basis of *use*. Thus, many agricultural uses of microorganisms, plants, and animals are regulated by the U.S. Department of Agriculture (USDA); foods, drugs, cosmetics, and biologics are regulated by the Food and Drug Administration (FDA); microorganisms used as pesticides are regulated by EPA under the Federal Insecticide, Fungicide, and Rodenticide Act (FIFRA); and uses of microorganisms not covered by other existing authorities are regulated by EPA under the Toxic Substances Control Act (TSCA) if the use involves a "commercial purpose." Examples of uses covered by TSCA include use of microorganisms in metal mining, degrading toxic wastes, conversion of biomass for energy, production of proteins and enzymes for nonpharmaceutical purposes, and nonpesticidal agricultural applications such as nitrogen fixation.

Four of the agencies of the Coordinated Framework have responsibilities that are spelled out by a statute that agency administers: the agencies are EPA, USDA, FDA, and the Occupational Safety and Health Administration (OSHA). These responsibilities may be directed at research and development (R&D) activities and at commercial activities (e.g., sale, importation, manufacture). Table 2 shows the various responsibilities of these agencies within the Coordinated Framework.

When the Coordinated Framework was developed, it was recognized that the network of oversight provided by the FDA, USDA, OSHA, and EPA might not extend to R&D endeavors that might present issues similar to R&D endeavors regulated by statute. In order to provide a "level playing field" so that researchers are treated similarly whether in academia or industry, and to address the public's concern, the Coordinated Framework provides that three agencies not possessing statutory responsibilities oversee some R&D activities under guidelines deriving from the contracts associated with grants and contracts. Indeed, the National Institutes of Health (NIH), under its "Guidelines for Research Involving Recombinant DNA Molecules" (7)

TABLE 2   Responsibility for Oversight of Commercial Products of Biotechnology

Product area	Responsible agency
Products produced in contained systems:	
Food and food additives	FDA[a], USDA
Human drugs, medical devices, and biologicals	FDA
Animal drugs	FDA
Animal biologicals	USDA
Other products	EPA
Products released in the environment:	
Plants and animals	USDA[a], FDA[b]
Pesticides	EPA[a], USDA
All other uses involving microorganisms	EPA[c], USDA[c]

[a]Lead agency.
[b]FDA has responsibility when the product is used for food.
[c]Depending on whether the product incorporates genetic information from pathogens, how the product is constructed (combining DNA from source organisms from different genera or not), and how it is used (agricultural use or not), either the EPA or the USDA will be the lead agency.
SOURCE: Ref. 11.

has, since 1986, provided oversight of certain research activities by funding only institutions complying with the guidelines.

The Coordinated Framework indicated that, in addition to NIH, two other agencies, Science and Education of the USDA and the National Science Foundation (NSF), would have guidelines operating under a similar premise (i.e., that the agency can set the conditions under which it will fund research projects). These agencies play an important role if both commercial and purely academic research are to be treated equally under the Coordinated Framework.

Table 3 shows the interweaving of responsibilities over research activities that was developed as part of the Coordinated Framework. Since several agencies might potentially have oversight responsibilities for a specific research activity, the Coordinated Framework indicated that the "lead" agency (i.e., the agency with primary responsibility) would be determined by (1) whether the research is regulated under a statute administered by FDA, USDA, OSHA, or EPA (for example, research involving field testing of pesticides is subject to FIFRA) or (2) which agency is funding the research.

TABLE 3   Research Jurisdiction

Research area	Responsible agency
Contained research: no intentional release to the environment:	
Federally funded	Funding agency,[a] FDA[b]
Nonfederally funded	NIH or S&E[c] voluntary review, APHIS[d], FDA[b]
Release to the environment:[e]	
Federally funded	Depending on several factors[e] review will be conducted by funding agency APHIS[d], EPA, S&E[c] voluntary review
Nonfederally funded	Depending on several factors[e] review will be conducted by APHIS[d], EPA, S&E[c] voluntary review

[a]Review and approval of research protocols conducted by NIH, and S&E or NSF.

[b]FDA has responsibility for foods and/or food additives, human drugs, medical devices, biologics, animal drugs.

[c]Science and Education, U.S. Department of Agriculture. Recently USDA has agreed that the NIH Guidelines would be applied.

[d]APHIS issues permits for the importation and domestic shipment of certain plants and animals, plant pests, and animal pathogens, and for the shipment or release in the environment of regulated articles.

[e]The agency responsible for review will be determined by whether the research involves genetic information from pathogens, how the research organism is constructed (combining DNA from source organisms from different genera or not), and whether the research is for commercial purposes.

SOURCE: June 26, 1986 *Federal Register* [OSTP]

The NIH guidelines address primarily biomedical experiments involving use of recombinant DNA techniques. The guidelines of the Science and Education portion of USDA would apply to agricultural research on plants, animals, and microorganisms. The NSF guidelines would apply to research projects funded by NSF.

The several agencies that form the Coordinated Framework have agreed to seek to operate their programs in a coordinated fashion.

The EPA's goals are (1) to protect human health and the environment, (2) to meet the public's need to be assured that testing and use of living products do not present unacceptable risk, (3) to expedite reviews of submissions, and (4) to respond to a rapidly developing technology. The EPA is pursuing these goals by developing a research program directed toward generating information on microbial ecology, and by implementing a sound regulatory program. The regulatory program consists of formal regulations for biotechnology and a review system for evaluating submissions.

EPA's Regulatory Program for Biotechnology

EPA's regulatory program in biotechnology is pursuant to its authorities under the law regulating pesticides, FIFRA, and the law regulating chemicals, TSCA. While these statutes were not written specifically for biotechnology, these two laws have been interpreted as investing EPA with authority to do so.

Federal Insecticide, Fungicide, and Rodenticide Act (FIFRA)

The Federal Insecticide, Fungicide, and Rodenticide Act (FIFRA) creates a statutory framework under which the EPA, through a registration process, regulates the development, sale, distribution, and use of pesticides, regardless of how these pesticides are made or their mode of action.

FIFRA states that a pesticide can be registered for use only if the pesticide will not cause unreasonable adverse effects to humans or the environment. The "unreasonable adverse effects" test involves a weighing of the risks and benefits of use of the pesticide. In order to demonstrate that a pesticide will not cause unreasonable adverse effects, an applicant seeking to register the product must submit or cite data on subjects such as product composition, toxicity, environmental fate, and effects on nontarget organisms.

FIFRA covers natural and genetically altered microorganisms that are used for pesticide purposes. Over the past 15 years, the EPA has reviewed and registered a number of microbial pesticides. The EPA's authority to approve pesticides for testing or use is a powerful regulatory tool. FIFRA places on the registrant or applicant the burden of proof that the benefits of use of the product outweigh the risks. Under FIFRA, the agency can require the registrant or applicant to test to acquire information necessary to evaluate potential risk.

Some of the information needed to support registration must be developed under actual field conditions, and such field testing is authorized under FIFRA through the "experimental use permit" (EUP) process. The results of such early testing are a critical element in determining when and under what circumstances a pesticide may be registered for use.

Toxic Substances Control Act (TSCA)

The Toxic Substances Control Act (TSCA) authorizes the EPA to acquire information on chemical substances and mixtures of chemical substances in order to identify and regulate potential hazards and exposures. TSCA applies to the manufacturing, processing, importation, distribution, use, and disposal of all chemicals in commerce, or intended for entry into commerce, that are not specifically covered by other regulatory authorities (e.g., substances other than foods, drugs, cosmetics, and pesticides). Under TSCA, the EPA can require testing of any chemical substance that may present an unreasonable risk to human health or the environment or which is produced in substantial quantities and may result in substantial environmental release or substantial human exposure. TSCA is concerned with all exposure media (air, water, soil, sediment, biota).

TSCA's applicability to the regulation of microbial biotechnology products is based on the interpretation that microbes are chemical substances under TSCA; an interpretation that was embodied in the development of the initial chemical inventory in 1979. The basis of this interpretation is that all substances, living and nonliving, have a chemical foundation at the most fundamental molecular level. As a result of this interpretation, microorganisms (except for those in excluded-use categories) are subject to all provisions of TSCA.

The heart of TSCA is Section 5, which implements the act's goal of identifying potentially hazardous new substances before they enter commerce. Section 5 requires that manufacturers and importers of "new" chemical substances intended for commercial use submit a notice to EPA. EPA has 90 days to prohibit or regulate a "new" substance, otherwise the submitter may proceed. The review period can be extended to 180 days for good cause. The submitter must include in the notice information as specified in the notification form and which is either "known or reasonably ascertainable" at the time of submission. They must also submit test data that are in their possession and under their control at the time of submission.

One important facet of TSCA should be emphasized: Unlike FIFRA, which applies to all pesticides, TSCA applies only to products developed for "commercial purposes." Investigators involved in purely academic research, thus, are not subject to regulation under TSCA[1]. They may, however, be subject to the guidelines of other agencies forming part of the Coordinated Framework.

TSCA and FIFRA applied to microorganisms

As can be seen in the above discussion, FIFRA and TSCA are different statutes with different goals (i.e., registration of pesticides versus identification and, if warranted, regulation of potential hazards and exposures of chemical substances and mixtures), different authorities, and different procedures. The agency is dedicated, however, to ensuring that these two statutes are administered for biotechnology in a coordinated manner and that the approaches taken under the two statutes are as similar as possible.

On what types of microorganisms, then, is the EPA focusing regulatory emphasis? The EPA, under FIFRA and TSCA, intends (11) to focus regulatory emphasis on three categories of microorganisms:

1. Microorganisms with "new" characteristics or that are new to the environment in which they are intended to be used (and whose behavior is therefore less predictable)

2. Microorganisms, such as pathogens, that have potential for causing adverse effects in other organisms

3. Microorganisms that are used in the environment (and therefore have the potential for widespread exposure).

The agency defines microorganisms with "new" characteristics as those microorganisms that have been formed through deliberate hu-

---

[1]EPA is in the process of developing guidance on what constitutes "research for commercial purposes." EPA is currently proposing to recognize several funding arrangements with university or other research institutions as "noncommercial" because these arrangements are free from intent to obtain an immediate or eventual commercial advantage. These include an outright gift by a commercial entity to a research institution if there is no limitation on the purpose for which the funds are used or on the use to be made of the results of the research; or a grant by a commercial entity to a research institution in a specific field with no limitations on the use to be made of the results. However, EPA is currently proposing to consider "commercial" a grant by a commercial entity to an institution to perform research in a specific field in which the commercial entity holds patent rights, and research performed under a contract in which a commercial entity holds patent or licensing rights. See footnote 2 under for the address of the TSCA Assistance Office, at which further information can be obtained.

man intervention by combining genetic material from dissimilar source material. The EPA defines "dissimilar" sources as organisms from different taxonomic genera; the organisms resulting from such a combination are called "intergeneric."

In developing this definition, the agency made several assumptions. These are: Combinations of genetic material from microorganisms from different genera are more likely to result in new traits than combinations of genes from microorganisms within the same genus; while genetic exchanges occur naturally and somewhat commonly among microorganisms, they are more likely to occur in nature within a single genus than across many different genera (1,13,14); and genus designations provide a practical criterion for administrative purposes. The "intergeneric" concept provides a definition of "new" under TSCA, and can be applied under FIFRA to describe a category of microbial pesticides.

The EPA also described in the 1986 Coordinated Framework (11) how it would address "new to the environment in which it will be used": both genetically engineered and geographically nonindigenous microorganisms could introduce new traits into their recipient ecosystems. The agency has, as it has been developing its program for regulation of biotechnology products, been further developing this concept.

Finally, the agency chose to focus attention on microorganisms that are released to the environment because such releases may involve large numbers and frequent applications of microorganisms that may have the ability, and indeed may be specifically designed, to survive, persist, and increase in number. The EPA feels it is prudent to evaluate these microorganisms before they are released to the environment because of the uncertainties associated with the behavior of microorganisms exhibiting new traits and the wide variety of expected products, new uses, and new exposures. Accordingly, EPA's regulatory approach encompasses procedures designed to provide sufficient oversight of the initial stages of environmental testing of microbial products of biotechnology.

The EPA described its policy for biotechnology under TSCA and FIFRA in the June 26, 1986 *Federal Register* (11), and requested voluntary compliance with those portions of the program for which formal regulations had to be written.

Under TSCA, the EPA must develop a full set of regulations applying to microorganisms. That means that the agency must develop regulations relating both to R&D activities, specifically including initial field testing, and large-scale commercial activities.

Since existing regulations under FIFRA apply to microbial as well as to chemical pesticides, the only modification needed is an amendment to the EUP regulations explaining how the EPA views the initial small-scale field trials.

The agency is currently in the process of developing formal regulations[2] under both TSCA and FIFRA.

Role of science advisory committee in regulatory program

Because of the complexity of the issues associated with regulation of biotechnology, the EPA is seeking the best scientific advice available. In *Federal Register* notices published on December 31, 1984 (9), November 14, 1985 (10), and June 26, 1986 (11), the EPA indicated its intention to form a science advisory committee for biotechnology. The July 2, 1986, *Federal Register* (3) announced the formation of such a committee, the EPA Biotechnology Science Advisory Committee (BSAC).

The committee's primary functions are to provide peer review of Agency assessments of specific product submissions under TSCA, FIFRA, and other agency statutes, and scientific advice on the EPA's biotechnology program. The BSAC, which is advisory to the EPA Administrator, is composed of eleven voting members: nine scientists and two representatives of the public. Members are named to provide the range of expertise required to assess the scientific and technical issues pertinent to BSAC responsibilities.

The BSAC charter ensures that three EPA science advisory committees, the BSAC, the Science Advisory Board (SAB), and the FIFRA Science Advisory Panel (SAP), are aware of issues of mutual interest. This is accomplished by having at least one full member from the SAB and the SAP appointed to the BSAC as a full member. This arrangement consolidates advice in the area of biotechnology using the BSAC as a focal point, while allowing the SAP and SAB to continue providing assistance within the context of their traditional advisory roles.

BSAC can be supplemented as needed by consultants to extend the range of expertise of the standing committee. The charter of the committee also authorizes the formation of subcommittees to perform specific functions consistent with the BSAC charter.

---

[2]For technical information on the regulatory program contact:

Under TSCA: TSCA Assistance Office, Office of Toxic Substances, U.S. Environmental Protection Agency, 401 M Street S.W., Washington, D.C. 20460. Telephone: (202) 554-1404.

Under FIFRA: Science Analysis and Coordination Staff, Environmental Fate and Effects Division, Office of Pesticide Programs, U.S. Environmental Protection Agency, 401 M Street S.W., Washington, D.C. 20460. Office location and telephone number: Room 1128, CM #2, 1921 Jefferson Davis Highway, Arlington, Virginia 22202, (703) 557-8127.

The EPA encourages open public discussion of the issues, and meetings of the BSAC and its subcommittees are open to the public except for those portions of meetings closed for cause, such as those dealing with material subject to statutory confidentiality requirements.

The BSAC has been very active in advising the EPA on science issues arising from the EPA's efforts to write formal regulations to fully implement the regulatory aspect of its program for biotechnology. For example, the committee's scientific advice was sought on terms important to regulations for biotechnology (e.g., release to the environment).

The BSAC's scientific advice has also been sought on other science issues—e.g., the benefits and risks associated with use of "antibiotic resistance genes" as markers in limited small-scale testing and for large-scale uses.

In addition, the BSAC has performed peer reviews of the agency's assessments of several submissions involving field trials.

Review process for submissions under TSCA and FIFRA

Since 1984, the EPA has reviewed approximately 50 small-scale tests of products of biotechnology, many of which involved release to the environment. A review process for such submissions has been developed (see footnote 2) that the EPA believes is sound, credible, and allows the public to feel confident that potential human or environmental impacts have been addressed. To date, the review procedures have involved case-by-case evaluation.

When a proposal for small-scale testing is received, TSCA or FIFRA staff groups evaluate the submission and develop a coordinated scientific position. During this process, hazard and exposure are addressed; potential problems, issues, or significant unanswered questions are identified; and the likelihood of significant risk from the proposed test is assessed.

Under both TSCA and FIFRA, intraagency workgroups then comment on the positions developed by staff. If appropriate, the submission and the EPA's scientific position are sent to other federal agencies for comment. Appropriate state regulatory agencies are also contacted in order to alert them to the submission, to discuss the EPA's assessment, and to ensure that the federal and state positions are as consistent as possible. For some submissions, visits to the test sites are conducted in order to evaluate actual field conditions.

In order to obtain an independent peer review of the EPA's scientific position, to address specific scientific questions raised by staff, and to identify any additional data that may be needed to complete the risk assessment, the submission and the EPA's scientific evaluation may be sent to the BSAC. (Prior to establishment of the BSAC, the SAP performed this function for microbial pesticide submissions.)

Public comment is considered an important aspect of the reviews, and for many proposals the public is provided several opportunities to comment during the review process.

At the conclusion of a review, the EPA determines whether the microorganism may be released into the environment. Under both TSCA and FIFRA, this decision is based on an assessment of both the potential risks and the potential benefits of the proposed use of the microorganism. Should the analysis indicate that use of the microorganism may pose unreasonable risks, the EPA has authority under either statute to impose restrictions on its use. These restrictions may represent "risk management," which seeks to reduce risk to "reasonable" levels through controls on manufacture, use, and disposal, and through mitigation, monitoring,[3] and other activities.

In some cases, microorganisms that are subject to TSCA or FIFRA are also subject to the Federal Plant Pest Act[4] (5) or other statutes administered by the Animal and Plant Health Inspection Service (APHIS) of the USDA (see Table 1). In such situations, APHIS and EPA conduct coordinated reviews. The two agencies cooperate closely, alerting each other to submissions, and sharing expertise and information. This close cooperation benefits both submitters and the agencies.

General observations on data submission

As can be seen from the preceding description, the EPA performs a risk assessment using data pertinent to the submission. The process the agency uses is a systematic scientific evaluation of hazard and ex-

---

[3]To date most biotechnology submissions that address environmental release have included some field-sampling activities. The EPA has sought to develop, with the submitters, a monitoring approach that includes gathering of data on dispersal and fate of the test microorganisms and the introduced genetic material. The goal of this approach is to produce data on the behavior of the microorganism in the environment, and thereby increase the data bases used to predict microorganism behavior. It is also to determine whether the agency's prediction on the risk associated with the trial has been accurate, and to permit remedial action to be taken at as early a date as possible should that prove necessary.

[4]The FPPA and the Plant Quarantine Act (12) give USDA regulatory authority over the movement into and through the United States of plants, plant products, plant pests, and any article or product that may contain a plant pest at the time of movement. Products developed through biotechnology that meet the standards of these laws would be covered by USDA regulations. "Plant pest" is defined to include virtually any form of living organism that can directly or indirectly harm plants or products of plants. Through a permit system, USDA restricts the transport of plant pests when their movement could result in the movement of plant diseases or pests.

posure data. A "hazard assessment" attempts to identify, and if possible, quantify any potential adverse impacts of the substance or product on organisms and processes. An "exposure assessment" addresses the distribution of the substance or product in relation to affected species. Exposure assessments emphasize the fate of the organism in the environment and the amounts that might reach people or other organisms.

These two assessments are combined to formulate conclusions about the potential for risk. In the biotechnology context two primary areas of risk concern exist: health and environmental. Health effects for both human and nonhuman organisms would include processes such as infection and intoxication. Environmental effects focus on processes affecting biological organization at the level of populations, communities, and ecosystems.

When evaluating risk, certain types of information are pertinent when the object of the assessment is a microorganism. These include: the identity of the parental organisms, particularly the characteristics and behavior of the recipient parental microorganism including natural habitat; the genetic modification, if any, and the intended function of the introduced genetic material; the behavior of the test microorganism including the potential for transfer and exchange of genetic material; and a description of the proposed site (i.e., the ecosystem), into which the microorganism will be released, including a comparison to the natural habitat of the parental recipient microorganism. It should be noted, however, that each assessment is, at this time, pursued on its own issues and these issues will vary depending on a number of variables including the behavior of the test microorganism, the genetic modification, if any, and the proposed release site (ecosystem).

For information on the types of data that might be pertinent in preparing submissions, the EPA generally refers submitters to several documents. These include:

1. The NIH "Points to Consider for Environmental Testing of Microorganisms" (6)

2. A detailed guidance document entitled "Points to Consider in the Preparation and Submission of PMNs for Microorganisms" prepared by EPA[5]

3. The "EPA Pesticide Assessment Guidelines: Subdivision M—Biorational Pesticides"

4. An NSF report entitled "The Suitability and Applicability of Risk

---

[5]Available from the TSCA Assistance Office, Office of Toxic Substances, U.S. Environmental Protection Agency, 401 M Street S.W., Washington, D.C. 20460. Telephone (800) 424-9065.

Assessment Methods for Environmental Applications of Biotechnology"

5. The booklet entitled "Recombinant DNA Safety Considerations" by the Organization for Economic Cooperation and Development (8)

Case history—a proposed field trial under TSCA

Table 4 presents a case history to illustrate some of the types of questions posed in a risk assessment. The case history involves a submission from Monsanto Company. In June 1987, Monsanto Company submitted to EPA a request to test in a limited field trial a genetically engineered microorganism. The microorganism, a fluorescent pseudomonad (*Pseudomonas aureofaciens*), was engineered to contain the genes from *Escherichia coli* K-12 that code for the production of lactose permease and β-galactosidase (*lacZY*). These genes permit the organism to grow on minimal lactose media and to cleave the chromogenic substrate X-gal, producing blue rather than the normal yellow-white colonies. The presence of this selectable color marker facilitates greatly the identification and/or reisolation of the engineered strain.

The two objectives of the field trial were to (1) verify that the *lacZY* insert will be an adequate marker to monitor the survival and location of the microorganism under actual field conditions and (2) evaluate the performance and survival of the organism. The field test design included three plantings of two crops (winter wheat and soybeans). The seeds of the first planting, winter wheat, were inoculated with either the modified microorganism or the parental pseudomonad strain. The subsequent two plantings, no-till soybeans in the summer of 1988 and minimum-till winter wheat in the fall of 1988, were not inoculated. The test microorganism was, however, monitored.

EPA completed its review of this submission in October 1987, and the company initiated the test in November 1987. Several progress reports have been submitted to EPA by Monsanto Company. The 6-month report showed that the modified microorganism effectively colonized roots, and that its population declined. Migration of the microorganism from the root zone was limited. A 9-month progress report showed that the modified microorganism effectively colonized the roots, the population continued to decline, and migration was limited. Monsanto also submitted 12- and 15-month progress reports showing the same pattern of behavior. Monsanto, in the summer of 1988, requested permission to scale down the monitoring effort because the continual decline in test microorganism numbers led to a very low population density. EPA agreed to a less aggressive monitoring effort.

Table 5 presents a chronology of events associated with this field trial.

TABLE 4 Case Study of a Risk Assessment

Issues considered in risk assessment	Agency evaluation
Dissemination from site	Dissemination from site expected albeit in very low numbers. Significant exposure not likely because of location and topography of test site, measures planned by company to limit movement from site, and/or application method and frequency.
Horizontal transfer of genetic material	Transport of inserted material to other organisms is expected to be very low. Introduced material inserted in chromosome. Transfer expected to be under control of natural gene-transfer mechanisms for chromosomal material.
Persistence	Survival and establishment at low levels predicted. No competitive advantage predicted based on (1) studies documenting the low level of survival of parental strain, (2) greenhouse studies indicating no difference in survival ability between test and parental strain, (3) studies showing no effect on plant growth, (4) information indicating general lack of ecological significance of added ability to absorb and metabolize lactose.
Environmental effects	No deleterious environmental effects predicted. Parental strain well-studied and never shown to cause adverse effects. Inserted genes not expected to confer competitive advantage or disadvantage: (1) inserted genes confer ability to metabolize lactose, few other sugars; (2) few sites would favor test organism (i.e., commonly available substrates would be limiting but lactose abundant).
Effect of genetic modification	Modification not expected to increase pathogenicity, infectivity, or toxicity to humans, animals, plants, or other organisms.
	Modification not expected to affect host range.
	No modification to introduce or increase penetrance of antibiotic resistance from that present in parent.
	Modification allows test microorganism to metabolize a limited number of sugars (mannose, xylose). Lactose permease is highly specific and only those compounds possessing an unsubstituted galactosidic residue in either α or β linkage will permeate membrane.
Ecological effects	No modification knowingly made to affect competitiveness or otherwise affect populations of other organisms.
	No modification to affect dispersal.
	No residual secondary effects expected.
Economic considerations	Water quality testing. Issue: potential for interference by lactose-metabolizing pseudomonad in water quality tests for coliforms. (Two procedures widely used for coliform testing depend on ability to metabolize lactose.) EPA predicts test microorganism not likely to interfere. Results of tests requested by EPA confirm prediction.
	Shelf life of dairy products. Issue: potential for reducing shelf life. (*Pseudomonas fluorescens* strains are a major cause of postpasteurization contamination reducing shelf life.) Agency predicts no effect on the dairy industry from this limited controlled test. Results of tests requested by EPA show test microorganism will not affect shelf life.

TABLE 5   Summary of Chronology of Events Associated with Case Study[a]

Date	Chronology of events
June 1987	Submission received.
August 1987	BSAC subcommittee meets to review submission and EPA risk assessment.
September 1987	Preliminary decision to approve test; review suspended to develop consent order.
October 1987	Review completed.
November 2, 1987	Test initiated.
January 1988	Clemson University requests permission to conduct similar test independent of Monsanto.
February 1988	Monsanto submits 3-month progress report to EPA.
March 1988	Monsanto requests modification to consent order to permit distribution of microorganism to noncommercial institutions.
April 1988	Monsanto submits data on milk testing.
April 1988	Monsanto submits 6-month progress report.
May 1988	Modification granted; Clemson may proceed.
August 1988	Monsanto submits 9-month progress report.
August 1988	Monsanto requests modification of monitoring protocols because of low population density of test organism.
September 1988	Modification to monitoring protocol granted.
November 1988	Monsanto submits 12-month progress report.
February 1989	Monsanto submits 15-month progress report.

[a]Reviewed under "Toxic Substances Control Act (TSCA)."

Observations on submission reviews

To date, the EPA has reviewed approximately 50 submissions, some of which have been "second-generation" variations of previously approved tests. While it is still too early to speak in general terms or draw definitive conclusions about the reviews, certain observations can be made:

1. Evaluation of the proposed tests often requires different types of background studies, data bases, and scientific expertise than are required for review of traditional chemical products. Similarly, monitoring for living microorganisms in the environment is frequently more complicated.

2. The issues giving rise to the greatest concerns are often not related

to whether the microorganism performs its intended function, but rather to other, nonintended impacts that might occur.

3. Many (if not most) of the risk concerns about these microorganisms have been related to their potential impacts on the environment and nonhuman, nontarget species; *to date*, potential impacts on human health have not been a significant concern.

4. Monitoring data from some of the tests have confirmed that under actual use conditions, physical containment of the test microorganisms may not be feasible or possible. Accordingly, issues related to spread of the test organisms from the test site are appropriately considered in most risk assessments.

5. The small-scale tests reviewed thus far have been found to pose no significant risks. However, large-scale or commercial use of some of these microorganisms may raise more significant concerns, particularly regarding environmental and ecological impacts, than those associated with small-scale tests.

As to nonscientific observations, it has become increasingly clear that public involvement is an important element of most test programs involving environmental release. Without such involvement, public fear of the unknown could derail the whole process and delay the testing through extensive administrative and judicial maneuvers. The impact of such administrative and judicial activity can be seen by contrasting the first and a more recent review of experimental use permit applications for two genetically engineered microbial pesticides. In the first, involving the testing of the Ice minus (INA$^-$) *Pseudomonas syringae* by Advanced Genetic Sciences Inc., the time elapsed between receipt of the application and the test was almost 2 years (Table 6). In between were two lawsuits brought in federal and state courts to stop the test, and numerous state and local administrative proceedings.

The more recent application involved the testing, by Crop Genetics International, of a plant endophytic bacteria into which the gene coding for the *Bacillus thuringiensis* toxin had been inserted. This second application, submitted after more extensive preliminary public outreach and presumably after the public had gained more confidence in the EPA's review process, moved from submission to field test in approximately 5 months (Table 7). There were no lawsuits or supplemental administrative proceedings involved with this review. It appears that the more the public knows and feels a part of and comfortable with the process, the less likely it is to obstruct the test. Hence, many submitters have elected to shield from public scrutiny only the barest essentials about their microorganisms and proposed tests.

TABLE 6 Chronology of Advanced Genetic Sciences (AGS) Field Trials[a]

Date	Chronology of events
November 1984	EPA receives notification
January 1975	EPA completes preliminary risk assessment; SAP meets to review submission and EPA risk assessment.
February 1985	EPA requests additional information of AGS and informs company EUP required.
July 1985	EPA receives EUP applications.
August 1985	EPA announces receipt of EUPs in *Federal Register*, and public comment period begins.
September 1985	Public comment period closes.
September–October 1985	SAP and intra-agency reviews of EUPs received by EPA.
November 1985	EPA grants EUPs; testing may begin.
November 1985	EPA sued by Foundation on Economic Trends.
February 1986	AGS outdoor testing (rooftop) disclosed.
March 1986	EPA audits AGS and suspends EUP; plant pathogenicity testing to be repeated.
July 1986	EPA reviews and approves AGS plant pathogenicity testing.
December 1986	AGS proposes and describes test sites.
February 1987	EPA approves test sites and reinstates EUPs.
April 1987	AGS sued and injunction requested in state court. Request denied; strawberry plot vandalized.
April 24, 1987	INA$^-$ bacteria applied to test site in Contra Costa County, California; EPA and California conduct monitoring.
June 1987	Test concludes; AGS monitoring continues.
August 1987	AGS submits summary of results of test.
September 1987	AGS submits amended EUP application to conduct second small test at same location.
November 1987	EPA approves amended EUP.
December 1987	AGS begins second test (fall and spring applications).
February 1988	Second application of INA$^-$.
December 1988	Progress report from company received by EPA.

[a]Reviewed under "Federal Insecticide, Fungicide, and Rodenticide Act (FIFRA)."

TABLE 7   Chronology of Crop Genetics International Field Trial[a]

Date	Chronology of events
December 1987	EPA receives EUP application.
January 1988	EPA announces receipt of EUP in *Federal Register* and public comment period begins.
March 1988	Public comment received; preliminary risk assessment completed.
April 1988	BSAC subcommittee evaluates submission and EPA risk assessment.
May 1988	EPA grants EUP.
June 1988	CGI initiates field trial.
January 1989	CGI requests extension, expansion of EUP; submits data accumulated in field trial in support of request.

[a]Reviewed under "Federal Insecticide, Fungicide, and Rodenticide Act (FIFRA)."

EPA's Research Program for Biotechnology

Twenty-five percent of EPA's budget is devoted to research. Although only a small portion of EPA's total research budget is allotted to biotechnology issues, the agency has initiated a research program directed specifically at issues associated with use of microorganisms in biotechnology.

These research activities support the science needs of the regulatory program. Six areas of research have been identified as essential to this support activity. These are:

1. Development of methods for the detection and enumeration of microorganisms in complex environmental samples. Microbial, serological, biochemical, physiological, and genetic methods are being tested and refined to maximize sensitivity, specificity, and stability in a variety of habitats including terrestrial, aquatic, air, sewage, plants, and animal and human gut. Innovative detection and analytical methods are being developed and assessed.

2. Development of data and predictive models for transport from the point of application or release to other locations. The mechanisms and dynamics of transport between and within various environmental components (e.g., air, soil, groundwater, water, plants, insects, animals, and humans) are being studied. Where appropriate, this research may extend to the development of mathematical modeling frameworks for predicting transport and consequent exposure.

3. Determination of potential for survival, growth, and colonization in

a variety of conditions and habitats in the environment. The pertinent properties of microorganisms and relevant environmental factors that influence survival and colonization are being studied and described.

4. Assessment of the stability and transfer frequency of introduced genetic material in the intra- and extracellular environment. Environmental, genetic, and organismal factors that affect gene stability and the rates of gene exchange in the environment are being studied in a variety of environmental situations.

5. Detection of adverse environmental response (e.g., ecological effects, toxicity, change in host range). This research focuses on developing conceptual and experimental framework(s) that may be used to identify and describe functional attributes of natural microbial communities, as well as perturbations in these communities. Research also addresses the genetic and molecular basis of infectivity, pathogenicity, and host range.

6. Methods of controlling risk. Criteria for evaluating containment and monitoring strategies are being developed, and a variety of mitigation and risk reduction strategies, including the development of genetic constructs for conditional lethal control of survival and gene exchange, and procedures for physical and chemical decontamination, are being investigated. Some portion of research in this area also addresses issues dealing with process equipment design, decontamination technology, and worker exposure.

Each of these areas is currently being investigated by EPA research laboratories. The strategy for program development includes the establishment of in-house scientific staff instrumental in building the necessary data base. Concurrently, staff scientists share responsibility for developing a complementary extramural program and fostering interactive information exchange with extramural scientists in pertinent disciplines such as genetics, biochemistry, ecology, and microbiology.

The strategy requires regular independent peer review of the research program to guide the focus toward the needs of EPA regulatory offices while maintaining a high standard of scientific quality.

The agency views its research program in biotechnology as a high priority; the panoply of microorganisms and uses under EPA regulatory purview necessitates a fundamental understanding of microbial ecology. While these research efforts were specifically conceived as support for EPA's regulatory program, basic information in microbial ecology is being generated. This information, while it should prove useful in risk assessment, should also be of value to those seeking to better understand microorganisms and their functions in the environment, as well as to those seeking to develop useful products.

Summary

Over the last few years, the EPA has met several goals with regard to regulation of microorganisms subject to TSCA or FIFRA. First, the agency has established a research program, which will assist the agency in devising and maintaining a sound regulatory process. Second, the agency has established a systematic approach for identifying and assessing potential human or environmental risks. Third, a peer review mechanism that draws upon independent scientists expert in diverse areas of relevant knowledge has been established and is proving useful. Fourth, the agency has devised an open process that allows for public participation. Finally, under the agency's regulatory process a number of field trials have been approved and conducted. The results of these field trials have confirmed EPA's judgments of no significant risk.

References

1. Campbell, A. 1978. Tests for gene flow between eucaryotes and procaryotes. *J. Infect. Dis.* 137:681–685.
2. U.S. Environmental Protection Agency. 1982. Pesticide Assessment Guidelines: Subdivision M—Biorational Pesticides. National Technical Information Service, Springfield, Virginia. Currently under revision. For information on the revised version contact the Science Analysis and Coordination Staff (see footnote 2).
3. U.S. Environmental Protection Agency. 1986. Establishment of the Biotechnology Science Advisory Committee. *Federal Register* 51:24221–24222 (July 2).
4. Federal Insecticide, Fungicide, and Rodenticide Act. 25 June 1947, and as amended; U.S. Code Title 7, Sec. 136–136y.
5. Federal Plant Pest Act. 23 May 1957, and as amended. U.S. Code Title 7, Secs. 147a, 149, and 150aa–150jj.
6. Milewski, E. A. 1985. Field testing of microorganisms modified by recombinant DNA techniques: applications, issues and development of "points to consider" document. *Recombinant DNA Technical Bulletin* 8:102–108.
7. National Institutes of Health. 1986. Guidelines for Research Involving Recombinant DNA Molecules; Notice. *Federal Register* 51:16958–16985 (May 7).
8. Organization for Economic Cooperation and Development. 1986. Recombinant DNA Safety Considerations. Paris, France.
9. Office of Science and Technology Policy. 1984. Proposal for a Coordinated Framework for Regulation of Biotechnology; Notice. *Federal Register* 49:50856–50907 (December 31).
10. Office of Science and Technology Policy. 1985. Coordinated Framework for Regulation of Biotechnology; Establishment of the Biotechnology Science Coordinating Committee; Notice. *Federal Register* 50:47174–47195 (November 14).
11. Office of Science and Technology Policy. 1986. Coordinated Framework for Regulation of Biotechnology; Announcement of Policy and Notice for Public Comment. *Federal Register* 51:23302–23393 (June 26).
12. Plant Quarantine Act. 20 August 1912, and as amended. US Code Title 7, Secs. 151–167.

13. Reanney, D. C., Gowland, P. C., and Slater, J. H. 1983. Genetic interactions among microbial communities. In *Microbes in Their Natural Environments*, 34th Symposium of Society of General Microbiology, J. H. Slater, R. Whittenbury, and J. W. T. Winpenny (eds.). Cambridge University Press, Cambridge, U.K., pp. 379–421.
14. Sanderson, K. E. 1976. Genetic relatedness in the family of Enterobacteriaceae. *Annu. Rev. Microbiol.* 30:327–349.
15. Toxic Substances Control Act. 11 October 1976 and as amended. U.S. Code Title 15, Secs. 2610–1629.

Index

ABOUT THE EDITORS

JAMES P. NAKAS is Professor of Microbiology at SUNY College of Environmental Science and Forestry, Syracuse. He conducts research dealing with bioconversion of renewable resources.

CHARLES HAGEDORN is Professor of Agronomy at Virginia Polytechnic Institute and State University where his research centers on biological disease control.